Particle Accelerator Design:
Computer Programs

PARTICLE ACCELERATOR DESIGN: COMPUTER PROGRAMS

John S. Colonias

Lawrence Berkeley Laboratory
Berkeley, California

ACADEMIC PRESS   New York and London   1974

A Subsidiary of Harcourt Brace Jovanovich, Publishers

Copyright © 1974, by Academic Press, Inc.
ALL RIGHTS RESERVED.
NO PART OF THIS PUBLICATION MAY BE REPRODUCED OR TRANS-
MITTED IN ANY FORM OR BY ANY MEANS, ELECTRONIC OR MECHAN-
ICAL, INCLUDING PHOTOCOPY, RECORDING, OR ANY INFORMATION
STORAGE AND RETRIEVAL SYSTEM, WITHOUT PERMISSION IN WRIT-
ING FROM THE PUBLISHER. REPRODUCTION IN WHOLE OR IN PART
FOR ANY PURPOSE OF THE UNITED STATES GOVERNMENT IS PER-
MITTED.

ACADEMIC PRESS, INC.
111 Fifth Avenue, New York, New York 10003

United Kingdom Edition published by
ACADEMIC PRESS, INC. (LONDON) LTD.
24/28 Oval Road, London NW1

Library of Congress Cataloging in Publication Data

Colonias, John S
　　Particle accelerator design: computer programs.

　　Includes bibliographical references.
　　1.　Particle accelerators–Design and construction–
Computer programs.　　I.　Title.
QC787.P3C64　　　　539.7'3　　　　72-9988
ISBN 0–12–181550–1

PRINTED IN THE UNITED STATES OF AMERICA

Contents

Foreword, ix

Preface, xiii

Acknowledgments, xiv

PART A | PROGRAMS FOR MAGNETIC AND ELECTRIC FIELDS

1. Introduction, 1

Chapter I **Two-Dimensional Magnetostatic Programs**

2. Program TRIM, 15
3. Program LINDA, 39
4. Program NUTCRACKER, 63
5. Program MAREC, 74
6. Evaluations: TRIM, LINDA, NUTCRACKER, and MAREC, 86
7. Program GRACY, 91
8. University of Colorado Magnet Program, 93
9. Program COILS, 94

Chapter II **Calculation of Electrostatic Fields**

10. Program JASON, 102

vi   Contents

Chapter III Three-Dimensional Magnetostatic Programs

11. Introduction, 114
12. Lawrence Berkeley Laboratory Magnet Program, 115
13. Carnegie Mellon Magnet Program, 118
14. University of Nevada Magnet Program, 118
15. CERN Magnet Program, 119
16. Program MAFCO, 119
17. Program FORCE, 129

PART B | ORBIT CALCULATION PROGRAMS

18. Introduction, 135

Chapter IV Programs Employing Matrix Formalism

19. Introduction, 138
20. Program TRANSPORT, 140
21. Program OPTIK, 164
22. Program TRAMP, 176
23. Program 4P, 188
24. Naval Research Laboratory Beam Transport Programs, 188
25. Program BOPTIC, 189
26. Program BEATCH, 189
27. Program SYTRAN, 190
28. CERN Beam Transport Programs, 190

Chapter V Programs Employing Integration of Equations of Motion

29. Introduction, 193
30. Program SOTRM, 194
31. Program TRAJECTORY, 203
32. Program GOC3D, 210
33. Program MAGOP, 223
34. Program MAFCOIII, 227
35. Program CYDE, 233
36. Program PINWHEEL, 246
37. Program SYNCH, 254
38. Program AGS, 264

PART C | LINEAR ACCELERATOR PROGRAMS

39. Introduction, 269

Chapter VI **Programs for Linac Cavities**

40. Program CURE, 270
41. Program JESSY, 279
42. Program MESSYMESH, 280
43. Program AZTEC, 281

APPENDIXES

I **Summary of Computer Programs That Solve the Nonlinear Partial Differential Equation for Magnetostatics in 2 Dimensions**, 288

II **Summary of Computer Programs That Solve the Partial Differential Equations for Magnetostatics in 3 Dimensions**, 291

III **Summary of Computer Programs That Use the Integral Equation Method**, 292

References, 293

Author Index, 301

Subject Index, 304

Foreword

The art of high-energy particle accelerator design has evolved greatly since the period around 1930 when the Cockcroft–Walton and Van de Graaff generators, the linear resonance-accelerator, and the cyclotron were developed for the study of nuclear reactions. Major steps in the evolution of cyclic particle accelerators include the discovery of phase stability and the use of alternating-gradient focusing, with the result that current devices are as large as 2 km in diameter and are capable of producing particles with an energy of several hundred GeV. Included in these developments are stored beams of sufficient intensity and quality that elementary-particle reactions can be productively studied at the intersection of two oppositely directed beams, thus making available a tremendously greater center-of-mass energy than would be realized from the particles of a single beam impinging onto a stationary target.

The present state of accelerator technology requires that the design of the instrument and facilities be optimized. For beyond considerations of reliability and versatility, one must respect that the construction and operation expenditures are substantial, the lead times for setting up the larger facilities amount to several years, and many experimental teams will depend on the scheduled availability of a particular accelerator facility. The experimental devices external to the accelerator and the beam-transport lines thereto involve concepts and components similar to those that are associated with the accelerator itself and must be planned with comparable care.

The progress of the accelerator art understandably has necessitated an increased understanding of the physical phenomena that govern accelerator performance and an increased sophistication in the quantitative evaluation of these phenomena. Back-of-envelope calculations, rule-of-thumb designs, and cut-and-try developments thus no longer are adequate approaches to many accelerator questions. Although model work in some cases can be useful—for example in checking and in providing manufacturing experience relating to a provisional final design—this frequently is an impractical or time-consuming route to the resolution of design problems.

It is current practice for the accelerator designer to rely heavily on extensive calculations for many features of his design, and these calculations, in order to be specific, typically are completed only after considerable numerical work. It thus has been fortunate that, beginning perhaps in the mid-1950's, accelerator designers were able to benefit from the concurrent development of automatic stored-program digital computers.

Automatic digital computation of course is of great assistance to the accelerator theorist both for investigating new concepts and in proceeding with a specific design. The most common and basic problems to which digital computations have been applied in accelerator design are, however, those of the types treated in this book—namely boundary-value (or eigenvalue) problems for static or dynamic electromagnetic fields and the single-particle dynamics of particles moving in a prescribed field. Flexible and efficient programs have been developed over the past several years, especially in accelerator and plasma-physics laboratories, and are now available to the accelerator designer for the numerical solution of such problems. Clearly these programs also have application in other fields of physical science and technology where similar (or mathematically identical) problems arise, and they indeed have been so used.

The basis on which certain of the computational programs have been constructed and the means for employing them most efficiently may have been aided in some cases by sophisticated theoretical numerical analysis and, in others, may have been guided chiefly by experience. A definitive treatment of such questions is not attempted in this book, which instead is intended primarily to outline in an explicit way (in some cases aided by flow charts) the characteristics of certain working programs that are available and to describe the options available in their use. Texts and monographs, symposia proceedings, and review articles are available concerning accelerator technology and the field of plasma physics, but a correspondingly explicit description of computational techniques applicable to these fields unfortunately has been essentially restricted to isolated reports concerning individual programs. We therefore are indebted to Mr. Colonias for undertaking the collection and exposition of the material contained in the present work, of which a substantial part has been developed through his own effort and that of his immediate associates. It may be hoped that this material not only will be informative for the physical scientist or engineer who will have applications for the programs discussed, but that also the numerical analyst may recognize areas where a fuller understanding or new approaches can be developed.

The reader will note that options available in some of the programs described here include the use of cathode-ray (CRT) display and the possibility of interactive computation. Such features can make a computer facility become a particularly powerful tool for a physical scientist or engineer, who, through use of these options, can obtain promptly a clear appreciation of the results of a computation and, based on these results, instruct the computer how to proceed. Although by no means a substitute for thought, the computer thus becomes a very effective tool in providing orientation with respect to a new problem (sometimes revealing the possibility of qualitatively unanticipated results), as well as furnishing accurate numerical results relative to any specific case. In this spirit the program SYNCH, described in Chapter V, permits one to "construct" an accelerator from specified magnetic elements, instruct the computer to optimize certain parameters (if desired) in accord with specified criteria, and to examine visually the orbit or beam-envelope characteristics that would result. The programs described in this book, and certainly others that will be developed in the future, thus may be regarded as illustrating an important trend to which Dr. Kowarski referred in the following words during the September 22nd evening meeting of the 1971 International Conference on High-Energy Accelerators:

> ...Now we are at the beginning of a new kind of extension by machine: the computer comes to supplement the theoretician's brain. We cannot foresee what this... kind of creativity in physics will bring, but we may expect that, just as Ernest Lawrence's contribution was decisive to the development of the nuclear machines, the name of John von Neumann will be remembered in connection with the origins of computational physics.

L. JACKSON LASLETT
Berkeley, 1972

Preface

Rapid progress in numerical methods, coupled with the development of advanced computer systems with large memories and nanosecond speeds, have made possible the formulation of the sophisticated computer programs that have become indispensable in the modern nuclear laboratory. However, the magnitude and complexity of the problems involved and the existence of an enormous amount of literature make it very difficult for one to obtain an understanding of the capabilities and limitations of a particular program; dozens of articles written by specialists must be read. Even then, the user may find himself unable to utilize one or the other program because of insufficient operational detail. Thus, he embarks upon the formidable task of writing "another program," adding another series of papers in the already existing large number of duplicate programs.

The aim of this book is to give a concise account of some of the most important computer programs applicable to accelerator design, assemble these programs under one cover, and present the research worker with information by which he may determine a program's effectiveness.

The arrangement of the book and the selection of the programs have been the result of the personal experience and professional career of the author for over fifteen years in the digital computing field, the last six of which have been spent in charge of computer applications for accelerator design at Lawrence Berkeley Laboratory.

The programs reflect what the author feels to be necessary for a generalized discussion and include mathematical development of programs, operating procedures, program organization, and application techniques. The book is concerned primarily with the description of these programs rather than with a comprehensive and exhaustive treatment of theoretical details. Most programs are followed by illustrative examples indicating the manner by which the program is used, and each chapter includes a number of suggested references keyed to the programs. At strategic points comparison tables are given, which describe the vital statistics of each program, such as computer language used, core size, time for solution, availability, etc.

In assembling this book I have delved liberally into the theoretical material published by accelerator scientists and engineers. I have also tried to be meticulous in acknowledging the sources of all material covered; any omissions are purely accidental.

My discussion of the programs is in the nature of sampling—rather than reviewing—with a definite bias toward topics with which I have been personally involved.

The book is divided into three parts. Part A covers computer programs relating to the calculation of two-dimensional magnetic and electric fields, as well as progress made in the area of three-dimensional calculations. Part B covers programs relating to the calculation of orbits of particles in a magnetic and/or electric field. Here, we deal with programs capable of analyzing various properties of cyclotrons, various problems relating to transverse orbit motion in synchrotron design, as well as with programs relative to the design of beam transport systems. Part C covers some representative programs useful in the design of linear accelerator-type cavities.

Acknowledgments

No book is ever written without the author having received a considerable amount of generous assistance from others. I wish to thank Dr. Elon Close for assisting me in preparing the introductions to this book and for his participation with Drs. Arthur Paul, Steven Sackett, and Paul Concus in undertaking the tedious task of reading and criticizing the content of the book from the point of view of the specialist in the field.

I am particularly indebted to my friend and colleague, Bruce Burkart, whose untiring efforts in criticizing, correcting, and supporting this book made its publication possible.

The editorial help of Mary Wildensten and the typing of the completed manuscript by Virginia Franks is also gratefully acknowledged.

Next, I would like to thank the many individuals, institutions, publishers, and organizations who have generously permitted the use of illustrations. In particular, I thank the University of California, Lawrence Berkeley Laboratory, and the Atomic Energy Commission for supporting my efforts in making this publication possible.

Finally, I thank my wife Becky and my children John-John and Elisabeth for their patience and understanding.

Part A | **PROGRAMS FOR MAGNETIC AND ELECTRIC FIELDS**

1. Introduction

Part A of this book deals with programs that calculate static magnetic and electric fields for particle accelerators. The calculation of such fields for essentially arbitrary geometries with various current and charge distributions has been of prime interest for many years, and much effort has been expended on the solution of such problems as they arise from applications of electromagnetic theory. Furthermore, the development of particle accelerators has created a need for easily available solutions to sets of problems, any one of which is of enough complexity to require an expenditure of considerable time and effort to arrive at a usable solution. The desire for usable engineering-type solutions is intense in this field, since it is the electric and magnetic field intensities that determine the motion of the particles for which these accelerators and their associated transport and experimental systems are designed.

Although in general, a solution for any particular problem can be arrived at by what we shall call classical methods (some of which are briefly mentioned below), only since the availability of the modern high-speed digital computer and numerical algorithms has it become possible to furnish solutions easily and economically to a wide class of problems. With such solutions available, it then becomes possible to do numerical design parameter studies before equipment is built.

Interest in the dynamics of charged particles in electric and magnetic fields is not limited to particle accelerators; other applications include cosmic rays, electron microscopes, field spectrometers, and the common television cathode-ray tube. Nor is interest in magnetic and electric field calculations of a limited nature restricted to only those in the accelerator field even though these programs have been developed principally by people associated with particle accelerators. While these programs may be considered simply as numerical algorithms for obtaining solutions to Maxwell's equations, they have inherent in them certain peculiarities and limitations.

In the sections that follow, we give a brief review of typical problems solved and the methods of solution. Following this, we describe some of the programs themselves. No attempt is made to be complete or exhaustive, our goal being to furnish a reasonable background for the programs which will be introduced to the reader. It is only by actually using such programs that one can begin to understand and master them. It is hoped, however, that the material presented will provide the reader with a sufficient view of what the programs do and how they are similar or different.

a. Statement of the Problem to Be Solved

We assume that there is a region in which the source of the electromagnetic field is a distribution of electric charge and current, and we further assume that the fields are static. We then wish to find solutions to Maxwell's equations, which can be written as

$$\nabla \times \mathbf{E} = 0, \quad \nabla \times \mathbf{H} = \mathbf{J}, \tag{1.1}$$

where $\mathbf{E}$, $\mathbf{H}$, and $\mathbf{J}$ are, respectively, the electric field intensity, magnetic field intensity, and current density. Our region, or universe, can be all space; or in the cases with boundary conditions, part of the total space; or in the cases with symmetry, part of the total region of interest. There may be different materials within the universe.

Associated with the intensities, $\mathbf{E}$ and $\mathbf{H}$, are the electric displacement $\mathbf{D}$ and magnetic induction $\mathbf{B}$. These quantities are related in that $\mathbf{D} = \mathbf{D}(\mathbf{E})$ and $\mathbf{B} = \mathbf{B}(\mathbf{H})$ are functions of the intensities. We also know that $\mathbf{J} = \mathbf{J}(\mathbf{E})$ for the current density. We shall restrict ourselves to the simple functional relations

$$\mathbf{D} = \varepsilon \mathbf{E}, \quad \mathbf{B} = \mu \mathbf{H}, \quad \mathbf{J} = \sigma \mathbf{E}, \tag{1.2}$$

where ε and μ are the inductive capacities of the material present and σ is the conductivity of the material. Equations (1.1) and (1.2) then furnish five relations between the quantities $\mathbf{E}$, $\mathbf{D}$, $\mathbf{H}$, $\mathbf{B}$, and $\mathbf{J}$.

Although ε and μ may be tensors and $\mathbf{E}$ and $\mathbf{D}$ may have different directions, in practice this occurs so rarely that only one of the programs, JASON, allows such to be the case. In the other programs ε and μ are assumed to be scalar quantities depending on the field intensities and spatial coordinates. These inductive capacities are often considered constant over different regions of the problem.

A typical problem might be a two-dimensional cross section of a quadrupole (see Fig. 1). The z-axis is out of the paper, and we wish to find the field in the transverse x, y plane. Regions R_1 and R_2 can have different inductive capacities and the universe is comprised of the union of R_1 and R_2 with boundaries Γ_{12} and Γ. In practice, the symmetry of this particular configuration would be used to advantage and a solution would be obtained in only one quadrant.

b. Method of Solution

Classically, obtaining the solution to Eqs. (1.1) and (1.2) for a simple problem such as the ideal quadrupole of Fig. 1 presents no difficulty. The analytic solution is known and furnishes a direct means for calculating the values of the field vectors, provided Γ_{12} is a hyperbola. This, then, is the catch. What if Γ_{12} is a circle, or is only part of a hyperbola? These simple modifications complicate obtaining a usable solution; thus, we are naturally lead to the use of numerical methods and we shall return to these shortly, but we first want to mention what we consider to be classical approaches.

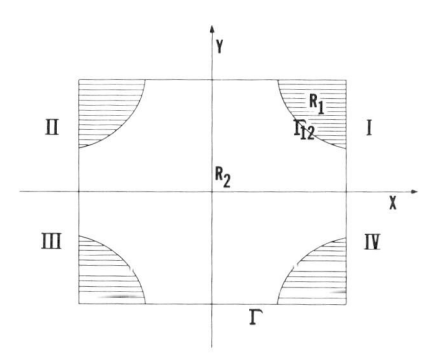

FIG. 1. Generalized two-dimensional cross section of a quadrupole.

The most direct, though not necessarily the most fruitful approach to obtaining the solution of a given problem is to solve analytically for the field at a given point. For geometries with suitable simplicity and symmetry, this can be done, as in the case of the ideal quadrupole of Fig. 1. Now, if the analytical answer is transformed into an algorithm that easily calculates the field intensities as a function of the problem parameters, we have a more useful tool. Further, if the results are applicable to a sufficiently large class of problems, we have a useful program, and in this case we shall consider the numerical function evaluations by a high-speed digital computer to be a modern method.

Another approach is to actually build the apparatus or a scale model and measure the field. We tend to ignore this as a solution to the problem, but classically this method, when coupled with elementary field calculations, was an accepted way of designing magnets. The field can be measured in many ways: electrolytic paper, tanks, wire filaments, magnetic probes, etc. In the final analysis, measurement is the definitive way to obtain the field even though it may be tedious, expensive, time-consuming, and sometimes difficult to establish the accuracy of the measurement. Analog computers, although of interest, will not be considered here.

Another approach is to solve numerically for the fields. This can be done by (1) graphical techniques, (2) engineering-type approximations to obtain solvable equivalent circuits and geometries, and (3) discretizing the problem and solving the resultant sets of equations. The first two techniques are quite useful, but do not readily yield the necessary precision that can be obtained using the techniques based on extensions of (3).

The discretization of the equations defining the analytic solution of the problem leads us to the "modern methods"; but, unless the resultant equations can be solved in a reasonably efficient manner (and this is not possible by hand) these methods, too, will be of no practical value. The use of modern computers possessing large memories and high speed central processors is essential for the successful solution by discretization.

To summarize, we can say that we are concerned here with programs that either efficiently evaluate analytic solutions or discretize the problem, i.e., substitute a discrete problem for a problem defined on a continuum, and then efficiently solve the resultant sets of equations that serve to define the approximate solution of the continuous problem.

c. Mathematical Formulation

The solution of Eqs. (1.1) and (1.2) by a computer is usually carried out in two main steps. The first of these is to define in suitable detail the problem to be solved mathematically, for example the solution of a two-dimensional problem where the boundary values are specified. This specification of the problem also implies the possible reformulation of Eqs. (1.1) and (1.2) into a form that is more suitable, e.g., solving for a potential rather than a field intensity vector. The second step is to take this defined problem and either obtain an analytic solution, which is used to evaluate the numerical solution, or discretize the problem and obtain equations, which, when solved numerically, give the desired solution. The "program" that is eventually built solves the problem either by actually evaluating the furnished analytic function or by numerically solving the set of discretized equations. We shall define the problem only in enough detail to convey the idea of what is being done; we shall not show how to solve the equations or evaluate the functions, since this can be a formidable task.

The first thing to notice is that although what is desired is a solution to Eqs. (1.1) and (1.2), none of the programs actually solve this problem. Instead, the properties of electric and magnetic field are used to transform these equations to other partial differential equations or to integral equation representations which are then solved. Along with basic vector identities, the following derived properties of the field are needed

$$\nabla \cdot \mathbf{D} = \varrho, \quad \nabla \cdot \mathbf{B} = 0, \quad \nabla \cdot \mathbf{J} = 0, \qquad (1.3)$$

where $\mathbf{D}$, $\mathbf{B}$, $\mathbf{J}$ are as previously defined and ϱ is the electrostatic charge density. The details of such derivations can be found in standard works dealing with electromagnetic theory [1]. For our purpose we take these as known results and use them as needed.

When considering the electric field, it is frequently the electrostatic potential ϕ that is solved for, and the field is obtained as $\mathbf{E} = -\nabla\phi$ where ϕ satisfies Poisson's equation

$$\nabla \cdot (\varepsilon \nabla \phi) = -\varrho, \qquad (1.4)$$

with suitable boundary conditions. For example, the value of the potential can be specified on conducting surfaces.

For magnetic fields one can also use a magnetic scalar potential, provided suitable care is taken to insure that the potential defines an irrotational field. In regions where the current density $\mathbf{J}$ is zero, this poses no problem; and some of the programs use such a magnetic scalar potential to define the fields as $\mathbf{H} = -\nabla\phi$ where again we have

$$\nabla \cdot (\mu \nabla \phi) = 0. \qquad (1.5)$$

However, in current-carrying regions, since $\mathbf{H}$ is itself not irrotational, it is necessary to somehow appropriately define an auxiliary vector $\mathbf{M}$ so that $\mathbf{H} - \mathbf{M}$ is irrotational. This leads to solving for a potential ϕ that defines the field as $\mathbf{H} = -\nabla\phi + \mathbf{M}$ where the potential satisfies

$$\nabla \cdot (\mu \nabla \phi) = \nabla \cdot (\mu \mathbf{M}). \qquad (1.6)$$

In cases where the inductive capacity is constant, Eq. (1.6) simplifies; if μ is a function of the field intensity, then the situation can get quite complicated.

The magnetic field can also be defined in terms of a vector potential, $\mathbf{A}$, where we have $\mathbf{B} = \nabla \times \mathbf{A}$. In this case we are led to solving for the vector potential, $\mathbf{A}$. There are no particular problems associated with the vector potential in current-carrying regions and thus in these regions it is usually this quantity that is solved for, where **A** satisfies

$$\nabla \times [(1/\mu) \nabla \times \mathbf{A}] = \mathbf{J}, \tag{1.7}$$

with suitable boundary conditions. Once **A** is obtained, **B** is readily found by taking $\nabla \times \mathbf{A}$.

In the two-dimensional case where A and J each have only one nonzero component, we can write Eq. (1.7) as

$$\nabla \cdot [(1/\mu)\nabla A] = -J.$$

The vector potential can be used throughout the universe to define the magnetic field; program TRIM does this, as does NUTCRACKER. On the other hand, a combination of magnetic vector and scalar potentials can be used where each defines the field in different regions of the universe, LINDA and MAREC have chosen this approach.

In the programs that are represented here, the mathematical model is based, not upon field Eqs. (1.1) and (1.2), but upon the equations satisfied by the potentials: either the scalar potential ϕ or the vector potential **A**. Thus the equations that are analytically solved or discretized are essentially Poisson's equation—or in the three-dimensional vector–potential case, the more complicated expressions given in Eq. (1.7).

d. Numerical Solution

Obtaining a numerical solution to a particular problem is a lot of work and can require literally thousands of arithmetic operations. This is precisely why it is essential to use modern high-speed computers.

In programs such as COILS, where the solution is being generated by evaluating the analytic solution to part or all of the problem, the details of how this numerical function evaluation is carried out are beyond the scope of this work. Suffice it to say that the evaluation of the solution is principally a task of numerically evaluating well-defined integrals using standard techniques. Different programs that solve different classes of problems will, in general, have different analytic functions that are to be evaluated and these will probably require different numerical techniques. Thus, there is not necessarily any set way of numerically obtaining the solution. For each such program the general field of numerical analysis is drawn upon to furnish a suitable calculation process.

In the cases where the problem is discretized and then the solution is found, the situation is somewhat different. In that case the original problem, for example Eq. (1.7), is replaced by a set of equations that can be written as

$$\mathbf{Ax} = \mathbf{b}, \tag{1.8}$$

where in the nonlinear case of nonconstant inductive capacities, it will be found that the coefficient matrix, $\mathbf{A} = \mathbf{A}(\mathbf{x})$, is a function of the unknown solution vector $\mathbf{x}$.

The actual process of deriving the discrete problem of Eq. (1.8) from the continuous problem, as for example Eq. (1.7), is called discretization. This can be carried out in a variety of ways and the equations arrived at in Eq. (1.8) will possibly vary for different discretizations of the same problem. The programs presented here have, to some degree, each chosen a somewhat different discretization approach. The reasons behind these choices are varied and one approach is not necessarily "better" than another. Judgement on the appropriateness of the discretization must be made within the context of the program and according to the particular type of problem it is intended to solve. The reader is referred to the programs themselves and the literature cited in their descriptive write-ups.

What is of interest to us is that the system of Eq. (1.8), possibly nonlinear, is solved numerically. The vector $\mathbf{x}$, which is the exact solution to Eq. (1.8), is still only an approximation to the true solution of our problem. The numerically obtained solution $\bar{\mathbf{x}}$ is an approximation to $\mathbf{x}$ and numerically defines the solution on a set of discrete points (nodes) of a mesh (grid) into which the universe of the problem has been subdivided. It is, therefore, an approximate solution to a set of equations that constitute an approximate representation of the original problem.

The coefficient matrix $\mathbf{A}$ of the system of Eq. (1.8) will have certain properties that depend on the method of derivation and on the particular problem being solved. The vector $\mathbf{b}$ will depend on the characteristics of the problem being solved, for example, the boundary conditions and the sources of charge and current. For linear systems (systems with constant $\mathbf{A}$) with appropriate conditions on $\mathbf{A}$, there is a vast body of theory dealing with how to effectively choose the "best" scheme to solve Eq. (1.8). Unfortunately, it happens to be a frustrating fact that the matrix $\mathbf{A}$ corresponding to the problem treated, possibly may not satisfy all such conditions. Thus, it is usually not possible to apply the developed theory directly in order to choose the best method of solving Eq. (1.8). However, the algorithm used in such ideal cases can also be used for our not-so-ideal cases and can even be applied to the nonlinear cases when $\mathbf{A}$ is a function of the unknown $\mathbf{x}$. The major difficulty is simply that we cannot theoretically predict much about what will, or will not, happen when these algorithms are applied. It is not our purpose here to discuss such theoretical questions, although they are of great interest to those trying to find more efficient solution algorithms. We shall, however, discuss the scheme most frequently used in the programs of this chapter.

The so-called "direct method approach" to solving Eq. (1.8), effectively calculates, in a finite number of arithmetic steps, the inverse $\mathbf{A}^{-1}$ of $\mathbf{A}$ to obtain $\mathbf{x} = \mathbf{A}^{-1}\mathbf{b}$. We say "effectively" because the actual inverse is rarely, if ever, calculated; rather, an elimination scheme is used. These schemes are usually not feasible when applied to the whole matrix $\mathbf{A}$ since the size of the array causes storage problems and in nonlinear cases this inversion process must be repeated many times. However, for rectangular meshes, cases occur where this approach is feasible due to the special structure of the matrix imparted by the special characteristics of the problem geometry and solution scheme. Although these methods are not used in the programs presented to solve for $\mathbf{x}$, they are used to solve for part of the solution $\mathbf{x}$ directly such as for a whole row of values. These methods then are usually referred to as direct block methods, and we shall comment further on this below.

The more generally applicable method used to solve the system of Eqs. (1.8) is an iteration scheme. In this scheme a sequence of vectors $\mathbf{x}^{(k)}$ is generated in such a manner that as k becomes infinite, $\mathbf{x}^{(k)}$ converges to the solution $\mathbf{x}$. The programs subsequently described use such iterative schemes. Much of the work expended in developing these programs pertains to choosing an appropriate iteration scheme and, having chosen it, to accelerate the rate of convergence of the generated sequence $\mathbf{x}^{(k)}$ to the solution $\mathbf{x}$.

The most popular iteration method has been the Point Successive Overrelaxation (PSOR) scheme, or a combination of this with some acceleration technique. PSOR is well known [2] especially in the linear ideal case where $\mathbf{A}$ has all the required properties. In this case PSOR is an easily defined method that can be efficiently coded. Thus it is fast and economical when applied to the large systems arising in this type of application.

The scheme is easily described as follows: The matrix $\mathbf{A}$ is factored into its diagonal elements $\mathbf{D}$, the lower triangular part $\mathbf{L}$, and the upper triangular part $\mathbf{U}$. Thus,

$$\mathbf{A} = \mathbf{D} + \mathbf{L} + \mathbf{U}. \tag{1.9}$$

The Gauss–Seidel value for the $(n + 1)$th iterate is calculated as

$$\bar{\mathbf{x}}^{(n+1)} = -(\mathbf{D}^{-1}\mathbf{L})\mathbf{x}^{(n+1)} - (\mathbf{D}^{-1}\mathbf{U})\mathbf{x}^{(n)} + (\mathbf{D}^{-1})\mathbf{b}. \tag{1.10}$$

Then the $(n + 1)$th iterate in the sequence $\mathbf{x}^{(k)}$ is taken as

$$\mathbf{x}^{(n+1)} = \mathbf{x}^{(n)} + \omega[\bar{\mathbf{x}}^{(n+1)} - \mathbf{x}^{(n)}], \tag{1.11}$$

where the parameter ω, the relaxation factor of the scheme, is usually taken in the range $0 \leq \omega \leq 2$; typical values ranging from 1.0 to possibly as high as 1.9.

In the linear case (constant **A**) the generation of a solution using (1.10) and (1.11) presents little, if any, difficulty. A suitable starting value of $\mathbf{x}^{(0)}$, frequently 0, is chosen and then the iteration is carried out by scanning over the mesh in a predetermined manner to successively calculate the components $x_i^{(n+1)}$ of the $(n+1)$th iterate. The convergence of the sequence $\mathbf{x}^{(k)}$ to the solution **x** is usually fast and smooth.

In the nonlinear case, problems can arise. It is necessary to calculate the elements of the matrices $\mathbf{D}^{-1}\mathbf{L}$, $\mathbf{D}^{-1}\mathbf{U}$, $\mathbf{D}^{-1}$ at some value of **x**; typically $\mathbf{x}^{(n)}$ could be chosen. However, the use of the actual values of these quantities can result in instabilities that cause a failure in the convergence of $\mathbf{x}^{(k)}$ to **x**. Thus these values are underrelaxed in the sense that one would, for example use

$$\mathbf{L}^{(n+1)} = \mathbf{L}^{(n)} + \varrho\{\mathbf{L}[\mathbf{x}^{(n+1)}] - \mathbf{L}^{(n)}\}, \tag{1.12}$$

where the convergence factor ϱ is less than 1.0. A typical value might be around 0.03.

The values of these relaxation factors are usually chosen from experience gained in developing and using these programs, since there is no well-developed mathematical theory that can tell us the best values.

The PSOR scheme when applied to nonlinear problems is sometimes called linearized PSOR. The problem solution is arrived at by a sequence of solutions, each of which represents an approximation to a locally linear problem.

For those cases in which the solution method leads to a matrix **A** with a block structure, it is possible to partition **A** into blocks $\mathbf{A}_i$ such that Eq. (1.8) can be solved as

$$\mathbf{A}_i \mathbf{x}_i = \mathbf{b}_i, \tag{1.13}$$

where $\mathbf{x}_i$ is the set of components of **x** corresponding to $\mathbf{A}_i$, and $\mathbf{b}_i$ is the sum of the source vector and the product of the off diagonal blocks for block i times the solution vector. For example, $\mathbf{x}_i$ can be a vector consisting of all the values of **x** in one row of the grid that has been used to discretize the universe. Equations (1.9)–(1.12) are directly applicable; we simply append the subscript i and we have the associated block SOR scheme that is used, for example, in program JASON. In such a scheme direct methods are used to effectively calculate $\mathbf{D}_i^{-1}$.

It is obvious that factoring **A** differently, would yield other schemes, such as the alternating direction implicit (ADI) method and many more. However, it is not obvious that factoring in this manner yields schemes that converge faster for the general class of nonlinear problems that we consider here. In fact, the simplicity of coding and the dependability of convergence that PSOR has displayed have so far made it the preferred method.

One of the programs mentioned below, THOR, uses a method called nonlinear PSOR. This scheme can easily be described as follows; we write

$$\mathbf{f} = \mathbf{Ax} - \mathbf{b}, \tag{1.14}$$

and then solve for $\mathbf{x}$ to make $\mathbf{f} = 0$. Newton's method is used to estimate the correct value of $\mathbf{x}$ and this leads to solving at the $(n+1)$th step the set of equations

$$\sum_j \frac{\partial f_i}{\partial x_j} [x_j^{(n+1)} - x_j^{(n)}] = -f_i. \tag{1.15}$$

This is solved iteratively by:

$$x_i^{(n+1)} = x_i^{(n)} - \frac{\omega f_i^{(n+1,n)}}{\partial f_i^{(n+1,n)}/\partial x_i}, \tag{1.16}$$

where the subscript refers to the solution vector component. That is, the value of the approximate solution at the ith point of the grid, and the superscript $(n+1, n)$ means that we use the current values $x_i^{(n+1)}$ if these are calculated, otherwise we use the previous values $x_i^{(n)}$.

Equation (1.16) is a "pointwise" method using a relaxation factor ω. It requires the recalculation of diagonal derivatives $\partial f_i/\partial x_i$ at each point at each iteration. This procedure is time consuming, however, it has the advantage of using only one relaxation parameter, and its convergence properties for sample test problems are such that it has been reported favorably [3].

In those programs that either use a combination of vector and scalar potentials or only solve systems of equations in a part of the universe, some means is employed as the iteration proceeds, to suitably match or adjust the boundary values at those interfaces that divide the problem into different representations of the solution. The method by which these adjustments are performed can be important factor in arriving at a good numerical solution and can have an effect on the specification of the boundary by the user of the program. The details of these calculations are, however, beyond the scope of the present work.

One of the chief concerns in the iterative solution of these problems is convergence of the sequence $\mathbf{x}^{(k)}$ to the numerical solution $\mathbf{x}$. To achieve this convergence, provisons are usually made to allow the user to selectively adjust such parameters as the relaxation factors and the initial guess at the solution. Also, some of the programs make use of known properties of the solution such as, the $\int \mathbf{H} \cdot d\mathbf{l}$ must be proportional to the current enclosed, to arrive at a global "block" correction that can be applied to all or many of the points to accelerate the convergence. The degree to which this is successful depends on many factors, one of which is the actual problem for which a solution is to be found.

In summary, for the ideal case in which Eq. (1.18) is a linear system, numerous methods of solution are available. For the practical case in which the problem is nonlinear, primarily the PSOR scheme has been used.

Perhaps, for a particular geometry other methods, e.g. block SOR, may be used, but for a general program applicable to diverse problem geometries, it is not presently evident that these other methods are significantly superior.

e. Solution Accuracy

All the programs are aimed at generating a numerical approximation of the analytic solution defined by Maxwell's equations applied to a particular problem. The accuracy with which such a solution can be obtained depends on many things. In this section we shall give some indication as to where some of the inaccuracies arise.

The first fundamental error is that introduced through the process of discretization, which implies that even if the discrete problem was solved exactly, the answer obtained would be only an approximation to the solution of the original problem. This error applies also to programs that solve the problem by numerical evaluation of the analytic function. For example in program COILS, which obtains the analytic solution and then superposes solutions to obtain the final solution, the error detected arises from the numerical methods used to evaluate the solution function.

When discretizing the problem (whether this be the differential equation or the integral equation equivalents) the discretization error usually depends on the mesh spacing. For example, in program TRIM, the vector potential is assumed to vary linearly over the small discretized regions; however, a detailed examination of the local error terms shows that they can depend upon the shape of the grid. Consequently, one obtains slightly different answers, depending on whether square, rectangular, or triangular meshes are used.

An additional error may also arise from partitioning the problem using different solution representations in different regions (for example, vector and/or scalar potentials) and then matching these different solutions at the region interfaces. This error is not entirely due to discretization since decreasing or eliminating the discretization error will not necessarily make this error vanish. Also, the solution scheme can involve the use of more than one solution method for different regions (for example, PSOR, or magnetic-field circuit approximations) and the approximation errors made when matching the different regions in the solution scheme will not necessarily disappear no matter how well the numerical solution is evaluated in each region.

Finally, there is the actual numerical solution of the resulting equations, boundary conditions, etc. The programs, such as COILS, that evaluate a function can be considered to be as accurate as one desires except that this evaluation becomes difficult, time consuming, and/or machine-precision limited.

With the iterative methods, there is a problem of knowing with what accuracy the solution is determined. Test cases, experience, and actual determination of measured values are the final guide lines. The iteration sequence $\mathbf{x}^{(k)}$ usually converges to a solution, but to say how close we are at a given iteration n can be a difficult task. It is usually obvious when the iteration sequence diverges.

There are guide lines for observing the convergence, such as monitoring the vector norms of the residuals (for example, the maximum residual over all points) or observing the average convergence rate (defined as a function that measures the difference in successive iterations of a quantity that is supposed to converge to a definite value as $\mathbf{x}^{(k)}$ converges to the solution $\mathbf{x}$). In practice, if the iterates are changing only in the sixth or seventh significant digit, then we can assume that either we have about five significant figures of the solution or else we are not making much progress per iteration. The standard trivial example of $f(x) = x^m$, for $m = 0, 1, 2, 3, \ldots$ shows that it is possible to have small residuals and yet still be relatively far from the correct root of $f(x) = 0.0$.

To estimate the error, one needs extra information about the numerically generated solution, or about the solution itself, and this is usually unavailable. So, in practice one develops a feeling for the problems being solved and can tell whether the answers are reasonable.

Finally, there are machine limits on the precision of the calculation. Most of the final results are obtained by performing calculations on the solution quantity selected; for example, if the vector potential is solved for and the field is desired then we have $\mathbf{B} = \nabla \times \mathbf{A}$. Thus, to obtain results with the necessary significant figures it is necessary that the machine precision be adequate, as with the CDC 6600 computer series with approximately 14 significant digits in single precision. The same codes, when run on machines with 7 to 8 significant digits, will produce different, less accurate answers and the error cannot be overcome by simply converging further, since once the solution is obtained to machine precision, further iterations simply result in an oscillating sequence.

Of the several sources of error, the user has most direct control over the discretization error that is a function of mesh size. (We are assuming here that machine-precision errors are negligible.) The accuracy of the solution can usually be increased by decreasing the mesh size, or by changing the density of the mesh points to effect a better distribution within the problem. Since the number of points, and thus the size of the system of equations, increases as the grid size decreases, the amount of computational effort is substantially increased and the time required to solve such problems can become prohibitive. Thus there are limits on grid density. Currently two-dimensional programs such as TRIM are effectively limited to 4000 or less points, while programs with two potentials, e.g., LINDA and MAREC, use 10,000 or more points.

The use of a rectangular grid versus a triangular grid is another factor of consideration. From the viewpoint of simplicity, as applied to arbitrary geometries, the triangular mesh has advantages, e.g., the user is easily permitted to distribute the density of points so that regions of greater interest can have more points, and presumably better answers, and the program can spend its time solving the important part of the problem. The use of an irregular mesh, such as that easily obtained in a triangular grid can result in a better approximation of the boundary values and an easier program treatment of these values since they can be considered as ordinary mesh point quantities, and no special interpolation or difference formulae are required. However, an irregular mesh can force a complicated interpolation elsewhere or distortion of the triangles to obtain these values as ordinary points; these complications can have various effects on the accuracy of the answer.

Rectangular grids allow one to fit certain geometries best (rectangular geometries) and can lead to more accurate answers. Curved boundaries require special boundary equations represented by special boundary approximations. They can also generate situations that violate conditions imposed by the use of the regular rectangular grids, and sometimes, it is more difficult to effectively distribute the point density. However, many of the programs use a rectangular mesh in spite of the complexities that they can introduce.

As a final comment on the numerical calculation of a solution using programs such as those mentioned in Part A, seldom do they solve for the actual quantities of interest. From a physical viewpoint it is usually the field intensity $\mathbf{E}$, or magnetic induction $\mathbf{B}$, or quantities derived from them, that are desired, and yet in many cases it is a potential that is solved for, for example, ϕ where $\mathbf{E} = -\nabla \phi$. Consequently, there is incorporated within the calculations an interpolation program that will (1) calculate the desired quantity from the numerical solution, and (2) furnish its value at some point in the universe, which in all likelihood is not a mesh point. We shall consider these "edit programs," as they are frequently called, as black boxes that give sufficiently accurate answers.

In summary, we would say that each program represents an algorithm that numerically solves a mathematical model of Maxwell's equations as applied to the class of problems it solves. The derivation of this numerical scheme has introduced certain assumptions and errors into the calculation. The numerical solution is itself carried out to a certain degree of completeness on machines of finite precision. The approximation is then subsequently edited to obtain the numerical value of the desired physically interesting quantities at specified points of the universe.

From the forgoing discussion we see that to describe and illustrate the programs and their differences we must indicate what the particular model is, what the numerical scheme is, what is solved for, and how this leads to the final desired results.

For a summary of the computer programs available to magnet designers, the reader is referred to the appendices at the end of this book. These appendices have been prepared by Dr. C. Trowbridge and his permission to reproduce them here is greatly appreciated.

I | Two-Dimensional Magnetostatic Programs

2. Program TRIM

a. Introduction

This program has been extensively used in the United States and abroad for the design of magnets. It is a general purpose, magnetostatic code capable of solving mathematical models of two-dimensional magnets and includes the effects of finite permeability of the iron. It uses a mesh composed of irregular triangles in which the mesh lines may be distorted to conform to irregular interfaces and boundaries. Additional features include:

1. The geometry may be specified by a variable triangular mesh of about 4000 points.
2. A magnet of any shape, either symmetric or asymmetric or possessing axial (cylindrical) symmetry, may be considered.
3. Any current distribution may be considered.
4. Energy stored in both air and iron regions may be calculated.
5. Nonzero boundary value problems may be treated.
6. Constant μ problems may be solved.

Perhaps the most useful characteristic of TRIM is its adaptability to practically any two-dimensional magnet problem. It is an easily defined mathematical model used to study entire magnets, and can include large borders of air or subsurface voids (wanted or unwanted).

Any mathematical model introduces some error from the necessary assumptions about the boundary conditions, but in TRIM these errors can be kept very small. For any particular problem or geometry, we have observed that the computed flux distribution depends upon the uniformity of the generated triangular mesh; therefore a sensible effort to avoid abrupt transitions from region to region should be exercised. With experience, one may avoid poor zoning and even improve upon a well-zoned problem. Uniform zoning is recommended in the air regions, principally by defining rectangular subregions, in particular on the median plane. New users of TRIM might be dismayed with first runs if they want results of the highest accuracy, but rezoning can produce astonishingly beneficial results. What can be accomplished by proper zoning is demonstrated by the simulation of the 184-inch Synchrocyclotron magnet at the Lawrence Berkeley Laboratory. This magnet has a cylindrical pole face and a rectangular return yoke, and since azimuthal and median plane symmetry was required, the magnet return yoke was adjusted for constant area in cylindrical geometry. Figures 2 and 3 show the generated mesh and indicate the flux distribution obtained. The computed magnetic field was better than 0.5 % in absolute value as compared with the measured field at the same excitation [4, 5]. Comparisons of computed and measured magnets indicate a probable upper limit of about 1% on the accuracy of calculated gradients with good zoning and using the maximum available mesh points. The reader is referred to an excellent description of the mathematical derivation of the algorithm by Winslow [6].

The program is written in Fortran (some output routines have been programmed in machine language) and is operational on the CDC 6600/7600, under the SCOPE or BKY operating systems. TRIM requires approximately $152K_8$ memory

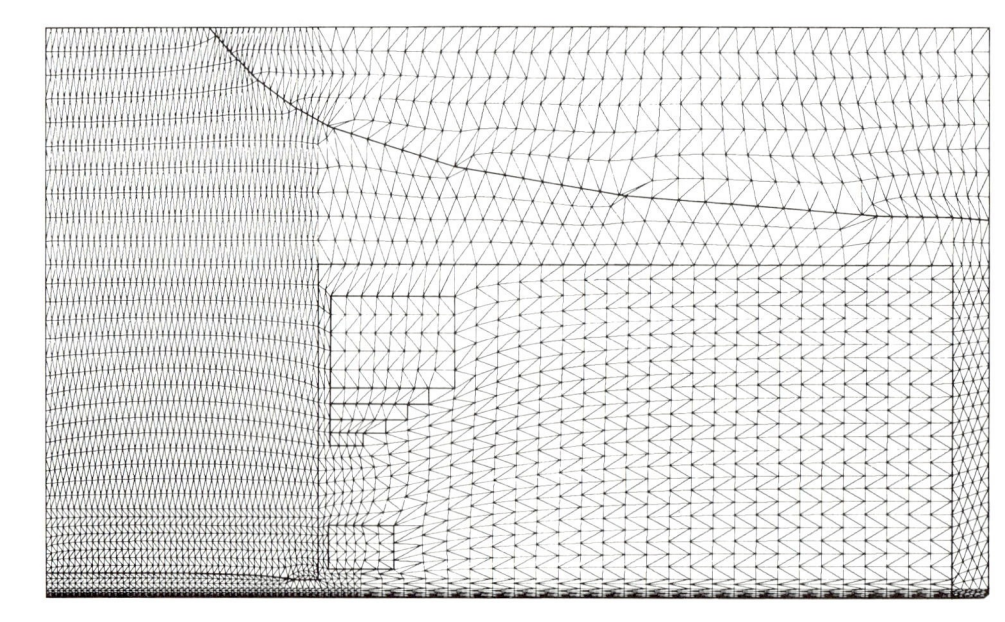

FIG. 2. Generated mesh for the 184-inch cyclotron magnet using program MESH.

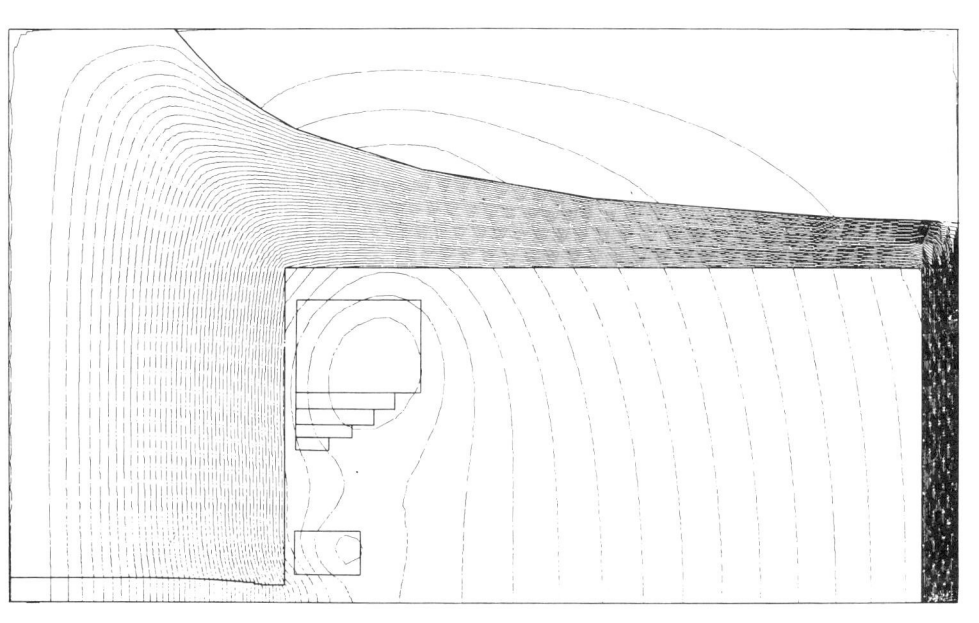

FIG. 3. A flux plot for the 184-inch cyclotron. It should be noted that the lines represent lines of constant rA since the magnet was simulated by a cylindrical model.

locations for a 4000-point problem. A typical problem solution run with 4000 points takes approximately 20 minutes on the CDC 6600 or about 3 minutes on the CDC 7600 computer. TRIM has also been converted to run on IBM 360-types computers at various laboratories.

b. Program Description

TRIM is divided into two separately executable programs: the mesh generator, MESH; and the magnetostatic program, FIELD. The purpose of MESH is to construct the irregular triangle mesh [7]. This is done by interpolating on specified (by input) boundary points, locating the internal mesh points of each region by a pseudo-equipotential method, and by assigning regional properties to each triangle. The mesh generator performs these functions by using four subroutines: INPUT, HSTAR, SETTLE, and GENOR (in a manner described in detail see Winslow [7]), and illustrated by a flow chart (see Fig. 4). The purpose of FIELD is to solve the non-linear Poisson's equation

$$\nabla \cdot (\lambda \nabla \psi) + S = 0, \tag{2.1}$$

where ψ or its normal derivative is specified on the boundary; λ is a function of ψ or its derivatives and S is a function of position. For magnetostatic prob-

18   I. Two-Dimensional Magnetostatic Programs

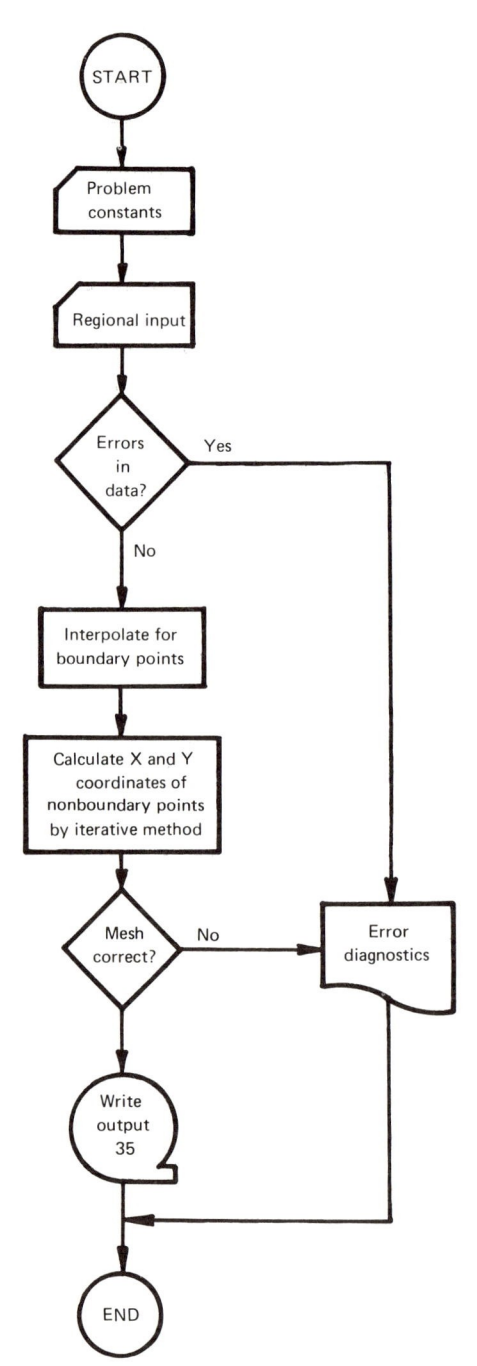

FIG. 4. Flow chart for program MESH.

lems the manner in which FIELD solves Eq. (2.1) is arrived at in the following fashion

$$\mathbf{B} = \mu \mathbf{H} = \nabla \times \mathbf{A}, \tag{2.2}$$

where μ is the permeability of the material.

The curl of the magnetic field intensity $\mathbf{H}$ is given by

$$\nabla \times \mathbf{H} = 4\pi \mathbf{J}, \tag{2.3}$$

where $\mathbf{J}$ is the current density. Substituting $\mathbf{H}$ from Eq. (2.2) to Eq. (2.3) we obtain

$$\nabla \times (1/\mu \nabla \times \mathbf{A}) = 4\pi \mathbf{J}. \tag{2.4}$$

Since $\mathbf{A}$ and $\mathbf{J}$ have only one nonzero component, for two-dimensional problems, Eq. (2.4) reduces to

$$\nabla \cdot (\gamma \nabla A) = -4\pi J, \tag{2.5}$$

where $\gamma = 1/\mu$.

The gradient of A, namely ∇A, is

$$\nabla A = \mathbf{i}(\partial A/\partial x) + \mathbf{j}(\partial A/\partial y), \tag{2.6}$$

and since γ depends on $|\nabla A|$, it is convenient to consider γ as a function of B^2 or

$$\gamma = f(B)^2 = f(|\nabla A|^2) = f[(\partial A/\partial x)^2 + (\partial A/\partial y)^2]. \tag{2.7}$$

The finite difference approximation in a triangular mesh is derived (see Winslow [7]) as

$$\sum_{i=1}^{6} w_i(A_i - A) + 4\pi J = 0, \tag{2.8}$$

where A is the vector potential at point C, A_i the vector potential at point C_i and the w_i's are geometric factors called couplings, which depend also on γ.

$$J = \sum_{i=1}^{6} \frac{J_{(i+1/2)}}{a_{(i+1/2)}} \tag{2.9}$$

is the total current through the mesh dodecagon surrounding the vertex C (see Fig. 5).

A more accurate value can be obtained from a different algorithm which has been recently implemented in TRIM by Winslow [8]. Briefly this algorithm involves the derivation of an expression for the appropriate area of the triangle to be used.

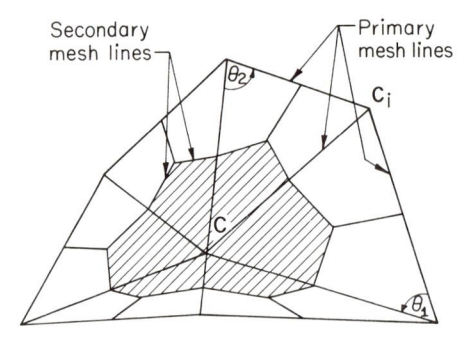

FIG. 5. One element of the triangular mesh showing primary and secondary mesh used to calculate currents.

Consider the acute triangle (see Fig. 6) of area A, where **a** and **b** are vectors representing two triangle sides.

The equations of the lines along bisector 1, and bisector 2 are

$$\mathbf{r}_1 = 0.5\mathbf{b} = s\downarrow\mathbf{b}\uparrow, \qquad (2.10)$$

$$\mathbf{r}_2 = 0.5\mathbf{a} = t\downarrow\mathbf{a}\uparrow, \qquad (2.11)$$

respectively, where $\downarrow\mathbf{a}\uparrow$, $\downarrow\mathbf{b}\uparrow$ represent the vectors **a**, **b** rotated 90° counterclockwise, and s, t are positive scalars at the intersection $\mathbf{r}_1 = \mathbf{r}_2$. Therefore we have

$$0.5\mathbf{b} + s\downarrow\mathbf{b}\uparrow = 0.5\mathbf{a} - t\downarrow\mathbf{a}\uparrow. \qquad (2.12)$$

Taking the scalar product of Eq. (2.12) with **a** we obtain

$$S = a^2 - \mathbf{a} \cdot \mathbf{b}/2\mathbf{a} \cdot \downarrow\mathbf{b}\uparrow = a^2 - \mathbf{a} \cdot \mathbf{b}/4A. \qquad (2.13)$$

Since $\mathbf{a} \cdot \downarrow\mathbf{b}\uparrow = 2A$, the vector **h** representing the height of triangle A_1 is

$$\mathbf{h} = \mathbf{r}_1 - (\mathbf{b}/2) = s\downarrow\mathbf{b}\uparrow = (a^2 - \mathbf{a} \cdot \mathbf{b}/4A)\downarrow\mathbf{b}\uparrow \qquad (2.14)$$

and

$$A_1 = \mathbf{h}\downarrow\mathbf{b}\uparrow/4 = [(a^2 - \mathbf{a} \cdot \mathbf{b})b^2]/16A. \qquad (2.15)$$

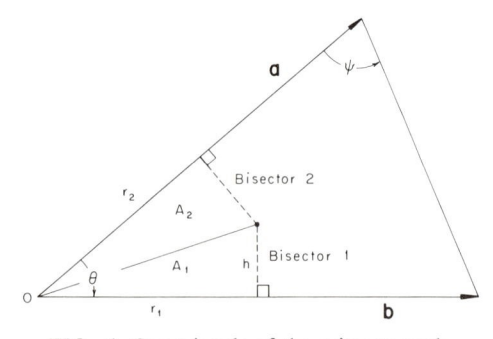

FIG. 6. One triangle of the primary mesh.

By symmetry

$$A_2 = [(a^2 - \mathbf{a} \cdot \mathbf{b})a^2]/16A, \tag{2.16}$$

so that for an acute triangle the fraction, $\bar{A}_i = (A_1 + A_2)/A$ assigned to the vertex labeled "0" in Fig. 6 is

$$\bar{A}_i = [2a^2b^2 - (a^2 + b^2)\mathbf{a} \cdot \mathbf{b}]/16A^2, \tag{2.17}$$

and a more accurate value of the current at node "0" is

$$J = \sum_{i=1}^{6} J_1 \cdot \bar{A}_i, \tag{2.18}$$

where J is the current density in triangle i. Since this method will fail to give the correct value if any of the triangles are obtuse, the program checks all angles of each triangle to determine whether the triangle is indeed obtuse; in that case $\bar{A}_i = 0.5A_i$ for the obtuse angle and $\bar{A}_i = 0.25A_i$ for the other two angles.

Diserens [9] has also noticed the original deficient algorithm in calculating currents for program TRIM, and has applied similar corrections by making the current allocation at a vertex, proportional to the angle of a triangle at that vertex.

The derivation of Poisson's equation for the solution in cylindrical coordinates follows the same reasoning applied to the rectangular case by observing.

$$\mathbf{B} = \mu\mathbf{H} = \mathbf{\nabla}_c \times \mathbf{A} = 1/r \, \mathbf{\nabla} \times (r\mathbf{A}), \tag{2.19}$$

where $\mathbf{\nabla}$ and $\mathbf{\nabla}_c$ represent the operator in Cartesian and cylindrical coordinates, respectively, and

$$\mathbf{\nabla} \times \mathbf{H} = 4\pi\mathbf{J}. \tag{2.20}$$

With proper substitution of (2.20), we obtain

$$\mathbf{\nabla} \cdot [\gamma/r\mathbf{\nabla}(rA)] = -4\pi J. \tag{2.21}$$

A comparison of Eqs. (2.5) and (2.21) shows that the two models differ from each other by γ/r replacing γ, and rA replacing A.

c. Program MESH

The triangular mesh which program MESH constructs is topologically regular and every interior mesh point is a common vertex for six triangles [10].

In Fig. 7, each horizontal line of mesh points is called a row and is numbered with index L as shown, running from L = 1 to L = LMAX at the top of the mesh.

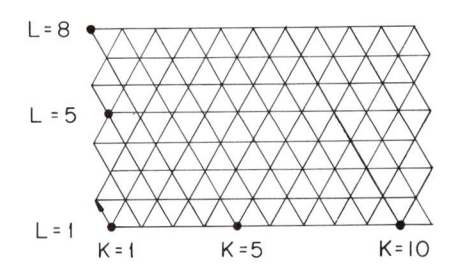

FIG. 7. Triangular mesh with defining parameters.

Within each row, mesh points are numbered with the index K, running from K = 1 at the left of each row to K = KMAX at the right. Note that the convention is made that the origin L = 1, K = 1 is a point at which three mesh lines meet. Also, by convention, lines of constant K always begin with the direction indicated by the arrow, that is, they are zigzag with the direction shown. The procedure for recording data for MESH is as follows:

1. Draw the universe of the problem with the desired dimensions.

2. Draw the logical diagram of the magnet inside the universe. On an isometric paper we draw a diagram which shows the magnet geometry as pictured in terms of mesh coordinates; we call this a "Logical Diagram" as shown in Fig. 8, corresponding to the CERN-PS magnet configuration; Fig. 9 shows the physical dimensions of this magnet.

3. Separate the problem into regions. Since a magnet consists of different materials (e.g., iron, conductors, etc.) the concept of regions is introduced, which besides its usefulness in separating various materials, also serves to force regular or otherwise special zoning in any part of the problem. In Fig. 8, it

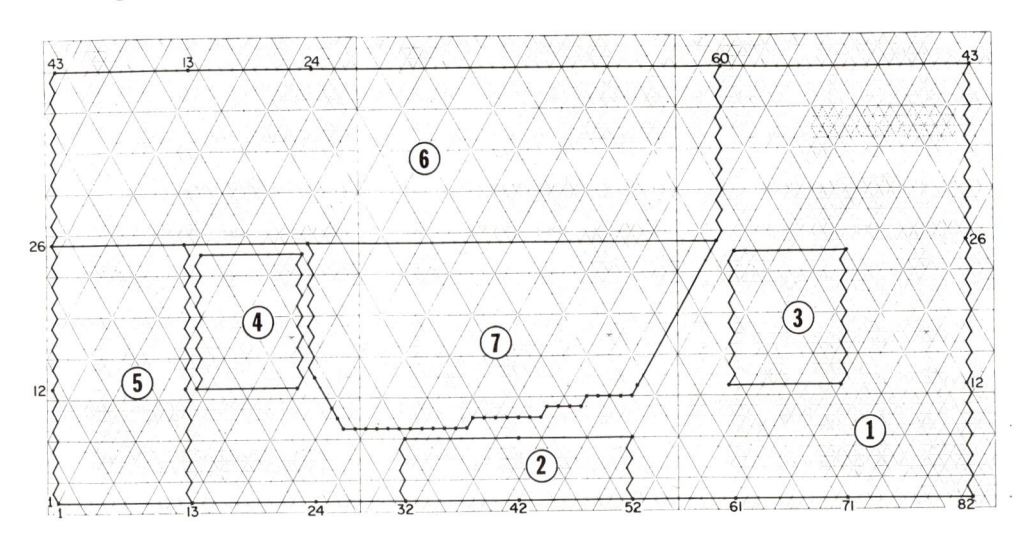

FIG. 8. The CERN-PS magnet logical diagram.

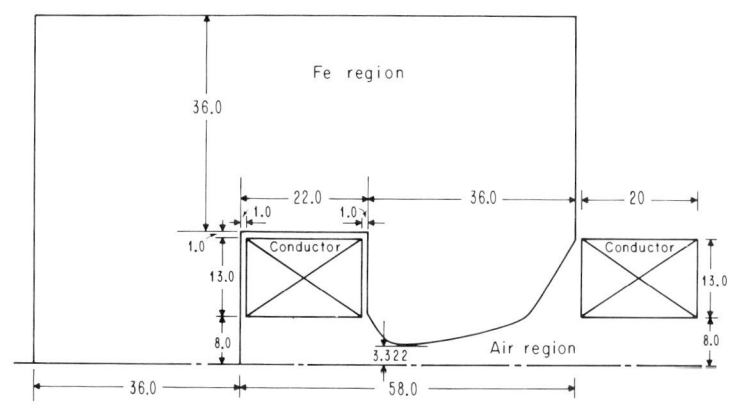

FIG. 9. The CERN-PS magnet, all units are in centimeters.

can be seen that the magnet is separated into seven regions. Program MESH superposes these regions in the order in which they are given in the input. Thus, all or part of a region which has once been specified, may be respecified later in the input. Care must be exercised to prevent mesh lines from crossing one another in impossible ways and producing triangles of negative or zero area. Such crossings can be avoided by twice recording points that belong to the common boundary of two regions. This will become more obvious as the input for the data for the test problem are analyzed.

Note, also, that the limited number of mesh points available necessitates the careful distribution of mesh lines. Namely, the designer should decide *a priori* which of the regions in his magnet are most important, i.e., where the most accuracy is desired, and then distribute the available mesh points accordingly. In magnet calculations for accelerator design, one is interested in the magnetic field and its gradient in a region defined within the limits of the vacuum chamber; the zoning in this region is very important. The CERN-PS magnet *was zoned* with this in mind.

4. Record the x, y coordinates corresponding to the mesh *indices*, in the following sequence:

a. L = index for horizontal mesh line
b. K = index for vertical mesh line
c. Y = y coordinate of physical dimension for this point
d. X = x coordinate of physical dimension for this point.

Although it is assumed that the points were taken counterclockwise, the direction is immaterial. Also, in closing the region, the first point should not be specified again. Therefore, referring to Fig. 8 for the logical diagram and to Fig. 9 for the actual dimensions, we record the coordinates of the four boundary points of region 4 to be those in Table 1.

Table 1
Boundary Coordinates for Region 4 of the Logical Diagram for the CERN-PS Magnet

Point	L mesh	K mesh	Y dimension (centimeters)	X dimension (centimeters)
1st	25	14	21.0	37.0
2nd	12	14	8.0	37.0
3rd	12	23	8.0	57.0
4th	25	23	21.0	57.0

Figure 10 shows the input data corresponding to the CERN-PS magnet. It should be observed that the last point of each region has the letter c following the last entry. This letter (peculiar to the Berkeley data reading routines) signifies the end of the regional boundary points, and provides a convenient indicator as to the number of points in each region.

Once the region boundary points have been specified, we proceed with the preparation of the region card. Information necessary for this card is as follows:

a. Region number
b. Flag describing the material of the region:
   0 = Regions to be skipped
   1 = Air
   2 = Iron
   3 = Iron (for example, SAE 1010)
   4 = Iron (for example, SAE 1020)
   5 = Iron (for example, Permeadur)
c. Region current in Ampere-turns (At)
d. Region current density in At/units^2
e. Sentinel indicating the type of triangles into which the region should be zoned:
   0 = equal weight relaxation
   1 = equilateral triangles
   2 = right triangles.

Note, if region At is specified, insert zero for At/units^2 and vice versa. Do not enter both At and current density for any current-carrying region.

In the same fashion, all regions are recorded and transferred onto punched cards. Any column from 1 to 80, may be used and any number of quantities

```
CERN-PS                                                          $ ALPHABETIC HEADING CARD $
*1+7. *2+43. *3+82. *4.    *5+1. *14+.95 *16+100. *19+.9987      $ PRBCON ARRAY $
*26+42. *29+10000. *35+.. *36+1. *37+1. *38+10.  S
  1  1   0.   0.   2          $ REGION CARD $
  1  1   0.   0.              $ BEGIN REGION BOUNDARY POINTS $
  1  13  0.   36.
  1  24  0.   58.
  1  32  0.   66.
  1  42  0.   76.
  1  52  0.   86.
  1  61  0.   95.
  1  71  0.   115.
  1  82  0.   150.
  12 82  8.   150.
  26 82  22.  150.
  43 82  58.  150.
  43 60  58.  94.
  43 24  58.  58.
  43 13  58.  36.
  43 1   58.  0.
  26 1   22.  0.
  12 1   8.   0.   C          $ LAST BOUNDARY POINT, 1ST REGION $
  2  1   0.   0.   2          $ REGION CARD FOR 2ND REGION $
  1  32  0.   66.
  1  42  0.   76.
  1  52  0.   86.
  7  52  3.   86.
  7  42  3.   76.
  7  32  3.   66.  C          $ LAST BOUNDARY POINT, 2ND REGION $
  3  1   40250.0  0.   2      $ REGION CARD FOR 3RD REGION (CURRENT) $
  12 61  8.   95.
  12 71  8.   115.
  25 71  21.  115.
  25 61  21.  95.  C          $ LAST BOUNDARY POINT, 3RD REGION $
  4  1   -40250.0  0.   2     $ REGION CARD FOR 4TH REGION (CURRENT) $
  25 14  21.  37.
  12 14  8.   37.
  12 23  8.   57.
  25 23  21.  57.  C          $ LAST BOUNDARY POINT, 4TH REGION $
  5  2   0.   0.   0          $ REGION CARD FOR 5TH REGION (IRON) $
  1  1   0.   0.
  1  13  0.   36.
  12 13  8.   36.
  26 13  22.  36.
  26 1   22.  0.
  12 1   8.   0.   C          $ LAST BOUNDARY POINT, 5TH REGION $
  6  2   0.   0.   0          $ REGION CARD FOR 6TH REGION (IRON) $
  26 1   22.  0.
  26 13  22.  36.
  26 24  22.  58.
  26 60  22.  94.
  43 60  58.  94.
  43 24  58.  58.
  43 13  58.  36.
  43 1   58.  0.   C          $ LAST BOUNDARY POINT, 6TH REGION $
  7  2   0.   0.   0          $ REGION CARD FOR 7TH REGION (IRON) $
  13 24  8.526  58.
  26 24  22.  58.
  26 60  22.  94.
  12 53  8.517  86.036
  11 52  8.045  85.2
  11 51  7.547  84.2
  11 50  7.105  83.2
  11 49  6.713  82.2
  11 48  6.361  81.2
  10 48  6.045  80.2
  10 47  5.827  79.8
  10 46  5.651  78.8
  10 45  5.4    77.8
  9  44  5.306  77.4
  9  43  5.17   76.8
  9  42  5.     76.
  9  41  4.802  75.
  9  40  4.62   74.
  9  39  4.451  73.
  9  38  4.324  72.2
  8  38  4.263  71.8
  8  37  4.147  71.
  8  36  4.01   70.
  8  35  3.882  69.
  8  34  3.762  68.
  8  33  3.663  67.2
  8  32  3.568  66.4
  8  31  3.455  65.4
  8  30  3.368  64.5
  8  29  3.298  63.6
  8  28  3.536  62.5
S 8  27  4.06   61.5
  9  26  4.94   60.5
  10 26  6.2    59.5  C        $ LAST BOUNDARY POINT, 7TH REGION $
```

FIG. 10. Program MESH input data for the CERN-PS magnet. Note the free-format specifications.

may be put on each card. However, it is easier to put one point per card so that an error may be more easily detected.

The final assembled deck for MESH will consist of the following cards:

1st card – Problem identification card (all 80 columns may be used for comments, such as Problem Name, Date, Account Number, etc.)

2nd card – Problem constants (see d. Problem Constants for MESH and FIELD Programs)

3rd card – Region card

4th & subsequent cards – Region boundary points (any number of cards) Region card (for next region), until all regions have been exhausted.

Figure 10 shows a complete listing for the CERN-PS magnet.

A successful MESH output consists of a listing of all input data, along with four quantities (the X and Y convergence rate, and the x and y overrelaxation parameters) which indicate the progress of convergence.

The message GENERATION COMPLETED indicates that the resulting mesh is suitable for submission to FIELD. It may (and usually does) still require some improvements if high quality results are desired, but it does, at least, satisfy the minimum requirements that no triangles of negative or zero area are present.

If error conditions have been detected, a listing of the locations of errors will be given. A careful review of the input data will usually reveal the error; if not, a CalComp plot certainly will. The mesh must be regenerated, if necessary with the modified input data, until the message GENERATION COMPLETED is printed in the output and CalComp plots indicate a mesh of sufficiently good quality.

d. Problem Constants for MESH and FIELD Programs

The first card after the Problem Identification card, consists of the program constants. A detailed description of these parameters is shown in Table 2.

These constants are stored in an array called PRBCON and they offer additional flexibility with which to process a variety of problems including symmetric and asymmetric magnets. If the user does not wish to change any of the constants shown, the standard values will prevail. The data for the test case show some of these constants entered in a format-free fashion as before (see Fig. 10). The remarks that follow clarify any questions that might arise; first, an asterisk (*) always precedes the location of the array number (this is characteristic of the Berkeley data reading routines); for example, *1 + 7.0 indicates that the first constant in Table 2, namely the number of regions, has been set to 7.0. In this fashion, any other constant may be set. If we wish to set also constant (6) and (7) we would write *6 + 1.0 + 1.0 or *6 + 1.0*7 + 1.0. The asterisk need not be repeated if sequential locations are used.

To indicate to the computer program that we do not intend to have any more problem constants, we insert the letter s (short for SKIP) at the end of the last entry.

After the data appearing in Fig. 10 have been transferred to cards and submitted to the computer, MESH will generate the irregular triangular mesh and produce a magnetic tape to be used in the second component of TRIM, program FIELD.

e. Program FIELD

This component of TRIM (see Colonias [10]) solves the two-dimensional mathematical model previously described. A generalized flowchart of this algorithm is shown in Fig. 11.

In solving this model, FIELD makes use of the magnetic tape generated by MESH, which describes the magnet under consideration in terms of the defining geometry, and performs the necessary calculations to arrive at the desired magnetic field and gradient. This pattern would have to be duplicated every time one wished to change coil currents, convergence criteria, B-DESIRED, or any other of the problem constants listed in Table 2. Therefore, to make the operation of TRIM easier, and to allow for the solution of many different problems in one computer run, additional input data are necessary as follows:

1. *Enter Permeability Table(s).*

The B versus H characteristics for the particular iron used is entered via cards. The quantities necessary are the following: the number of tables (NTABLE), type of tables (TTABLE), and the values of B versus H. Specifically, the number of tables indicates to the program whether one, or more B versus H tables are used, in the following fashion.

If NTABLE = 0. Use is made of the standard table already in the program, in which case no entries of B versus H are needed.

If NTABLE = − 1. No B versus H tables needed (used for infinite μ problems).

If NTABLE = N. Where N = 1 to 6, indicating the number of tables used for each problem.

Table 2
TRIM Input Parameters and Specifications

Location	Array name	Description	Standard value
PRBCON(1)	NREG	Number of regions; maximum number, 40	(Specify)
(2)	LMAX	Maximum value of row index (y-axis); maximum, 150	(Specify)
(3)	KMAX	Maximum value of column index (x-axis); maximum, 150 $[(KMAX + 2) \times (LMAX + 4)]$ must never exceed 4000	(Specify)
(4)	MODE	If = -2.0, air solution only (infinite μ solution); if = -1.0, air solution followed by an all points solution; if = 0.0, all points solution only (finite μ solution)	0.0
(5) - (8)	IBUP, IBLO, IBRT, IBLF	Upper, lower, right, left boundary reflection sentinels; if = 1.0, true; if = 0.0, false	0.0
(9)	RHOX, RHOY	Initial overconvergence factor for mesh generation	1.6
(10)	RHOAIR	Initial overconvergence factor for points surrounded entirely by air and interface points	1.94
(11)	RHOFE	Overconvergence factor for points in iron	1.0
(12)	RHOCPI	Fraction of new couplings to use in iron	0.03125
(13)	RHOCPA	Fraction of new couplings to use in air	1.0
(14)	PACFAC	Stacking factor: fraction of total magnet volume that is iron	1.0
(15)	RESID	Convergence criterion: value of residual/length at which the problem has sufficiently converged (FIELD)	10^{-6}
(16)	IPRFQ	Print frequency: interval between iteration prints or between JFACT adjustments (in program FIELD)	20.0
(18)	CONVER	Conversion factor: if input is inches, user must enter *18+2.54	1.0
(19)	JFACT	Current multiplier (FIELD): actual current in coils is JFACT times input coil current	1.0
(22)	ETAMAX	Maximum rate of convergence	1.0
(23)	BETA	Underrelaxation of OMEGA	0.5
(24)	OMEGA	Overrelaxation parameter	0.005
(25)	RMAX	Maximum number of cycles after which automatic optimization resumes	25.0
(26)	KBZERO	K-mesh value where the desired B_0 on the midplane is found (FIELD)	1.0
(27)	LI	Edit row: mesh row at which edit is taken if problem fails to converge (FIELD)	1.0
(28)	EPSO	Convergence criterion for program MESH	10^{-5}
(29)	BDES	B-desired: desired field at K = KBZERO; useful when a desired field is specified; thus, FIELD will calculate the necessary current required by adjusting JFACT to produce the desired field; input in gauss	10^{15}
(30)	LIMTIM	Time limit (internal CDC 6600 clock system)	15 Seconds
(31)	MODEL	Set = 1.0 for cylindrical model, Z(L) axis of rotation	0.0
(32)	KBOUND	Set = 1.0 if boundary value problem is being solved	0.0
(33)	GAMX	Constant γ ($\gamma = 1/\mu$) problem: GAMX = 0.0, variable γ problem; others are γ = GAMX	0.002

Table 2 (*cont.*)

Location	Array name	Description	Standard value
PRBCON(34)	NORHO	Air relaxation constant: set = 1.0, air relaxation factor remains constant throughout the problem; otherwise set at standard value and adjust every RMAX iteration	
(35) - (38)	L1,L2, K1,K2	Lower, upper, left, right limit of edit (FIELD); L1 = 0.0, no edit	1.0,LMAX, 1.0,KMAX
(40)	-	If = 1.0, use 1/8 section of symmetric quadrupole; if = 0.0, use 1/4 section of symmetric quadrupole	0.0
PBMK(1)	LMAX	See PRBCON(2)	
(2)	KMAX	See PRBCON(3)	
(3) - (6)	IBUP,IBLO, IBRT,IBLF	See PRBCON(5) through PRBCON(8)	
(7) - (8)	RHOAIR,RHOFE	See PRBCON(10); see PRBCON(11)	
(9) - (10)	RHOCPA,RHOCPI	See PRBCON(13); see PRBCON(12)	
(11) - (12)	EPSOC,IPRFQ	See PRBCON(15); see PRBCON(16)	
(14) - (15)	ETAM,BETAM	See PRBCON(22); see PRBCON(23)	
(16) - (17)	OMEGA,IRMAX	See PRBCON(24); see PRBCON(25)	
(18)	FNM	Problem name: alphabetic heading (10 Hollerith) for problem identification	
(19) - (20)	PACFAC,XJFACT	See PRBCON(14); see PRBCON(19)	
(21) - (22)	KBZERO,LC	See PRBCON(26); see PRBCON(27)	
(23) - (24)	CONVER,BDES	See PRBCON(18); see PRBCON(29)	
(26) - (27)	MODEL,KBOUND	See PRBCON(31); see PRBCON(32)	
(28) - (29)	GAMX,NORHO	See PRBCON(33); see PRBCON(34)	
(43)	-	See PRBCON(40)	
TVAR(1)	CYCLE	Iteration counter	
(2) - (3)	AMIN,AMAX	Minimum, maximum vector potential in mesh	
(4)	DAMAX	Absolute value of maximum change that has occurred in vector potential on last cycle	
(5) - (6)	RESID,BMAX	See PRBCON(15); value computed in subroutine ENERGY after convergence	
(8)	EPSOC	Convergence sentinel; if = 1.0, problem converged; if = 0.0, not converged	
(12)	IPRFQ	Cycles to print, see PRBCON(16)	
(14)	NDUMP	Dump number	
(15)	MODE	See PRBCON(4)	
(20)	-	Negative triangle sentinel: if = 1.0, negative triangle(s) in mesh; if = 0.0, no negative triangle(s) in mesh	
(22)	LIMTIM	See PRBCON(30)	
(23) - (24)	L1,L2	See PRBCON(35); see PRBCON(36)	
(25) - (26)	K1,K2	See PRBCON(37); see PRBCON(38)	

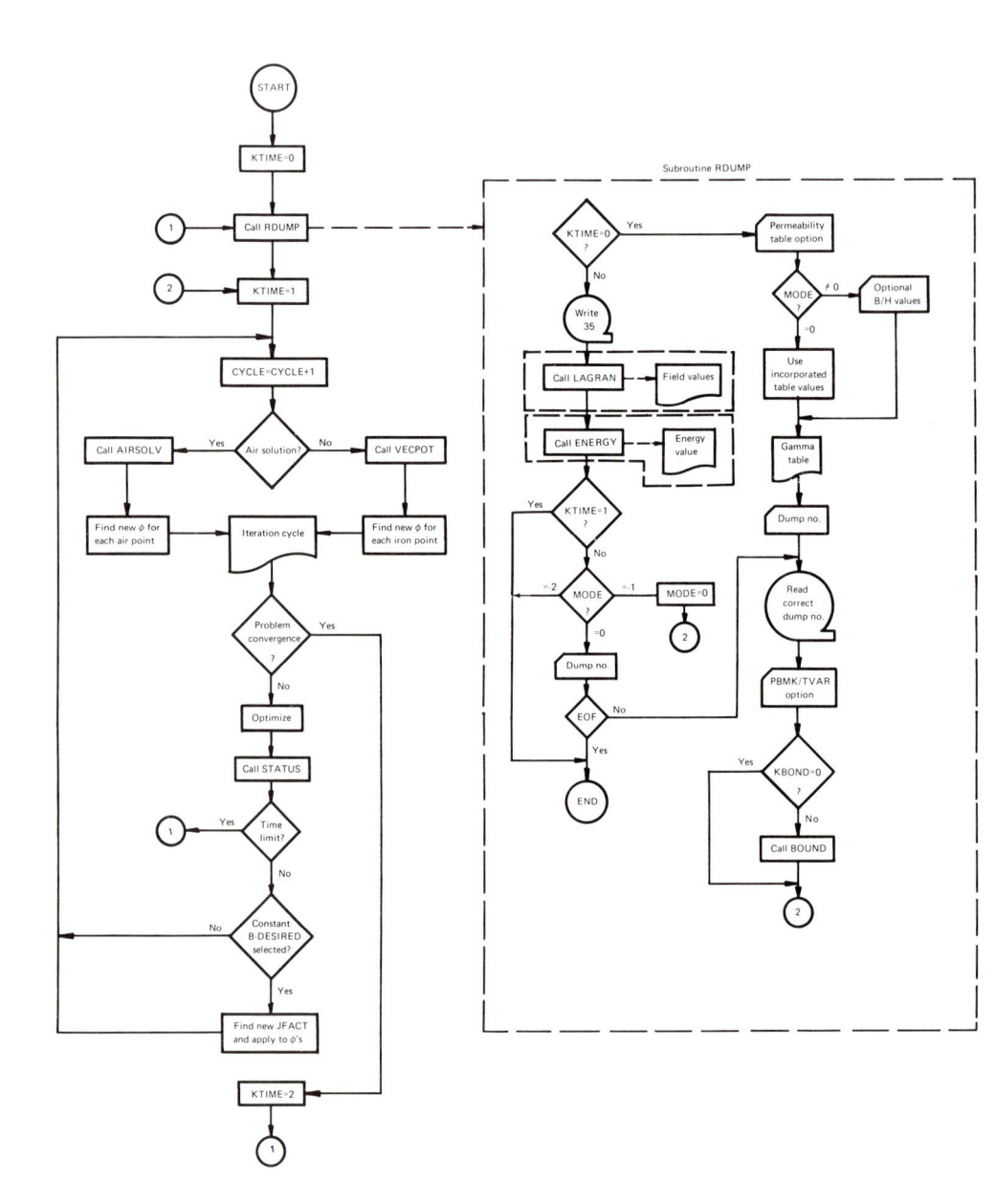

FIG. 11. Flow chart for program FIELD.

The "type of tables" entry allows the user to specify the form of the B versus H table as follows:

If TTABLE = 0. The B versus H table is entered in terms of B^2 (gauss^2) and γ ($\gamma = 1/\mu$) form.

If TTABLE = 1. The B versus H is entered as B (gauss), and H (oersteds) form.

The actual values of B, H, B^2 or γ are entered free format (see Table 3).

2. *Enter Magnetic Tape Block Number.*

This number refers to the number of successive blocks of information (dumps) that have already been written on the magnetic tape prepared by program MESH. If the magnetic tape has just been generated by program MESH and not yet used as input to FIELD, there will be only one block on this tape, namely the block that was written by MESH. This constitutes block number 1. If, however, one or more solutions have already been obtained (each solution corresponds to a separate block), the initial FIELD block number, will be higher, possibly as high as 5 or 10. The only limit to the number of blocks that a magnetic tape will hold is its length. A separate card in integer form will be necessary for this block number.

3. *Enter Problem Constant Variations.*

The last information required by program FIELD is the set of desired changes (or absence thereof) to the set of problem constants specified during the execution of program MESH.

While in MESH these constants were called PRBCON's, they are now called PBMK's and TVAR's. Table 2 gives a complete list of these additional constants.

Suppose we have generated a mesh for a magnet using MESH, and with coil excitation of 20,000 Ampere-turns (At). We would now like a solution at this excitation as well as at 40,000 and 60,000 At. The entries in Table 3 have been progressively modified to accommodate this data for the successive solutions by program FIELD. The bracketed insertions cite appropriate comments.

While the dump cards may be self-explanatory in Table 3 the PBMK and TVAR cards may be more difficult to understand. The important thing to remember is that the code will first try to read values for PBMK and then, when this array is satisfied (when fifty numbers have been read), it will read values for the TVAR array (also fifty numbers). If that many changes are not required for either array, one uses an S after the last entry to skip the remaining values in the array. For example, in Table 3, when Dump 3 is picked up we change the

Table 3

FIELD Sample Input Data[a]

	CARD COLUMN	
1	20	72

1	$ ONE PERMEABILITY TABLE WILL BE READ FROM CARDS $
1	$ TABLES ENTERED IN B VS. H FORM $

0.0   3.0E2   6.0E2   8.0E2   1.1E3   1.55E3   2.1E3   2.725E3   3.4E3   4.05E3
4.675E3   5.275E3   5.8E3   6.35E3   6.9E3   7.4E3   7.9E3   8.35E3   8.75E3
9.1E3   9.4E3   9.7E3   9.95E3   1.0175E4   1.035E4   1.06E4   1.075E4
9.1E3   9.4E3   9.7E3   9.95E3   1.0175E4   1.035E4   1.06E4   1.075E4
1.095E4   1.115E4   1.225E4   1.31E4   1.375E4   1.42E4   1.4525E4   1.4775E4
1.5E4   1.515E4   1.53E4   1.5425E4   1.555E4   1.5625E4   1.5725E4   1.58E4
1.585E4   1.5925E4   1.6E4   1.605E4   1.6125E4   1.6175E4   1.6225E4
1.6275E4   1.6425E4   1.6525E4   1.695E4   1.73E4   1.7625E4   1.787E4   1.81E4
1.83E4   1.85E4   1.885E4   1.9175E4   1.9475E4   1.97E4   1.995E4
C $ END B VALUES $

0.0   2.0E-1   4.0E-1   5.0E-1   6.0E-1   7.0E-1   8.0E-1   9.0E-1   1.0   1.1
1.2   1.3   1.4   1.5   1.6   1.7   1.8   1.9   2.0   2.1   2.2   2.3   2.4   2.5
2.6   2.7   2.8   2.9   3.0   4.0   5.0   6.0   7.0   8.0   9.0   10.0   11.0   12.0
13.0   14.0   15.0   16.0   17.0   18.0   19.0   20.0   21.0   22.0   23.0   24.0
25.0   28.0   30.0   40.0   50.0   60.0   70.0   80.0   90.0   100.0   120.0
140.0   160.0   180.0   200.0
C $ END H VALUES $

1	$ READ DUMP1, GENERATED BY MESH, FROM TAPE35 $
SS	$ SKIP PBMK AND TVAR ARRAYS OF DATA $

> At this point, the code will produce a solution at the first excitation level (20 000 AT), will store the solution on magnetic tape (TAPE35), and will proceed to read the next data card

2	$ READ DUMP2, FROM PREVIOUS SOLUTION $
*20+2.0	$ ENTER PBMK20, JFACT=2 $
SS	$ END PBMK AND TVAR ENTRIES $

> Now a solution with two times the original excitation level of (20 000 AT), or 20 000 × JFACT is written on TAPE35 as Dump 3.

3	$ READ DUMP3 $
*20+3.0	$ ENTER PBMK20, JFACT=3 $
S	$ END PBMK ENTRIES $
*24+5.0	$ ENTER TVAR24, NEW UPPER LIMIT OF EDIT=5 $
S	$ END TVAR ENTRIES $

> Again, a solution now Dump 4, with three times the original excitation or 20 000 × JFACT is written on TAPE35.

$ A NUMERIC ZERO ENTRY TERMINATES DATA $

---

[a] All entries are free-format. Data routine ignores comments between dollar signs ($).

20th array location of PBMK—that is, PBMK(20) = 3.0—and then terminated the rest of the PBMK array with an s. Next we changed the 24th array location of TVAR—that is, TVAR(24) = 5.0—and the s again means skip TVAR(25) to TVAR(50).

One final note regarding dumps—FIELD writes a dump only under two conditions: first, every time the problem has converged to a solution, and secondly, whenever the time estimate for the job is exhausted. Thus, one will never lose computer time due to a misjudgment of how long a particular solution will take to converge.

A sample of FIELD output begins by listing the permeability table(s) used, followed by a list containing the problem name and the operating parameters in use during this run (see Fig. 12). The total positive current, negative current, and current sum are also given, as well as the number of the dump from which execution is proceeding.

Next, the iteration printout follows the problem solution, at intervals specified by the user: every 20 iterations is standard, but may be changed at will, as may any other constant specified in Table 2.

The quantities shown have the following meaning:

a. CYCLE—Iteration counter
b. AMIN—Minimum value of vector potential (A) in mesh at cycle n
c. AMAX—Maximum value of vector potential (A) in mesh at cycle n
d. DAMAX(L, K)—The absolute value and mesh location of the maximum change that has occured in the vector potential on the last cycle
e. RESIDUAL/LENGTH—The convergence criterion; that is,

$$e = \left[\sum_i (A_i^{n+1} - A_i^n)^2 / \sum_i (A_i^{n+1})^2 \right]^{(1/2)}, \qquad (2.2)$$

where the summation is extended over all points

f. RHOAIR—The theoretically optimum value of the over-relaxation factor used for air points
g. JFACT—Current multiplier; this feature allows the user to vary the current from case to case
h. CPTIME—The amount of central processor time used thus far.

If the problem has converged, the program prints the energy stored in air, iron, total energy, and the maximum field in the magnet. An edit, as specified in PRBCON(35) through PRBCON(38), is then taken. In both cases (converged and nonconverged) the edit gives the quantities shown in Table 4 for each point in the edit region.

I. Two-Dimensional Magnetostatic Programs

```
            PERMEABILITY TABLE ENTRIES --- 1 TABLES

                  (B),FLUX DENSITY    (H),OERSTEDS       (B),SQUARED         GAMMA
TABLE 1    41 ENTRIES
                       0.                 0.                 0.             2.250000E-04
                   1.200000E+04       3.696000E+00       1.440000E+08       3.080000E-04
                   1.400000E+04       6.300000E+00       1.960000E+08       4.500000E-04
                   1.500000E+04       1.000500E+01       2.250000E+08       6.670000E-04
                   1.550000E+04       1.449250E+01       2.402500E+08       9.350000E-04
                   1.600000E+04       2.256000E+01       2.560000E+08       1.410000E-03
                       .                  .                  .                  .
                       .                  .                  .                  .
                       .                  .                  .                  .
                   9.272109E+04       7.132391E+04       8.597200E+09       7.692307E-01
                   1.069888E+05       8.559103E+04       1.144660E+10       8.000000E-01
                   1.283889E+05       1.069907E+05       1.648370E+10       8.333330E-01
                   2.353899E+05       2.139906E+05       5.540840E+10       9.090900E-01
                   3.162278E+49       3.162278E+49       1.000000E+99       1.000000E+00


THE STANDARD BUILT-IN TABLE ABOVE,RESEMBLES THE DATA,PAGE 211,USS ENGINEERING MANUAL,TITLED,NON-ORIENTED ELECTRICAL STEEL SHEETS.

HOT ROLLED CARBON STEEL PLATES
GRADE 1010-OVER .250 INCHES
DC MAGNETIZATION, AFTER ANNEALING AT 1500F.


TRIM PROBLEM CERN-PS

CYCLE     AMIN           AMAX           DAMAX(L,K)            RESIDUAL/LENGTH    RHOAIR        JFACT         CPTIME
   0     0.             0.             0.        (  0,  0)   1.00000E+00       1.94000E+00   9.98700E-01       51.788
 100    -1.550592E+05   2.85247E+04    1.40005E+03( 25, 12)   8.70864E-03       1.95393E+00   9.98700E-01      169.430
 200    -3.112795E+05   1.33706E+04    1.35331E+03(  1,  9)   4.74492E-03       1.97253E+00   9.98700E-01      287.282
 300    -3.846672E+05   6.67343E+03    6.34919E+02(  1,  8)   1.75427E-03       1.97483E+00   9.98700E-01      405.247
 400    -4.15047E+05    4.03348F+03    3.06329E+02(  1,  8)   6.83403E-04       1.97551E+00   9.98700E-01      523.095
 500    -4.26315E+05    3.06834E+03    1.16575E+02(  1,  7)   2.35318E-04       1.97465E+00   9.98700E-01      640.978
 600    -4.30259E+05    2.75157E+03    3.51686E+01(  1,  9)   7.62961E-05       1.97365E+00   9.98700E-01      758.366
 700    -4.34245E+05    2.65082E+03    1.43850E+01( 33, 42)   2.86051E-05       1.97277E+00   1.00466E+00      875.755
 800    -4.33146E+05    2.59861E+03    4.26739E+00( 27, 12)   8.96683E-06       1.97962E+00   1.00093E+00      993.831
 900    -4.32874E+05    2.57385E+03    1.89425E+00(  1, 11)   4.33647E-06       1.98001E+00   9.99575E-01     1111.785
1000    -4.32679E+05    2.56533E+03    5.95894E-01(  1, 10)   1.21958E-06       1.97625E+00   9.98895E-01     1229.725

          ----- CONVERGED IN 1010 CYCLES  --  DUMP NO. 2  ---  ELAPSED TIME  19.63 MINUTES


TRIM PROBLEM CERN-PS
CONVERGENCE CRITERION =   1.00E-06
STACKING FACTOR =   .9500
RHO IRON =    1.0000
RHO CP AIR =  1.0000
RHO CP IRON =  .0313
B-DESIRED =   1.00000E+04
SYMMETRY =    LINEAR MODEL
NORMAL DERIVATIVE CONDITION ON LOWER BOUNDARY

 + CURRENT =     40195.94
 - CURRENT =    -40195.94
     TOTAL =     7.683E-09

       ENERGY
       IN AIR  =   9.29547E+03
       IN IRON =   6.69768E+01
       TOTAL   =   9.36244E+03

       B-MAX   =   1.97201E+04


TRIM PROBLEM CERN-PS
CYCLE NUMBER IS  1010
DUMP NUMBER IS   2
LMAX IS  43
KMAX IS  82
MODE IS   0
JFACT IS  .9987
KBZERO IS 42

HARMONIC POLYNOMIAL FIT *** CARTESIAN COORDINATES

L   K    X(CM)    Y(CM)       A(G-CM)        BX(GAUSS)   BY(GAUSS)   BXX(G/CM)   BXY(G/CM)    BABS(GAUSS)
                             FITERR(O/O)                             BYY(G/CM)   K(1/CM)

1   1    0.       0.         0.              0.          1.17329E+04  0.          0.           1.17329E+04
                             7.28707E-02                              -0.
1   2    3.000    0.        -3.51989E+04     0.          1.17337E+04  0.          5.05563E-01  1.17337E+04
                             4.30403E-04                              -0.         5.05558E-05
1   3    6.000    0.        -7.04022E+04     0.          1.17356E+04  0.          8.22589E-01  1.17356E+04
                             3.72527E-04                              -0.         8.22580E-05
1   4    9.000    0.        -1.05612E+05     0.          1.17379E+04  0.          7.81747E-01  1.17379E+04
                             2.00191E-04                              -0.         7.81738E-05
1   5   12.000    0.        -1.40829E+05     0.          1.17394E+04  0.          1.76499E-01  1.17394E+04
                             6.92265E-05                              -0.         1.76498E-05
1   6   15.000    0.        -1.76048E+05     0.          1.17382E+04  0.         -1.19880E+00  1.17382E+04
                             4.33179E-04                              -0.        -1.19879E-04
1   7   18.000    0.        -2.11257E+05     0.          1.17319E+04  0.         -3.56224E+00  1.17319E+04
                             8.94220E-04                              -0.        -3.56220E-04
1   8   21.000    0.        -2.46437E+05     0.          1.17176E+04  0.         -7.13670E+00  1.17176E+04
                             1.45685E-03                              -0.        -7.13662E-04
1   9   24.000    0.        -2.81559E+05     0.          1.16916E+04  0.         -1.21350E+01  1.16916E+04
                             2.13391E-03                              -0.        -1.21349E-03
1  10   27.000    0.        -3.16581E+05     0.          1.16498E+04  0.         -1.87531E+01  1.16498E+04
                             2.98175E-03                              -0.        -1.87529E-03

END PRINT
```

FIG. 12. Partial output for the CERN-PS magnet derived from program FIELD.

Table 4

Program FIELD Output Column Description

Cylindrical coordinates	Rectangular coordinates	Heading description	Unit
L	L	L-mesh coordinate	–
K	K	K-mesh coordinate	–
Z	X	X-coordinate	Specified by input
R	Y	Y-coordinate	Specified by input
RA	A	Vector potential at (L,K)	gauss-centimeter
BZ	BX	Z-component or X-component of field	gauss
BR	BY	R-component or Y-component of field	gauss
BZZ	BXX	$\partial B_Z/\partial Z$ or $\partial B_X/\partial X$	gauss/centimeter
BZR	BXY	$\partial B_Z/\partial R$ or $\partial B_X/\partial X$	gauss/centimeter
BRR	BYY	$\partial B_R/\partial R$ or $\partial B_Y/\partial Y$	gauss/centimeter
BABS	BABS	Magnitude of the field	gauss
k	k	$k = (1/B_0)(d|B|/dX)$ or $(1/B_0)(d|B|/dR)$	centimeters^{-1}
%	%	Fitting error	percent

f. Program TRIP

The ability of TRIM to calculate magnetic fields for two-dimensional magnets of arbitrary geometry is greatly enhanced by the use of auxiliary programs such as TRIP, ZIG, and TRED. TRIP (see Colonias [10]) allows the user to produce CalComp plots of the generated mesh and of the resulting flux distribution. These plots provide a visual means of evaluating the quality (the minimum number of distorted triangles) of the generated mesh, as well as the quality of the magnetic field produced.

Data necessary for this program are:

1. YMIN—Minimum value of y coordinate (usually zero)
2. YMAX—Maximum value of y coordinate
3. XMIN—Minimum value of x coordinate (usually zero)
4. XMAX—Maximum value of x coordinate
5. SCALE FACTOR—Usually set to (YMAX-YMIN), although a larger value may be used; for example, if the universe = 20.0 inches and SCALE FACTOR = 40.0 inches, the plot will only use half the available space in the y direction; the plot uses 10.0 inches full scale so that SCALE FACTOR may be used to plot the universe, in some exact ratio

6. TSENT—Triangle sentinel: TSENT = 1.0, plots triangles TSENT = 0.0, does not plot triangles

7. NPHI—Number of flux lines to be plotted: if NPHI = −1.0, the next card to be read contains information pertaining to the constant A lines to be plotted. For example, if the card contained 50.0, 100.0, 120.0 C then the line $A = 50.0, 100.0, 120.0$ would have been plotted. (The letter C is peculiar to the data-reading routine and is required.) If NPHI = −2.0, the next card to be read contains information pertaining to the starting value of A desired and the increment to be stepped. For example, if the card contains 50.0, 10.0 C then beginning with $A = 50.0$ and at intervals of 10.0 all flux lines will be plotted until AMAX have been reached. If NPHI = −3.0, the next card to be read contains information pertaining to the starting value of A, the ending value of A, and the desired interval. For example, if the card contains 200.0, 400.0, 100.0 C then three flux lines will be plotted at intervals of 100.0 beginning at $A = 200.0$ and ending at $A = 400.0$

8. DMPNUM—Dump number from which plotting is to be done. The example of the three cards which follow show the useage of these constants:

```
0.0   147.32   0.0   381.0   147.32   1.0   0.0    1.0
0.0   147.32   0.0   381.0   147.32   0.0   50.0   2.0
S
```

The first card describes the dimension of the universe in the units used originally (0.0, 147.32, 0.0, 381.0), followed by SCALE FACTOR = YMAX − YMIN = 147.32; next is the flag 1.0 indicating that the triangular mesh plot is required, without flux lines (0.0), and that this plot should be done from the first dump of Tape 35. The second card contains the same dimensional and scaling information, but it requires a flux plot of 50.0 lines from dump number 2.0. The

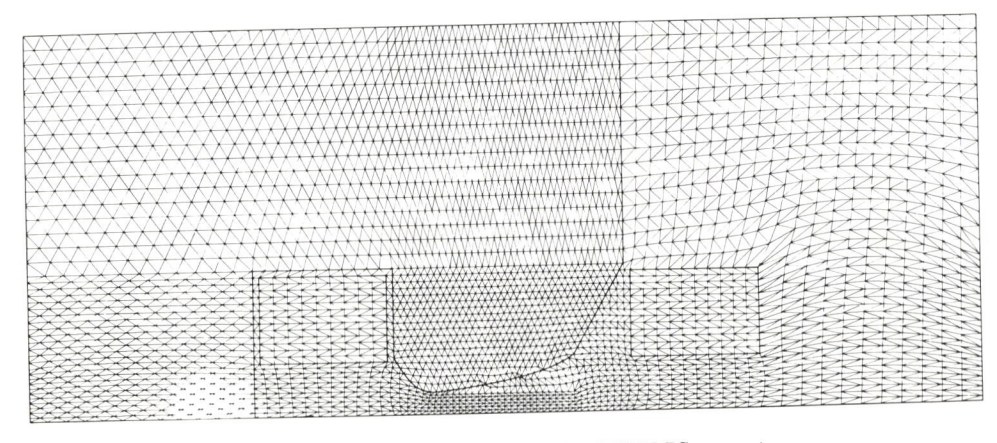

FIG. 13. Generated mesh for the CERN-PS magnet.

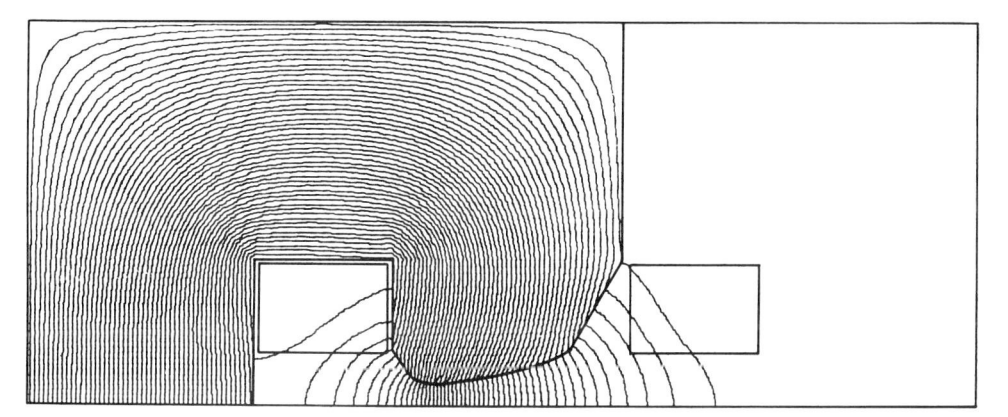

FIG. 14. Flux distribution for the CERN-PS magnet.

third card contains an s which is a sentinel indicating the end of data. These cards have produced the CalComp plots for the CERN-PS magnet shown in Figs. 13 and 14.

g. Program ZIG

This program [11, 12] allows the user to observe on a cathode ray tube the mesh as it is generated by program MESH, and to introduce changes on the mesh interactively with a light-pen. This program is very useful for correcting a mesh to insure that minimum distortion exists, or for altering the dimensions of the magnet being calculated. Input to this program is the output tape (Tape 35) generated by MESH. Figure 15 shows the flux distribution of an H-type window frame magnet on the CDC 252 model display console.

h. Program TRED

This program is provided to allow the user a flexible editing option separate from the built-in edit in FIELD. It is useful when a problem has converged to an acceptable solution and an edit is desired along more than just one row. The edit is obtained in the same manner as in FIELD; that is, a second degree harmonic surface is fitted to all nearest neighbors and its derivatives are evaluated. The output quantities are identical.

The following input is required to run TRED:

1. Dump number
2. Lower limit of edit (L1)
3. Upper limit of edit (L2)

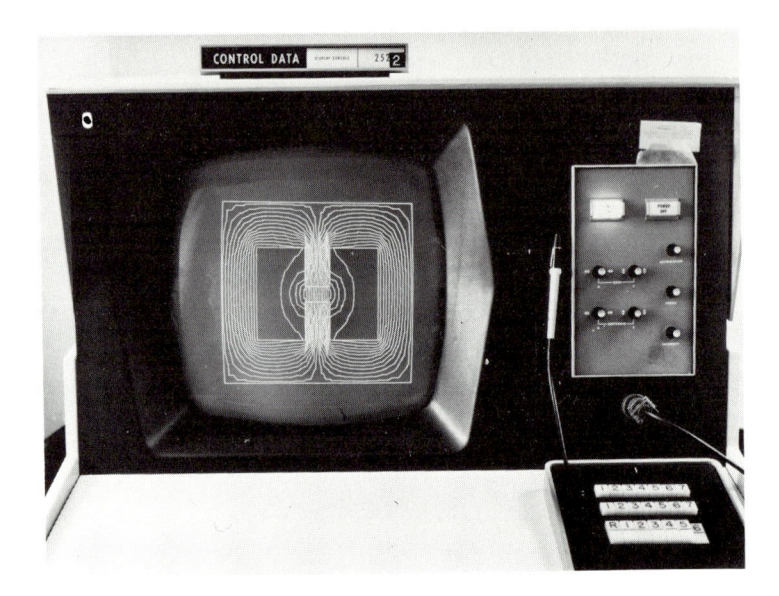

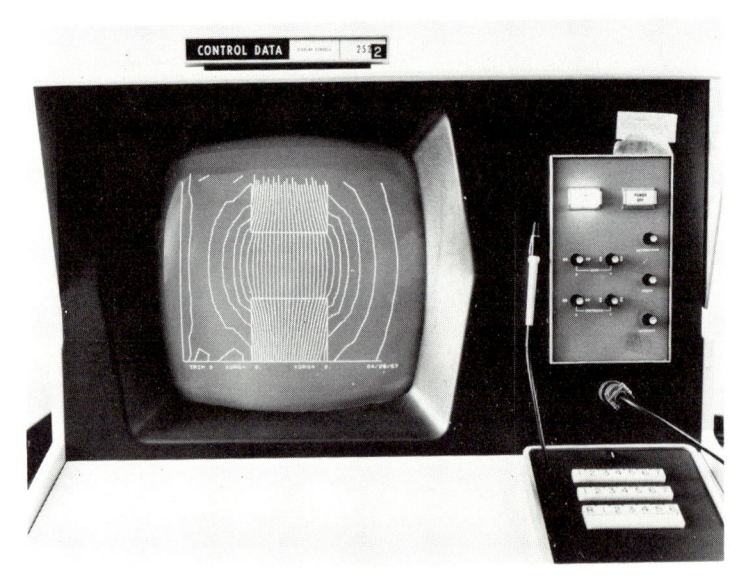

FIG. 15. Flux distribution generated on a CRT screen with the on-line version of TRIM. The lower picture shows the "zoomed" section of the pole tip.

4. Left limit of edit (K1)
5. Right limit of edit (K2)
6. Repeat the above steps for each region to be edited
7. A card with the dump number ≤ 0.

All data are integer numbers and are recorded on cards format free. The dump number must be on a card by itself, but the four edit limits may all be recorded on the same card. The output from TRED is identical to that shown in the lower part of Fig. 12.

3. Program LINDA

a. Introduction

LINDA [13] uses a combination of scalar and vector potential to model two-dimensional magnetostatic problems which have a simple general interface of iron of finite nonuniform permeability. The computer model consists of a rectangular problem space or universe with a single region of iron and one or more regions of air. There may be current conductors inside or outside the problem space, but currents in iron are not allowed. The problem space is divided into a mesh composed of uniform rectangles. The overall size of the mesh depends on the size of computer memory available. For a typical 2500-point mesh, $127K_8$ locations are required.

Perhaps the most outstanding characteristic of this code is its accuracy, allowing experimenters to perform perturbations on pole face profiles with very fine detail. The uniform mesh used by LINDA, in contrast to the nonuniform mesh of TRIM, could be considered a disadvantage since it does not offer variable point density as does TRIM. However, residual errors, presently existing in TRIM, are greatly reduced in LINDA.

LINDA is written entirely in Fortran and is operational on the CDC 6600/7600 system as well as on the IBM 360/75 and other IBM computers. Normally, the code does not use overlays; however, Brown [14] and Lin [15] have used LINDA with overlays successfully.

b. Program Description

LINDA uses the two-potential method (scalar and vector potential) used by Christian [16] and also used in SIBYL [17] and MAREC [18–20]. The total problem space is divided into two parts: the iron region, and the air and conductor region. The working potential in air is a modified scalar potential with automatic current cuts; the working potential in iron is the single component of the vector potential. These two regions communicate with each other through values of vector and scalar potentials on the interface. A complete solution requires cycling: solving alternatively the air and iron regions until the potentials on the interface approach final values as closely as may be desired.

The two-potential method, where an initial set of scalar potentials on the interface is changed once per cycle, is not applicable to all problems of magnet design, for example, a closed ring with circulating flux. However, some have been successfully modeled with an artificial air gap containing large extra currents. The degree of difficulty encountered by LINDA in solving magnet problems is a function partly of the level of flux density desired, but mostly of the relationship between the scalar potentials on the interface at the start and the end of a cycle. Small changes in local flux densities may make very large changes in H and in $\int H \, dl$ in the iron. Large portions of the interface of "difficult" magnets may produce changes of 20 percent in scalar potential from a one percent change in local flux. Since a change in flux at any point on the interface changes the scalar potential by some amount at all points on the interface, the overall convergence of all scalar potentials toward final values is truly unpredictable. LINDA will sense and make appropriate changes in the internal parameters (especially in the boundary mixing process) and will converge eventually for all but some extremely difficult magnet models—such as those having salient square corners, small air gaps and narrow return paths.

LINDA is a complicated program with many subroutines, and even though the preparation of data for a particular magnet is straightforward, the determination of errors in cases of an unsuccessful run becomes very difficult.

The program is divided into six major sections:

1. Setup and control—translating input (interface and current) from dimensional coordinates to mesh coordinates, testing of the interface, calculating the current cuts, preparing tables for use in subsequent problem-solving operations, and doing the necessary housekeeping.

2. Solving for the scalar potential in air.

3. Integrating the field in air to produce values of vector potential at every point on the interface.

4. Solving for the distribution of vector potential in iron to achieve $\int H \, dl \approx 0$ on every closed path in the iron, while changing the reluctivity in every mesh rectangle (with iron) to match the properties of the iron.

5. Integrating $\int H \, dl$ within the iron to produce values of scalar potential at every point on the interface.

6. Selecting values of scalar potential on the interface for the mesh cycle.

The flow chart (see Fig. 16) portrays in a general way the cycling process used in LINDA and Table 5 describes briefly the main subroutines used by this program.

3. Program LINDA 41

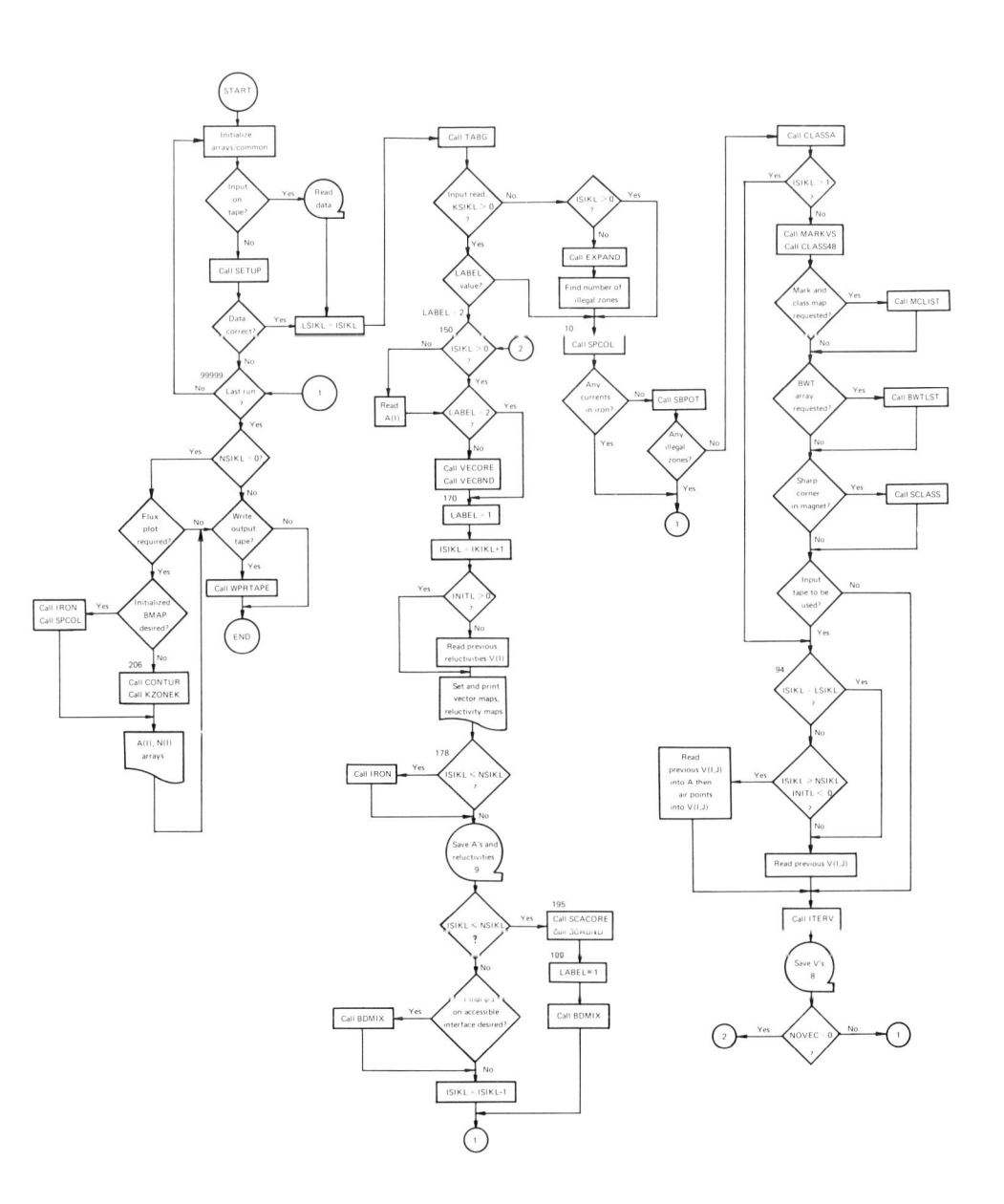

FIG. 16. LINDA flowchart.

Table 5

LINDA Subroutines and Functions

Name	Description
MAIN	Controls program and performs housekeeping functions
SETUP	Reads input data
ERROR	Prints error diagnostics
ONINT	Tests dimensions
EXPAND	Calculates closed boundary points and intersection mesh points
LEGAL	Tests legality of input boundary points
SPCOL	Calculates current distribution in coil section of magnet
SBPOT	Changes potentials to include sum of all current cuts
CLASSA	Determines class of each point, distance to each of the surrounding boundaries if any, makes an entry in the BWT table if necessary, and puts the mark and class of each point into the corresponding CV word
DIST	Computes the distance to the nearest boundary from any point P
MARK	Determines whether a point P is in iron, air, or on the interface between iron and air
CBWT	Sets up the BWT table that consists of the potentials and distances, in mesh units, to first boundary
IGBWT	Searches BWT table for errors
MARKUS	Classifies boundary points as to their location in relation to their neighbors
SCLASS	Finds special intersections in air which are near salient corners of the interface and calculates special factors
ITERV	Computes scalar potentials in air
EDIT, TRANS, XRAY, NTPO	Edit routines. Prints desired quantities (fields, gradients etc.)
VLIST	Prints V array
CVLIST	Prints CV array
MCLIST	Prints MARK and CLASS array
BWLIST	Prints BWT table
ALIST	Prints map of vector potentials throughout mesh
DATA, DATAI, RDNUM, RDBYTF	Control Data 6600/7600 system data-reading routines
LSTSQ	Least-square routine

3. Program LINDA

c. Program Restrictions

The complicated nature of LINDA necessitates the proper definition of the contemplated magnet geometry and restricts the use of the program to geometries which meet the following requirements:

1. *One region of iron.*

Separate pieces of iron are not allowed. The flux-normal borders have a scalar potential of zero, so these borders commonly provide a reference scalar equipotential for the scalar potentials in iron.

2. *Region of air.*

Separate regions of air are allowed for any model, but for most magnet types (MTYPE) only the fields in one air region contribute to the calculation of vector potential through the interface. The flux in two air regions is calculated and used when both air regions touch the same no-flux border at left or right and that mesh border contains the fiducial point for the vector potential. This can be done with MTYPE = 4, 8, 11. If the vector potential on the interface has not been computed at all points by VECBND or SETABD, the program will print a message, and will not try to reach a solution in iron.

3. *Fine structure restriction.*

This basic restriction applies to the interface for all problems, whether the evaluation includes the effects of finite iron or not. In any mesh rectangle zone there may be only one interface segment which is not on a mesh line. The interface may divide a zone into only two pieces, one in iron and one in air. LINDA tests for this restriction with FUNCTION LEGAL and will stop without computation if the interface is illegal.

4. *Types of magnets.*

As a vehicle by which magnet geometries can be simulated with the best possible definition, LINDA provides an array of topologically defined types of magnets which provide the necessary information about the location of the interface on mesh borders.

The program communicates this information through a variable called MTYPE which is defined in the XCTL(126) array position. It is specified by the user. (Topologically valid geometries have been programmed into LINDA and are described below.) In the figures that follow, the dotted line represents a flux normal boundary, a solid line a no-flux boundary, and the dashed line may be either.

    a. MTYPE = 0. Magnet iron does not touch any mesh border.

I. Two-Dimensional Magnetostatic Programs

b. MTYPE = 1 and MTYPE = 7. The iron portion of the magnet touches the lower flux-normal border and touches the top, side, or both no-flux borders (see Fig. 17). In these topologies the vector fiducial (the starting point for integrating scalar fields to get vector potentials) must be somewhere in the accessible air region; usually this is in the right air region when MTYPE = 1, and in the left air region when MTYPE = 7. All the flux between the ends of the accessible part of the interface is assumed to return across the lower flux-normal border in iron.

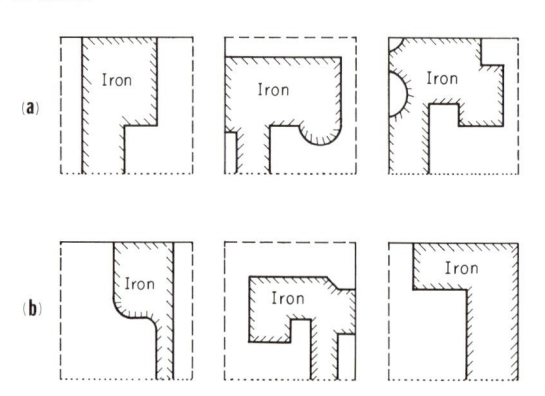

FIG. 17. General magnet outlines for (a) MTYPE = 1 and (b) MTYPE = 7. The solid line indicates zero flux boundary. The dotted line indicates flux normal boundary and the dashed line indicates equipotential surface.

c. MTYPE = 2 and MTYPE = 3. When MTYPE = 2, the two iron portions of the magnet touch the lower flux-normal border; when MTYPE = 3, one leg touches a left flux-normal border and the other touches a lower flux-normal border (see Fig. 18). In these topologies there is one accessible air region between the iron legs, and the vector fiducial must be in that region. The flux in the legs must be balanced during solving in iron by using the results of SUMDHL. Again, it is assumed that all flux entering the accessible interface returns through the iron legs across the flux-normal border(s).

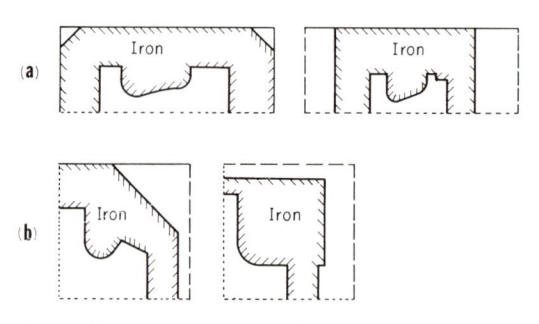

FIG. 18. General magnet outlines for (a) MTYPE = 2 and (b) MTYPE = 3.

d. MTYPE = 4 and MTYPE = 8. When MTYPE = 4, the iron portion touches a horizontal flux-normal border and a no-flux border at the right (see Fig. 19). The iron may also touch other no-flux borders. With MTYPE = 8, the iron portion touches a horizontal flux-normal border and a no-flux border at the left. The iron may touch other no-flux borders. These topologies were coded with the assumption that no flux escapes the iron into an outer air region. If the vector fiducial is on the appropriate vertical no-flux border, and if an outer air region also includes a part of that mesh border, then the flux in the outer (inaccessible) air will be calculated and used. Specifying the location of the vector fiducial in any other part of the accessible air region will prevent the calculation and use of flux in an outer air region, even if there is such a region in the problem space. If, on the other hand, the fiducial is specified in an inaccessible air region (usually the upper region), then the code will not proceed into the iron calculation portion of the program.

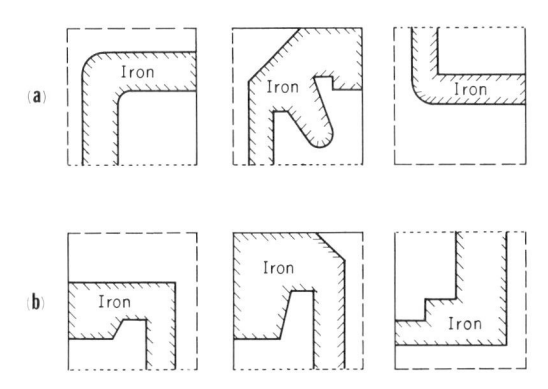

FIG. 19. General magnet outlines for (a) MTYPE = 4 and (b) MTYPE = 8.

e. MTYPE = 9. Here, the iron touches only flux-normal borders, and may touch any one or all such borders in one continuous line (see Fig. 20). There is only one air region. This topology can be used to replace part of the return-path iron by infinitely permeable iron.

f. MTYPE = 10. The iron touches only one no-flux border at the left or right (see Fig. 21). As in MTYPE = 0 models, the scalar fiducial must be input. It may be at any interior iron point, including mesh points on the no-flux border, provided the adjacent interior iron mesh point is not near a boundary point.

g. MTYPE = 11. This topology is a combination of MTYPE = 9 and MTYPE = 10 types (see Fig. 22). The iron touches only one vertical no-flux border, at left or right, and has a continuous interface on one or more flux-normal borders. When there are two or more separated air regions, the vector fiducial must be in air on the vertical no-flux border, and each air region must

46   I. Two-Dimensional Magnetostatic Programs

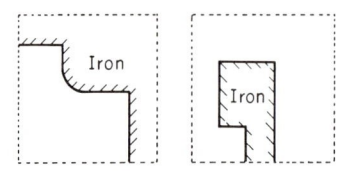

FIG. 20. General magnet outline for MTYPE = 9.

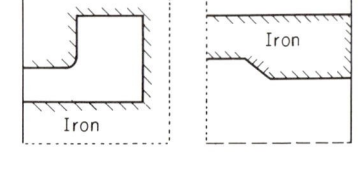

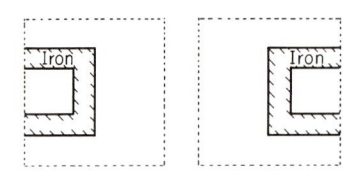

FIG. 21. General magnet outline for MTYPE = 10.

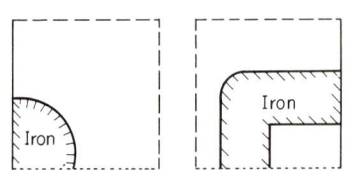

FIG. 22. General magnet outline for MTYPE = 11.

touch that border. With only one air region, the vector fiducial may be anywhere in the air of that region.

The location of the fiducials (whether vector or scalar) is communicated to the program through the following CTL array locations:

1. CTL(27) = XVFID (x coordinate of fiducial)
2. CTL(28) = YVFID (y coordinate of fiducial)
3. CTL(29) = VVFID (value of fiducial).

For types of MTYPE = 0 and MTYPE = 10 it is sufficient to specify the x and y coordinates of a point residing in iron and to enter zero for CTL(29).

In general, it is not necessary to specify the vector fiducial since the program will "hunt" for a legal point and set it to that location. However, there is a possibility that during this hunt the program will locate the fiducial in an inaccessible region, in which case the code will come to an abrupt stop. Only experience gained in operating this program will permit one to avoid such events.

d. Preparation of Input Data

Input data for a magnet geometry solution requires a considerable amount of effort and attention. Therefore, the instructions and sample data should be studied carefully before an attempt is made to use this program. It is also suggested that the user perform simple runs until a certain degree of familiarity with the code is achieved.

There are seven "banks" of data required for the simulation of a magnet, which include data pertaining to the geometry, coil configuration, boundary conditions, input/output options and various mandatory problem constants necessary for the execution of the program. Data may be stacked to process multiple runs. A description of these seven data banks follows:

1. Bank I. This is a one-integer flag for controlling the beginning option: if INTAPE $= 0$, there is no input tape; if INTAPE $\neq 0$, the code expects data from magnetic tape. This option is useful for continuation runs.

2. Bank II. The CTL (or XCTL) array of 200 locations has been reserved in memory for storing program constants. (A partial listing of XCTL array is shown in Table 6.). This array is initialized to zero at the beginning of each run, except when INTAPE $\neq 0$, in which case this array will contain values left from the previous case. When in doubt as to whether a CTL-value should be included, the motto "when in doubt, leave it out" should prevail.

3. Bank III. The ICTL bank of variables consists also of constants, which, even though listed in the CTL bank, have been renamed here. This inconsistency is only of historical importance and it should be ignored. The parameters of this bank are listed in Table 7.

4. Bank IV. The PRBCON array contains the mandatory program constants (some of which appear also in the CTL array). Table 8 lists these constants, their standard value and description.

5. Bank V. The arrays XB, YB, and VB contain the interface points. These points must form a closed curve, described in the order in which one would see them if he traversed the interface with the air always on his right. The last point must be the same as the first point and usually on a flux-normal border. The VB array must be inputed, usually 0.0.

6. Bank VI. The data contained in this bank called "CO Bank", pertains to the conductor geometry and excitation. LINDA allows the user to describe the conductors with either rectangular, parallelogram or polygonal coils. Each type of conductor is characterized by a slightly different set of input data as outlined below:

a. *Rectangular coils*

Rectangular coils require a group of five quantities per conductor:

$CO(1,K) = x$ coordinate of the left side of the Kth conductor
$CO(2,K) = x$ coordinate of the right side of the Kth conductor
$CO(3,K) = y$ coordinate of the bottom side of the Kth conductor
$CO(4,K) = y$ coordinate of the top side of the Kth conductor
$CO(5,K) = $ current in the Kth conductor (per unit current)

If $CO(5,K) \geq 1000.0$, then this entry is used as a flag to specify parallelogram coils.

Table 6
Partial List of XCTL Array

Array location	Variable	Description
1 - 4	XO,YO,XMAX,YMAX	Locations of problem boundaries in dimension units; same as PRBCON(1) through PRBCON(4), (see Table 8)
5 - 6	EL,H	Vertical and horizontal dimension unit per mesh interval
7	EPALP	Criterion for ALPHA recalculation
8	EPFIN	RMS convergence criterion
9 - 10	MCNT,MXCNT	Minimum and maximum number of iterations in air
11	ALPHA	Overrelaxation factor in air
12	CMSPER	Unit conversion factor (centimeter per dimension unit)
13	AMPTRN	Excitation factor (ampere-turns)
14	XBZERO	Location of edit center (dimension)
31	BLPHA	Relaxation factor for vector potentials in iron
32	BETAG	Weight factor for GAMMA recalculation
37	STAK	Stacking factor for laminated iron
38	GLIM	Amount by which GAMMA may change
46	CABDY	Weighting factor for a new ABDY from old values
47	CLIM	Factor that revises GLIM from iteration to iteration
58	ABDY	Value of vector potential on inaccessible interface
60[a]	BZERO	Desired B on midplane at (XBZERO,YO)
64	TOTNI	Total excitation
114	ILIST	Flag for extra printing
126[a]	MTYPE	Type of magnet geometry
139	ISIKL	Cycle number that program is on
140[a]	NSIKL	Number of the last cycle to be run
168	KABDY	Iteration number to begin $\int H\,dl$
178[a]	MODEI	If MODEI = 0, change currents[b]; if MODEI < 0, do not change currents[c]

[a] Variables most commonly used in each run.
[b] Current modified to produce the desired field set by KBZERO at XBZERO.
[c] Current specified by AMPTRN will be frozen to produce the corresponding solution at that excitation

3. Program LINDA   49

Table 7
ICTL List of Parameters

Array location	Name	Description
1	ID	Case number or run identification number
2	JANY	End sentinel: if = 0, no output tape written; if ≠ 0, write problem output tape
3	ILIST	Print-control flag, see Table 12
4	NOPRNT	Print-control suppression flag, see Table 13
5	KARDS	Card-output control flag, see Table 14
7	ICTL(7)	Sentinel: if ≠ 0, plot map of flux distribution; if = 0, flux map not printed; ICTL(7) used only when NSIKL = 0
9	ICTL(9)	Vector potential maps printed before or after any desired cycle
10	ICTL(10)	Reluctivity maps printed before or after any desired cycle. However, if ISIKL = 0 and INITL = 0 then, and only then, it prints a scalar-potential map. To choose a particular cycle to print reluctivities or vector-potential maps we proceed as follows: assume we want to print vector-potential map after cycle 4. Since this is before cycle 5, the 5th bit of ICTL(9) in binary code is 10000 which in turn is equivalent to 16 in decimal code; therefore ICTL(9) must be set to 16. In a similar manner a reluctivity map would be obtained with ICTL(10) set equal to 16

Table 8
PRBCON Parameters

Array location	Variable	Description	Standard value
1 - 4	X0,Y0,XMAX,YMAX	Locations of problem boundaries; normal input (inches)	(Specify)
5 - 8	IBDX0,JBDY0, IBDXF,JBDYF	Type of boundaries on the four sides of problem rectangle; enter zero for flux-normal boundary, +1 for no-flux boundary	(Specify)
9	EPALF	Criterion for ALPHA recalculation	0.005
10	EPFIN	Convergence criterion (rms)	10^{-7}
11	MCNT	Minimum number of iterations in air	-
12	MXCNT	Maximum number of iterations in air	300.0
13	ALPHA	Overrelaxation factor in air	1.5
14 - 16	Not used		
17	IMAX	Maximum value of I index (x direction)	(Specify)
18	JMAX	Maximum value of J index (y direction)	(Specify)
19	CMSPER	Conversion factor (centimeters per dimension unit)	2.54
20	AMPTRN	Excitation factor (ampere-turns)	1000.0
21	XBZERO	Location of edit center (dimension)	(Specify)
22	HSTAR	Desired mesh interval for extra edit (HINC = HSTAR)	-
23 - 24	H,EL	Horizontal and vertical dimension unit per mesh interval	-
25	XXORGP	Origin + 1000 for HINC = HSTAR edit	-

I. Two-Dimensional Magnetostatic Programs

b. *Parallelogram coils*

This type of conductor requires 10 entries per conductor all following:

$co(1,K)$ = x coordinate of the first point of the short side of the parallelogram (starting at any corner)
$co(2,K)$ = y coordinate of the point above
$co(3,K)$ = x coordinate of the 2nd point of the short side
$co(4,K)$ = y coordinate of the 2nd point of the short side
$co(5,K)$ = $\geq$ 1000 flag
$co(6,K)$ = x coordinate of adjacent long side
$co(7,K)$ = y coordinate of adjacent long side
$co(8,K)$ = subdivision of the short side (SUBS)
$co(9,K)$ = subdivision of the long side (SUBL)
$co(10,K)$ = total current per coil.

If SUBS and SUBL are zero, the code assumes 12 and 60 subdivisions respectively. In general the total area is subdivided into SUBS * SUBL equal areas. As an example, consider the coils of the CERN-PS magnet (See Table 9); if these coils were entered as parallelograms the data would appear as follows:

Coil No. 1] 37.0 8.0 37.0 21.0 1002.0   57.0 21.0 0.0 0.0 -1.0
Coil No. 2] 95.0 8.0 95.0 21.0 1002.0 115.0 21.0 0.0 0.0   1.0

c. *Polygonal coils*

If $co(5,K) \geq 2,000.0$ then this entry is used as a flag to indicate a polygonal coil. This type of conductor requires at least 10 entries in the co bank arranged five entries per card as following:

$\left.\begin{array}{l}co(1,K)\\co(2,K)\end{array}\right\}$ x and y coordinate of the first point of the polygon

$\left.\begin{array}{l}co(3,K)\\co(4,K)\end{array}\right\}$ x and y coordinate of the second point of the polygon

$co(5,K)$ current or current density if $co(5,K) \geq 7000$ then $co(5,K)$ contains the total current of the conductor as: current = 10000.0 $-$ $co(5,K)$ if $co(5,K) < 7000.0$ then $co(5,K)$ contains the current density of the conductor on: Current density = 5000.0 $-$ $co(5,K)$

$\left.\begin{array}{l}co(6,K)\\co(7,K)\end{array}\right\}$ x and y coordinate of the third point of the polygon

$\left.\begin{matrix} \text{CO}(8,\text{K}) \\ \text{CO}(9,\text{K}) \end{matrix}\right]$ x and y coordinate of the fourth point of the polygon

CO(10,K) number of input coordinate pairs.

This sequence is repeated until all polygon boundary points are exhausted, with every fifth entry (e.g., CO(15,K), CO(20,K), etc.) being entered as a dummy number. As an example, consider the coils of the CERN-PS magnet (see Table 9). If these coils were entered on polygons the data would appear as follows:

Coil No. 1 $\left.\begin{matrix} 37.0 & 21.0 & 57.0 & 21.0 & 9999.0 \\ 57.0 & 8.0 & 37.0 & 8.0 & 4.0 \end{matrix}\right]$

Coil No. 2 $\left.\begin{matrix} 95.0 & 21.0 & 115.0 & 21.0 & 10001.0 \\ 115.0 & 8.0 & 95.0 & 8.0 & 4.0. \end{matrix}\right]$

7. Bank VII. The GAMMA bank is the last bank of the required data that pertains to the particular characteristics of the magnetic material used in the problem solution. The user has the option to enter B and H instead of B^2 and γ ($\gamma = 1/\mu$) provided the first entries for B and H are both zero. For the benefit of the user, a 40-point table for decarbonized iron is included in the program; therefore, the user may skip reading this array by entering an s s s in the last card of input data. However, if one desired to enter his own permeability tables, they would be entered in three parts: first, enter B (gauss) or B^2 (gauss^2) values followed by an S; secondly, enter the number of B or B^2 points just entered, followed by an S; and thirdly, enter the H (oersteds) or γ values.

The description of the input data given thus far is sufficient to produce solutions for a variety of magnet geometries. Experience with the program will allow the designer a maximum use of its capabilities and the various options available. In the following section, we will apply the instructions given thus far to a specific illustrative example, so that the organization of the input data to a coherent input deck may become more obvious.

The necessary data shown in Table 9 will simulate the CERN-PS magnet with program LINDA.

The first card constitutes the title of the problem. Any number of cards may be used provided each line is enclosed in "$" signs. The second card (INTAPE = 0) signals the program to expect data from cards, since this is a new case. If INTAPE = + 1, then the code will expect this case to be a continuation run; it will read the specified magnetic tape and data from cards reflecting the changes required for this particular run. The letter s appearing at the end of each data bank signifies the end of data for that particular bank. (See program TRIM for further definition of s.)

52   I. Two-Dimensional Magnetostatic Programs

Table 9

Program LINDA Input Data for the CERN-PS Magnet

			Card Column				
1	10	20	30	40	50	60	70
$ CERN PROTON SYNCHROTRON MAGNET				$			
+0 S				$ INTAPE=0 $			
*37+0.95	*126+1	*140+4	S	$ XCTL ENTRIES	$		
*1+1000	*2+1	*3+2048	S	$ ICTL ENTRIES	$		
*1+0.0	0.0	150.0	58.0 1 0 1 1	$ PRBCON ENTRIES	$		
*19+1.0	40250.0	76.0	*23+1.0 1.0 S				
0.0 36.0 36.0 58.0 58.0 59.5 60.5 61.5 62.5 63.6 64.5 65.4 66.4 67.2							
68.0 69.0 70.0 71.0 71.8 72.2 73.0 74.0 75.0 76.0 76.8 77.4 77.8 78.8							
79.8 80.2 81.2 82.2 83.2 84.2 85.2 86.036 94.0 94.0 0.0 0.0 C $ END XB $							
0.0 0.0 22.0 22.0 8.526 6.2 4.94 4.06 3.536 3.298 3.368 3.455 3.568							
3.663 3.762 3.882 4.01 4.147 4.263 4.324 4.451 4.62 4.802 5.0 5.17 5.306							
5.4 5.651 5.927 6.045 6.361 6.713 7.105 7.547 8.046 8.517 22.0 58.0 58.0							
0.0 C			$ END YB	$			
+0.0R150 C			$ END VB	$			
37.0	57.0	8.0	21.0 -1.0	$ COIL 1	$		
95.0	115.0	8.0	21.0 +1.0 C	$ COIL 2	$		
S S S				$ USE STD. BUILT-IN PERM. TABLE		$	

The third card (CTL Bank II) contains the laminations stacking factor, the type of magnet and the number of cycles to run this problem.

The fourth card (ICTL Bank III) consists of the case number, a flag (*2 + 1) indicating that we desire to write the results on an output tape, and the last entry, which is the print control option (see Table 10).

The PRBCON array constants are shown in cards 5 and 6. The first four numbers in card 5 indicate the location of the problem boundaries, and the remaining numbers (1 0 1 1) simulate the required boundary conditions. Similarly the rest of the PRBCON array constants in the 6th card may be identified by making use of Table 8.

The next array of numbers contain the interface points. The first (XB) set, in cards 7 through 9, contains the x coordinate; the second (YB) set, in cards 10 through 13, contains the y coordinate. It is imperative that each set be termi-

Table 10

Print Control Options for ICTL(3)

ICTL(3) value	Subroutine	Description
1	MARKVS	Defines interface points (MARK and CLASS)
2	SCABND	Diagnostics describing least-squares fit used to compute gradient of scalar potential close to the boundary
4	CONTUR	CX, CY = tables of interface intersections on each mesh line. ICX(1,5) is the number of crossings or corners with YM = J, CX(2,J) = XM(K), CX(3,3) = AM(K), and ICX(4,J) = (K), etc.
8	KZONEK	Array of boundary point numbers; one for each zone that is not full of iron
16	VECBND	Vector potential A(I,J) in air at every cycle
16	CONTUR	Vector potential A(I,J) and initialization in iron (once)
16	IRON	Vector potential A(I,J) in iron and air at every cycle
32	IRON	Maps of fields and gradients when KOUNT = 1 (for all cycles)
64	SCABND	Scalar potential V(I,J) in iron every cycle
128	IRON	Reluctivity map every cycle
256	SBPOT	Scalar boundary with and without current cuts, when ISIKL = 0 or when VSFAC $\neq$ 0 on last cycle
512	SBPOT	Same as above but only on last cycle
1024	IRON	Vector potential A(I,J), IRON(NSIKL), AIR(NSIKL-1)
2048	SCABND	Scalar potential V(I,J) in iron
4096	VECBND	Vector potential, AIR(NSIKL), IRON(NSIKL)
8192	MAIN	MARK and CLASS map, when ISIKL = 0
16384	IRON	Relucitivity map (Gamma in iron) for last cycle

nated by the letter C, which flags the data-reading routine to count the number of entries. In case the *x*-array entries are not equal to the *y*-array entries, the program stops with an error diagnostic. The last entry in this bank of data, card 14, is the initial guess of the scalar potentials. In the absence of known values, the array is initialized to zero. The interpretation of this card (+ 0.0R150) is the following: zero indicates the value of the scalar potential, R indicates repeat, and the value 150 indicates the number of times to repeat the value of zero. Cards 15 and 16 contain the coordinates and currents respectively, in the coils. The interpretation of these numbers follows the description given previously in Bank VI, for the CO array. The last card in the deck contains three S entries, indicating to the code to skip reading the three sections of the GAMMA bank, since the internal table will be used.

There are two ways to specify coil excitation in LINDA; first, by entering XCTL(60), in which case the code will pay no attention to PRBCON(20) since it will calculate its own value of AMPTRN. (If both BZERO and AMPTRN are specified then the dominant parameter is BZERO.) The second way to specify current is by simply entering the proper excitation in AMPTRN, after first assuring that

54   I. Two-Dimensional Magnetostatic Programs

BZERO has been set to zero. Also, MODEI is used to ask whether we want to keep the same current in later cycles (iron cycles) so that the calculated field BZERO remains constant. In this case, a positive number in MODEI will continually change the current to maintain this condition. However, there are magnets in which the symmetry plane is not defined in the lower part of the universe, in which case BZERO cannot be specified. Therefore, we must enter AMPTRN and MODEI (a negative number) implying to the code to freeze the current specified by AMPTRN and proceed to calculate the resulting BZERO.

It should be emphasized that in preparing data the user must distinguish inetgers and floating-point number usage. The following rule must be adhered to: all input data, in any bank, whose array location number is greater than 100, for example, *126 + 1, must be entered as an integer value; also, all array location values less than 100, for example, *60 + 1.2, must be entered as a floating-point number.

e. Output Description

Output from LINDA is controlled by the ICTL(3), ICTL(4), and ICTL(5) flags described in Tables 10, 11, and 12. Figure 23 shows the resulting field distribution in the median plane at the end of the fourth cycle. The field (BY) and the gradient of the field (DB/DX) are printed along with other quantities at each mesh point (K). Figure 24 shows a partial listing of the magnitude of the magnetic field (B) (in kilogauss) in iron at every mesh point.

In addition to providing a tabulation of midplane gradients and fields, LINDA provides other useful information about the magnet, such as magnetic

Table 11

Print Suppression Options for ICTL(4)

ICTL(4) value	Subroutine responsible for suppression	Output suppressed
1	BDMIX	Boundary printout (VS,VM,DV,WT1 etc.) for all but the last cycle
2	BDMIX	Boundary printout for all cycles
4	ITERV	KCOUNT, ALPHA, SRMS, F, TIME, after each pass through air points
8	IRON	Gradient of vector potential in iron, all but the last cycle
16	IRON	Gradient of vector potential in iron, all cycles
32	IRON	CALCG parameters (NEEDB, BETA, GLIM, etc.) after each call
64	IRON	ITERA parameters (ITVEC, BLPHA, S, etc.) after each call
128	SUMHDL	$\int H\, dl$ on interior iron path; DELA, RABDY, each call
256	TABG	BSD, GAMMA table, B, H, DGMMA/DB, DH/DB
512	ITERV	Duplicate calls of edit, ISIKL = 0, and ISIKL = NSIKL
1024	VECBND	Vector potential on boundary, all but last cycle and when ISIKL = 0

3. Program LINDA   55

Table 12

Output Options for ICTL(5)

ICTL(5) value	Subroutine	Quantities punched on cards
1	EDIT	X(L), BY(L) on midplane, with H = X(L+1) - X(L)
2	EDIT	X(L), BY(L) on midplane, with HSTAR = X(L+1) - X(L)
4	BDMIX	VB at every mesh intersection with accessible interface, one per card; corresponding XB and YB are punched with "$" signs
8	BDMIX	VB at every mesh intersection with accessible interface, 8 values per card; corresponding XB, 0, YD, 0 have 8 values per card

```
ISIKL    4,   KASE   1000.  CMSPER 1.000,  AMPTRN -57409.72   TOTNI -64201.25    AMPFAC  1.1183
X0=   0.        XMAX= 150.000  H= 1.0000   XBZERO=   76.000   HSTAR= 0.           AVGDV= -.000145
                                                                        PER CENT
   K     X        DELX       BY(TESLAS)    DB/DX        K(1/M)      B/BMID       G/GMID        BMID= 1.442000     GMID= -.041466
   1    0.       -76.000     0.            0.           0.          0.           0.
   2    1.000    -75.000     0.            0.           0.          0.           0.
   3    2.000    -74.000     0.            0.           0.          0.           0.
   4    3.000    -73.000     0.            0.           0.          0.           0.
   5    4.000    -72.000     0.            0.           0.          0.           0.
                                               .
                                               .
                                               .
  64   63.000   -13.000      1.819277      .102624      7.1168     126.163459   -247.4891
  65   64.000   -12.000      1.895570      .050175      3.4796     131.454202   -121.0025
  66   65.000   -11.000      1.922187      .006992       .4849     133.300055    -16.8612
  67   66.000   -10.000      1.914566     -.019408     -1.3459     132.771540     46.8050
  68   67.000    -9.000      1.887345     -.033570     -2.3280     130.883824     80.9568
  69   68.000    -8.000      1.849733     -.041057     -2.8473     128.275492     99.0142
  70   69.000    -7.000      1.806360     -.045453     -3.1521     125.267674    109.6140
  71   70.000    -6.000      1.759376     -.048368     -3.3542     122.009442    116.6445
  72   71.000    -5.000      1.709927     -.050436     -3.4976     118.580232    121.6308
  73   72.000    -4.000      1.658687     -.051983     -3.6049     115.026809    125.3626
  74   73.000    -3.000      1.606072     -.053207     -3.6898     111.378116    128.3133
  75   74.000    -2.000      1.552339     -.054242     -3.7616     107.651770    130.8107
  76   75.000    -1.000      1.497623     -.055181     -3.8267     103.857363    133.0733
  77   76.000    0.          1.442000     -.056059     -3.8876     100.000000    135.1920
  78   77.000    1.000       1.385530     -.056869     -3.9437      96.083940    137.1447
  79   78.000    2.000       1.328305     -.057556     -3.9914      92.115476    138.8024
  80   79.000    3.000       1.270490     -.058022     -4.0237      88.106100    139.9254
  81   80.000    4.000       1.212351     -.058195     -4.0357      84.074242    140.3436
  82   81.000    5.000       1.154156     -.058191     -4.0355      80.038527    140.3339
  83   82.000    6.000       1.095973     -.058197     -4.0359      76.003668    140.3478
  84   83.000    7.000       1.037763     -.058222     -4.0376      71.966938    140.4091
  85   84.000    8.000        .979576     -.058113     -4.0300      67.931779    140.1452
  86   85.000    9.000        .921651     -.057660     -3.9986      63.914761    139.0536
  87   86.000   10.000        .864450     -.056874     -3.9302      59.946570    136.6743
  88   87.000   11.000        .808514     -.055037     -3.8167      56.068916    132.7261
  89   88.000   12.000        .754564     -.052757     -3.6586      52.327597    127.2283
  90   89.000   13.000        .703165     -.049971     -3.4654      48.763138    120.5104
                                               .
                                               .
                                               .
 140  139.000   63.000       -.007684     -.000124      -.0086      -.532881      .2988
 141  140.000   64.000       -.007796     -.000101      -.0070      -.540650      .2432
 142  141.000   65.000       -.007887     -.000082      -.0057      -.546956      .1969
 143  142.000   66.000       -.007961     -.000066      -.0046      -.552047      .1584
 144  143.000   67.000       -.008019     -.000052      -.0036      -.556126      .1263
 145  144.000   68.000       -.008066     -.000041      -.0029      -.559361      .0995
 146  145.000   69.000       -.008102     -.000032      -.0022      -.561888      .0769
 147  146.000   70.000       -.008130     -.000024      -.0017      -.563816      .0577
 148  147.000   71.000       -.008151     -.000017      -.0012      -.565232      .0411
 149  148.000   72.000       -.008165     -.000011      -.0008      -.566199      .0264
 150  149.000   73.000       -.008173     -.000005      -.0004      -.566762      .0129
 151  150.000   74.000       -.008175      0.            0.         -.566947      0.
END OF EDIT   455.185

TIME AT LOCATION 3   455.277
```

FIG. 23. Partial output from program LINDA showing quantities of interest on the median plane.

```
MAGNITUDE OF B IN IRON OF EVERY ZONE. (KILOGAUSS)  STAK=  .95  KASE   1000  CYCLE  4.  BCOMP= 1.442000                412.383

        61     62     63     64     65     66     67     68     69     70     71     72     73     74     75     76     77     78     79     80

60   0.     0.     0.     0.     0.     0.     0.     0.     0.     0.     0.     0.     0.     0.     0.     0.     0.     0.     0.     0.
59   0.     0.     0.     0.     0.     0.     0.     0.     0.     0.     0.     0.     0.     0.     0.     0.     0.     0.     0.     0.
58  17.12717.02716.91516.79016.65216.49616.32416.13515.92415.69215.43315.14114.81614.46014.06313.62513.15712.65212.09111.475
57  17.12717.02716.91516.79016.65116.49616.32416.13415.92315.69015.43115.13914.81414.45814.06413.62713.16012.65812.10511.496
56  17.12817.02816.91616.79116.65116.49616.32316.13215.92115.68715.42815.13614.81114.45614.06413.63013.16512.66912.13211.537
55  17.13017.03016.91716.79116.65116.49516.32116.13015.91815.68315.42315.13114.80714.45214.06613.63413.17412.68712.17211.599
54  17.13217.03116.91816.79216.65116.49416.31916.12615.91315.67715.41715.12514.80114.44714.06813.64113.18612.71112.22411.682
53  17.13517.03416.92016.79316.65116.49316.31616.12215.90815.67015.40915.11814.79414.44214.07013.65013.20212.74212.28511.786
 .                                                                                          .
 .                                                                                          .
 .                                                                                          .
12  20.52320.41720.31820.22120.12120.02319.91519.80619.69419.57719.45719.33519.21219.08918.96718.85118.74818.66918.60818.572
11  20.65120.56120.47720.39320.30720.21720.11920.01919.90319.78719.66419.53419.39919.26019.11518.96218.80218.65318.54118.436
10  20.57420.56620.55120.51620.46520.39920.32220.23320.12920.02119.89519.76519.62919.48619.33719.17919.00418.79418.57018.405
 9  20.03120.33520.49220.56620.58620.56920.52620.45720.36920.26620.14820.02519.88119.73719.58719.43419.28019.10718.87418.522
 8  19.96420.28420.54820.67820.75120.77620.75720.70720.63420.54220.43020.30020.16020.01619.84919.68819.53219.39419.27219.103
 7  21.21121.23921.18421.20421.18821.15221.11121.05120.96920.86220.74420.61720.47720.31520.14819.97619.77719.60019.46519.447
 6  22.10322.21522.18822.06421.94921.82821.68521.54221.40121.26921.12220.98220.82220.65720.48520.29620.10119.86219.59019.396
 5   0.    22.92223.07023.10122.92822.67922.43722.18321.96121.77021.58521.39421.18420.97620.75220.52720.291 0.     0.     0.
 4   0.     0.    24.01923.91823.85923.69723.18022.79022.40622.09921.845 0.     0.     0.     0.     0.     0.     0.     0.     0.
 3   0.     0.     0.     0.     0.     0.     0.     0.     0.     0.     0.     0.     0.     0.     0.     0.     0.     0.     0.     0.
 2   0.     0.     0.     0.     0.     0.     0.     0.     0.     0.     0.     0.     0.     0.     0.     0.     0.     0.     0.     0.
 1   0.     0.     0.     0.     0.     0.     0.     0.     0.     0.     0.     0.     0.     0.     0.     0.     0.     0.     0.     0.

        81     82     83     84     85     86     87     88     89     90     91     92     93     94     95     96     97     98     99    100

60   0.     0.     0.     0.     0.     0.     0.     0.     0.     0.     0.     0.     0.     0.     0.     0.     0.     0.     0.     0.
59   0.     0.     0.     0.     0.     0.     0.     0.     0.     0.     0.     0.     0.     0.     0.     0.     0.     0.     0.     0.
58  10.83510.180 9.511 8.826 8.127 7.412 6.682 5.937 5.179 4.408 3.626 2.836 2.042 1.258  .554 0.     0.     0.     0.     0.
57  10.85910.208 9.544 8.865 8.173 7.467 6.748 6.018 5.278 4.532 3.785 3.046 2.335 1.702 1.280 0.     0.     0.     0.     0.
 .                                                                                          .
 .                                                                                          .
 .                                                                                          .
26  16.68516.64516.60616.57016.53516.50416.46916.43816.41116.38816.37116.35716.34916.34716.353 0.     0.     0.     0.     0.
25  16.79316.73316.67416.61716.56316.51316.45816.40316.35916.32516.29816.28016.26416.25216.242 0.     0.     0.     0.     0.
24  16.95416.87916.80116.72116.63816.55416.46416.36516.26616.17416.10016.04216.01016.00316.017 0.     0.     0.     0.     0.
23  17.18017.10217.02116.93516.84316.74516.63816.52216.37916.22416.05815.84215.59315.26614.846 0.     0.     0.     0.     0.
22  17.43317.36017.28517.20717.12317.03316.93616.83116.71016.57816.41916.22016.00715.65315.119 0.     0.     0.     0.     0.
21  17.68017.61517.55017.48117.40517.32617.24417.15417.05816.95116.83016.69216.53716.34816.134 0.     0.     0.     0.     0.
20  17.90917.85217.79417.73217.66817.60117.53217.45617.37217.28217.18517.08416.96316.812 0.     0.     0.     0.     0.     0.
19  18.11418.06518.01517.96217.90717.85017.79017.72517.65617.58417.50717.40417.31817.209 0.     0.     0.     0.     0.     0.
18  18.29518.25318.21018.16618.12118.07418.02617.97317.91517.85117.78617.72617.617 0.     0.     0.     0.     0.     0.     0.
17  18.44718.41318.37918.34518.31018.27318.23518.19218.14818.10018.04017.97017.913 0.     0.     0.     0.     0.     0.     0.
16  18.56918.54318.52018.49718.47218.44618.41918.39118.35718.32018.29218.221 0.     0.     0.     0.     0.     0.     0.     0.
15  18.65418.63818.62418.61218.60118.58818.57518.56018.54718.52518.481 0.     0.     0.     0.     0.     0.     0.     0.     0.
14  18.69218.68618.68518.68718.69018.69718.70118.70218.69618.70218.669 0.     0.     0.     0.     0.     0.     0.     0.     0.
13  18.66818.67218.68518.70518.73118.75218.78218.81218.84318.820 0.     0.     0.     0.     0.     0.     0.     0.     0.     0.
12  18.55818.56618.59018.62918.68018.75518.80418.86118.91918.951 0.     0.     0.     0.     0.     0.     0.     0.     0.     0.
11  18.36418.33018.34218.39018.47218.55118.73518.88018.928 0.     0.     0.     0.     0.     0.     0.     0.     0.     0.     0.
10  18.23418.09117.97717.92217.93818.07618.08018.421 0.     0.     0.     0.     0.     0.     0.     0.     0.     0.     0.     0.
 9  18.29918.08717.86817.61717.37417.06217.28017.633 0.     0.     0.     0.     0.     0.     0.     0.     0.     0.     0.     0.
 8  18.50218.20718.01017.80217.49717.11616.449 0.     0.     0.     0.     0.     0.     0.     0.     0.     0.     0.     0.     0.
 7  19.64218.28418.08717.963 0.     0.     0.     0.     0.     0.     0.     0.     0.     0.     0.     0.     0.     0.     0.     0.
 6  19.578 0.     0.     0.     0.     0.     0.     0.     0.     0.     0.     0.     0.     0.     0.     0.     0.     0.     0.     0.
 5   0.     0.     0.     0.     0.     0.     0.     0.     0.     0.     0.     0.     0.     0.     0.     0.     0.     0.     0.     0.
 4   0.     0.     0.     0.     0.     0.     0.     0.     0.     0.     0.     0.     0.     0.     0.     0.     0.     0.     0.     0.
 3   0.     0.     0.     0.     0.     0.     0.     0.     0.     0.     0.     0.     0.     0.     0.     0.     0.     0.     0.     0.
 2   0.     0.     0.     0.     0.     0.     0.     0.     0.     0.     0.     0.     0.     0.     0.     0.     0.     0.     0.     0.
 1   0.     0.     0.     0.     0.     0.     0.     0.     0.     0.     0.     0.     0.     0.     0.     0.     0.     0.     0.     0.
```

FIG. 24. Partial output from program LINDA showing the magnitude of the field in every zone. These printouts may be assembled to produce a large flux map of the entire magnet.

efficiency, the distribution of magnetomotive force or potential along the pole, the flux density in the iron, and the total flux linkage of the coils. These quantities may be printed by choice of the proper ICTL(3) flag or may be ignored by use of the ICTL(4) flag.

f. Program PLUTO

This program is used to plot equipotential and flux lines obtained from calculations performed by the two-dimensional magnetostatic computer program LINDA. Program PLUTO is written in Fortran language. It occupies $100K_8$ words of memory and is operational on the CDC 6600/7600 computer systems. The code is organized in a flexible way to allow interaction of various sets of input which produce a variety of plots as demonstrated in several examples. In program PLUTO the functions of each subroutine are briefly:

Program PLUTO is the main control program, which allows the initiation of various options to perform the desired results (options are described under Preparation of Input Data).

Subroutine SETUP reads data from File PRTAPE, identifies and matches parameters from LINDA to be used by PLUTO.

Subroutine SIZEUP reads dimensions of the area to be plotted (X1, X2, Y1, Y2) and the scaling factor (XF), and computes quantities associated with the size of the plot.

Subroutine LIMITS determines the number of flux lines to plot (if not given as input) and arranges pertinent quantities to describe the contour map.

Subroutine CONMAP the main plotting routines makes use of various other subroutines to plot flux lines, equipotential lines, etc.

1. *Preparation of Input Data.* The input required to produce a plot depends largely on the type of plot desired. In this section the various options available to the user are developed and in the next section examples of their usage are given.

The first card in the input deck is called INDMP. This flag allows the user to utilize the following functions:

If INDMP ≤ 0, job is terminated

If INDMP $\geq$ N, N $>$ 0, the program will look for file N on PRTAPE and process it.

The next set of three cards (except for option KPLOT = 4) consists of the parameters necessary to furnish information relating to the quantities plotted, as outlined below.

1st card. Contains the parameters KPLOT, NOL, IX. KPLOT has seven possible values:

KPLOT = -1, indicates termination of file being read. In other words, "we are through," go back to beginning to read next file depending on the value of INDMP.

KPLOT = 0, plot mesh. Actually, tick marks are plotted at mesh line intersections.

KPLOT = 1, plot scalars in iron plus the magnet outline and coils.

KPLOT = 3, plot ellipse.

KPLOT = 4, plot magnet and coil outline.

KPLOT = 5, plot tick marks and numbered scales at left and bottom.

58   I. Two-Dimensional Magnetostatic Programs

Parameter NOL has four values associated with contour lines to be plotted:

NOL = 1, all contour lines are input.

NOL = 2, enter number of desired contour lines.

NOL = 3, enter one contour line and interval.

NOL = 4, enter range and interval, i.e., from where to where (dimensions) and at what interval plots are desired. When KPLOT = 0, NOL defines the length of line in fifths of mesh unit; when KPLOT = 3, NOL defines the number of times to trace lines.

The third parameter, IX, used in subroutine CONMAP, is renamed as LAY and it may have any of the four values shown as follows:

LAY = − 1, new plot; calls subroutine BOUNDS to plot magnet boundary and write labels; coils are not plotted.

LAY = 0, new plot; calls subroutine BOUNDS to plot magnet boundary, write labels, and plot coil outline.

LAY = 1, overlay previous plot; write values in border.

LAY = 2, overlay previous plot; do not write values.

2nd card. This card contains information relating to the area to be plotted and the desired scale. In particular, the four coordinates of the plot size are entered, followed by the scaling. These quantities and their relation to the problem universe are shown in Fig. 25. The scale factor (XF) indicates how many dimensions units are needed to make 10 inches or 25.4 cm of chart paper. The order by which these parameters are entered is x1, x2, y1, y2, XF. The units of the dimensional coordinates (x, y) are the same units with which the particular run of LINDA was made. The choice of the scale factor (XF) may be better understood by the following example.

Consider Fig. 25 in which the large universe is Y1 = 0.0 inches and Y2 = 59.0 inches. To produce a plot of full CalComp chart paper (PP$_{inches}$), we set

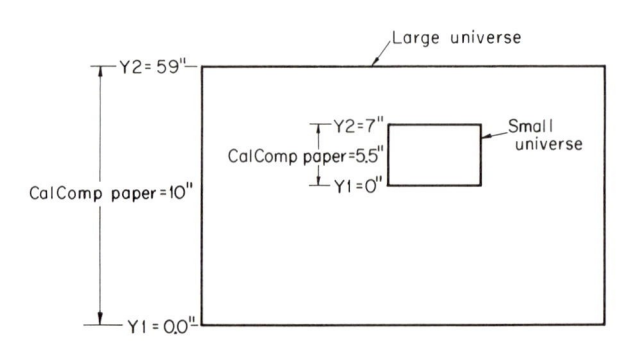

FIG. 25. Scaling universe for program PLUTO.

$PP_{inches} = 10$, and find the scaling factor (XF) by

$$XF = (Y2 - Y1)*10/PP_{inches} \qquad (3.1)$$

In the same manner, XF for the small universe may be calculated. Choosing a scaling factor of 12.7 inches, the resulting plot would occupy 5.5 inches of chart paper. Figures 26 and 27 are examples of the CERN-PS magnet showing different scaling factors.

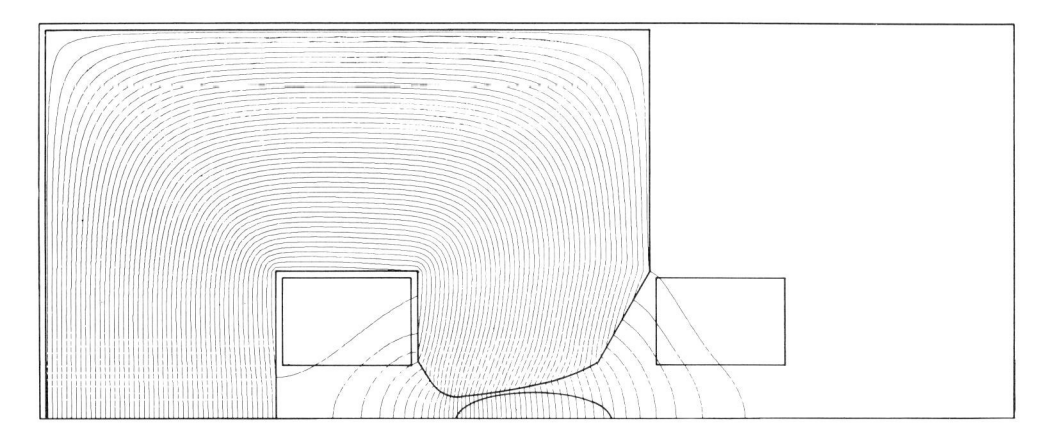

FIG. 26. The flux distribution for the CERN-PS magnet from program PLUTO.

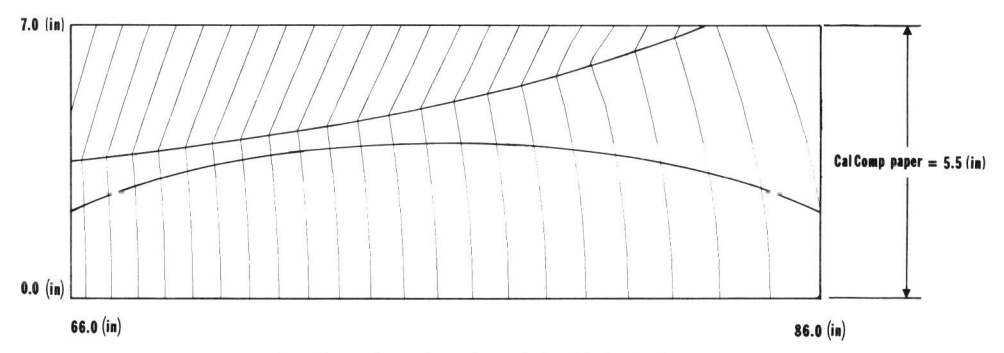

FIG. 27. An enlarged section of the CERN-PS magnet.

3rd card. The parameters necessary for the last card in this set depend on the value of KPLOT and NOL. In the following examples some commonly used arrangements are shown so that the potential user may recognize the pattern and arrange new cases at his discretion.

a. Case 1 – KPLOT = 0 (NOL may have any value). The parameters of the third card are XA, XB, YA, YB, ISTEP; the first four parameters correspond to

60   I. Two-Dimensional Magnetostatic Programs

X1, X2, Y1, Y2, described previously, while ISTEP specifies the interval of plot, e.g., ISTEP = 2 will represent plotting every other mesh line, while ISTEP = 1 —plotting every mesh line.

  b. Case 2 – KPLOT = 2, NOL = 2. The parameter on the third card is NOC, denoting the number of contour lines to be plotted.

  c. Case 3 – KPLOT = 3, NOL = 3. The parameters on the third card are XC, YC, A, B; where XC = x coordinate of center of ellipse, YC = y coordinate of center of ellipse, A = length of semi-major axis, and B = length of semi-minor axis.

  d. Case 4 – KPLOT = 4. No third card is required.

  e. Case 5 – KPLOT = 5. This option permits the insertion of numbered scales and tick marks in the left and bottom borders of the magnet. The parameters on the third card are DEL and INC, where DEL = frequency of tick marks, and INC = frequency of the labeling tick marks.

  2. *Illustrative Examples.* The set of input cards in Table 13 is required to plot the flux distribution for the CERN-PS magnet (see Fig. 27) calculated by program LINDA. The first card in Table 13 indicates the magnetic tape/file (PRTAPE) from which PLUTO will extract the information needed for plotting.

Table 13

PLUTO Input Data for a Flux Distribution Plot of the CERN-PS Magnet

	Card Column	
1	30	72
$ PLUTO INPUT DATA FOR THE CERN-PS MAGNET $		
1	$ INDMP=1, LOOK FOR 1ST FILE SOLUTION ON TAPE $	
2 2 0	$ KPLOT,NOL,IX- PLOT FLUX LINES, NEW PLOT $	
0.0 0.0 0.0 0.0 0.0	$ X1,X2,Y1,Y2,XF- PLOT WHOLE UNIVERSE $	
50	$ NOC- NUMBER OF FLUX LINES TO BE PLOTTED $	
5 0 1	$ KPLOT,NOL,IX- PLOT TICKMARKS IN SAME PLOT $	
0.0 0.0 0.0 0.0 0.0	$ X1,X2,Y1,Y2,XF- PLOT WHOLE UNIVERSE $	
1.0 5	$ DEL,INC- INCREMENT + FREQUENCY OF TICKMARKS $	
3 3 1	$ KPLOT,NOL,IX- PLOT ELLIPSE IN SAME PLOT $	
S S S S S	$ SKIP X1,X2,Y1,Y2,XF $	
76.0 0.0 12.0 7.0	$ XC,YC,A,B- COORDINATES OF ELLIPSE CENTER + AXES $	
-1 S S	$ NO MORE SOLUTIONS TO BE PLOTTED $	
-1	$ EXIT $	

The next three cards specify the plotting of flux lines (KPLOT = 2) on a new plot (IX = 0) throughout the universe of the problem (x1 = x2 = y1 = y2 = 0). Had we specified a smaller portion of the universe, then x1, x2, y1, and y2, would have been the coordinates of the portion of the universe to be plotted. The use of the letter s allows the user to omit repeating information specified earlier; that is, the use of each s directs the program to utilize the previously specified value. In a similar fashion the other sets of cards are interpreted.

In another example, Fig. 27 shows the flux distribution for a portion of the CERN-PS magnet located under the pole tip. Interpretation of the data in Table 14 follows the same reasoning as the previous example. In Table 14, the s's omit the repetition of the values 66.0, 86.0, 0.0, 7.0, 12.7, specified in the previous set. In the same manner any quantity may be omitted by replacing it with s.

Table 14

PLUTO Input Data for an Enlarged Portion under the Pole Tip of the CERN-PS Magnet

					Card Column	
1					30	72
$	PLUTO INPUT DATA FOR A ENLARGED PORTION UNDER THE POLE TIP					$
$	OF THE CERN-PS MAGNET	$				
1					$ FILE 1 OF CERN-PS SOLUTION ON MAG. TAPE $	
2 2 2					$ KPLOT,NOL,IX- PLOT FLUX LINES, NEW PLOT $	
66.0	86.0	0.0	7.0	12.7	$ X1,X2,Y1,Y2,XF- COORDINATES OF ENLARGEMENT $	
25					$ NOC- PLOT 25 FLUX LINES $	
3 3 1					$ KPLOT,NOL,IX- PLOT ELLIPSE IN SAME PLOT $	
S S S S S					$ SKIP X1,X2,Y1,Y2,XF $	
76.0	0.0	12.0	4.0		$ XC,YC,A,B- COORDINATES OF ELLIPSE CENTER + AXES $	
-1 S S					$ LAST PLOT $	
-1					$ EXIT $	

g. Program BMAP

This is an auxiliary program to make additional calculations on the results of computations made by program LINDA. Specifically, BMAP calculates components of flux density at mesh points in air, or iron, and the components of the force in zones (mesh rectangles) with current.

I. Two-Dimensional Magnetostatic Programs

1. *Preparation of Input Data.* The input required is as follows:

1st card. Contains the parameters IRST, LAST, ILIST, WRD.

IRST = number of first file on tape produced by LINDA.

LAST = number of last file on tape produced by LINDA.

ILIST = used as flag. When nonzero, prints a map of the MARK and CLASS of the entire problem.

WRD = hollerith description to name problem and magnet (optional). This card has a fixed format of [2(8X,I2),5A10].

2nd card. This card specifies the rectangle on which an edit is made. The parameters needed are the following: XMIN, YMIN, XMAX, YMAX, RHZ. The first four specify the corner points of the rectangle. RHZ is a four-bit number with each bit acting as a flag to initiate a map of values appropriate to the bit being activated. Let RHZ = KLMN (the four bits):

If N $\neq$ 0, print B_x component of field in specified rectangle

If M $\neq$ 0, print B_y component of field in specified rectangle

If L $\neq$ 0, print F_x component of force in current-carrying elements in the specified rectangle

If K $\neq$ 0, print the F_y component of force. This card also has a fixed format (5F10.5,3A10).

3rd card. This card is a blank card (no parameters) and terminates the job.

2. *Illustrative Example.* Figure 28 shows a partial printout for the CERN-PS magnet obtained by using the program BMAP with the data listed in Table 15.

Table 15

BMAP Input Data for a BX and BY EDIT of the CERN-PS Magnet

		Card Column			
1	10	20	30	40	50
$ BMAP	INPUT	FOR A BX	AND BY	EDIT OF THE	CERN-PS MAGNET $
	1	1		0 CERN-PS	MAGNET
35.0	0.0	120.0	15.0	11.0	

(A blank card must be inserted here to terminate program)

4. Program NUTCRACKER

```
BY IN KILOGAUSS IN RECTANGLE  35.00000 .LE. X .LE. 120.00000,  0. .LE. Y .LE. 15.00000
      KASE    1000    RCOMP 1.442000    AMPFAC 1.1182901   AMPTRN  -57409.72   TOTNI  -64200.72
```

Y \ X	75.000	76.000	77.000	78.000	79.000	80.000	81.000	82.000	83.000	84.000
15.000	0.	0.	0.	0.	0.	0.	0.	0.	0.	0.
14.000	0.	0.	0.	0.	0.	0.	0.	0.	0.	0.
13.000	0.	0.	0.	0.	0.	0.	0.	0.	0.	0.
12.000	0.	0.	0.	0.	0.	0.	0.	0.	0.	0.
11.000	0.	0.	0.	0.	0.	0.	0.	0.	0.	0.
10.000	0.	0.	0.	0.	0.	0.	0.	0.	0.	0.
9.000	0.	0.	0.	0.	0.	0.	0.	0.	0.	0.
8.000	0.	0.	0.	0.	0.	0.	0.	0.	0.	0.
7.000	0.	0.	0.	0.	0.	0.	0.	0.	0.	0.
6.000	0.	0.	0.	0.	0.	0.	0.	0.	10.50675	9.87703
5.000	0.	0.	0.	0.	0.	11.89231	11.51391	11.03268	10.44191	9.82613
4.000	0.	0.	13.92529	13.34951	12.75094	12.07043	11.52551	10.98154	10.40207	9.80020
3.000	15.01437	14.46046	13.89344	13.31570	12.72297	12.11295	11.53615	10.96590	10.38466	9.79177
2.000	14.99543	14.43857	13.87232	13.29686	12.71230	12.12226	11.54025	10.96126	10.37887	9.79189
1.000	14.98229	14.42567	13.86042	13.28711	12.70710	12.12355	11.54132	10.96002	10.37769	9.79423
0.	14.97775	14.42142	13.85654	13.28406	12.70544	12.12351	11.54149	10.95979	10.37763	9.79536

Y \ X	85.000	86.000	87.000	88.000	89.000	90.000	91.000	92.000	93.000	94.000
15.000	0.	0.	0.	0.	0.	0.	2.38830	2.50529	2.61406	2.72039
14.000	0.	0.	0.	0.	0.	2.55980	2.68266	2.78528	2.87686	2.96486
13.000	0.	0.	0.	0.	0.	2.89382	2.99726	3.07633	3.14323	3.20723
12.000	0.	0.	0.	0.	3.15097	3.26727	3.33621	3.37956	3.41249	3.44577
11.000	0.	0.	0.	0.	3.62943	3.68808	3.70074	3.69322	3.68141	3.67631
10.000	0.	0.	0.	4.16944	4.19849	4.15499	4.08546	4.01117	3.94361	3.89169
9.000	0.	0.	5.27629	5.05265	4.84021	4.64752	4.47504	4.32238	4.19018	4.08185
8.000	0.	0.	6.64279	5.92468	5.46219	5.11992	4.84481	4.61312	4.41288	4.23736
7.000	0.	8.60775	7.35901	6.54109	5.96393	5.52517	5.17116	4.87242	4.61087	4.37232
6.000	9.22537	8.40670	7.64442	6.91674	6.32727	5.84568	5.44224	5.09454	4.78585	4.50220
5.000	9.18539	8.50466	7.90523	7.15418	6.58272	6.08803	5.65757	5.27764	4.93579	4.62095
4.000	9.18081	8.54730	7.91367	7.31204	6.76141	6.26617	5.82237	5.42264	5.05873	4.72262
3.000	9.19034	8.58603	7.99011	7.41889	6.88470	6.39286	5.94311	5.53183	5.15387	4.80361
2.000	9.20269	8.61656	8.04182	7.48868	6.96563	6.47748	6.02536	5.60771	5.22130	4.86215
1.000	9.21199	8.63564	8.07192	7.52840	7.01168	6.52605	6.07315	5.65236	5.26148	4.89745
0.	9.21536	8.64212	8.08182	7.54132	7.02664	6.54189	6.08882	5.66709	5.27482	4.90924

FIG. 28. Partial output from program BMAP showing the flux distribution in the air region of the CERN-PS magnet.

4. Program NUTCRACKER

a. Introduction

Program NUTCRACKER [21, 22] was developed at the Stanford Linear Accelerator Center to provide solutions to problems arising in the design studies of iron-bound and iron-core magnets having simple geometries. The program is capable of static evaluation of two-dimensional magnets or for magnets possessing axial symmetry. It is best suited for magnet geometries with rectangular shape and should not be used with curved boundaries, since the square mesh necessitates that a curved boundary be specified by a series of square regions resulting in a decrease of the program's accuracy to unacceptable levels.

The program is written in Fortran IV for the IBM 360/91 and has been subsequently converted by the author to run on the CDC 6600/7600 computer systems. It requires approximately 290_{10} bytes in the IBM 360/91 and $160K_8$ words of memory in the CDC 6600/7600 system. Maximum mesh size is 171×151 points and a typical run takes approximately 20 minutes on the CDC 6600 computer system.

b. Program Description

NUTCRACKER makes use of the vector potential method in a fashion similar to that described in program TRIM. Briefly it makes use of Poisson's equation

$$\nabla \times \gamma \nabla \times \mathbf{A} = \mathbf{S}, \tag{4.1}$$

where $\gamma = 1/\mu$, taking the curl as indicated in Eq. (4.1) and noting that for two-dimensional problem $\mathbf{A}$ is independent of z we obtain

$$\nabla \times \gamma \nabla \times \mathbf{A} = -\left[\frac{\partial}{\partial x}\left(\gamma \frac{\partial A_z}{\partial x}\right) + \frac{\partial}{\partial y}\left(\gamma \frac{\partial A_z}{\partial y}\right)\right]\mathbf{k}, \tag{4.2}$$

which, when combined with the right side of Eq. (4.1) yields,

$$\frac{\partial}{\partial x}\left(\gamma \frac{\partial A_z}{\partial y}\right) + \frac{\partial}{\partial y}\left(\gamma \frac{\partial A_z}{\partial y}\right) = -S_z. \tag{4.3}$$

In a similar fashion we derive the equivalent equation in cylindrical coordinates

$$\frac{\partial}{\partial r}\left[\frac{\gamma}{r}\frac{\partial}{\partial r}(rA_\theta)\right] + \partial/\partial_z\left[\frac{\gamma}{r}\frac{\partial}{\partial_z}(rA_\theta)\right] = -S_\theta. \tag{4.4}$$

NUTCRACKER solves Eqs. (4.3) and (4.4) in two phases. The first phase involves the overrelaxation of the vector potential A throughout the available universe using finite difference equations to approximate Eqs. (4.3) and (4.4); the calculation of the ratio

$$C_n = \iint \mathbf{S} \cdot d\mathbf{s} \Big/ \int \mathbf{H} \cdot d\mathbf{l}, \tag{4.5}$$

for scaling the potential values in the available universe by C_n, if appropriate; and the underrelaxation of the permeability throughout the mesh. The second phase involves the successive overrelaxation of the vector potential at every point in the available universe and the simultaneous underrelaxation of every permeability corresponding to a specific potential value. Convergence is achieved when the residual at each mesh point is less than 0.1 percent of the mean potential value.

NUTCRACKER is organized to allow the user to solve for a variety of problems relating to magnet design. One important limitation of the program is its inability to cope with curved boundaries resulting from the square mesh utilized by the program. A second limitation is that the number of nodes that may be used in describing a magnet geometry is a function of the available computer memory and the large mesh that must be used for reliable results.

The flow of information in NUTCRACKER is shown in a simplified chart (see Fig. 29); the program begins by reading and checking the input data for valid conditions and prints appropriate error diagnostics. Table 16 describes briefly the main subroutines of NUTCRACKER. The problem geometry may possess

Table 16

NUTCRACKER Subroutines and Functions

Name	Description
MAIN	The main control program: reads, checks, and prints input data. A large number of error diagnostics guide user in correcting discrepancies, and proper definition and sequence of input
AMPERE	Calculates ampere-turns within a region by summing values of JH22 (K,L)/2
BX,BY,HX, HY,MX,MY	Calculates the field components of B, H, M using the appropriate difference equations. A determination is made as to whether the point lies on the boundary; if so, the problem symmetry is checked to determine correct expressions to be used
DRAW	Draws map of magnet using the following symbols: I = Iron, C = Coil, B = Both, . = Air
EQUI	Draws an equipotential plot with each equipotential line symbolized by a letter of the alphabet
IMPROVE	Calculates improved values for the SOR factor ω; calls subroutine RELAX to perform the overrelaxation
INTHDS	Performs calculation of line integral $\int H \cdot ds$ for each curve; integration is performed counter-clockwise
MJH	Finds the value of μ corresponding to H; binary search is used in this process; a linear interpolation between tabulated values is performed if needed. If H is outside the tabular range, a warning message is printed and the last acceptable value of μ at this point is used
PRINTD	Prints final results, relative permeabilities, vector potentials, flux density as well as magnet moments, if requested
READD	The array $\vec{A}$ (vector potential) and MJH (μ) are read from magnetic tape under the format specified in WRITED
RELAX	When called by the MAIN program, this routine sweeps the entire mesh starting with column K = 1 and proceeds until K = XMESH2
WRITED	Writes arrays $\vec{A}$ and MJH (μ) on magnetic tape

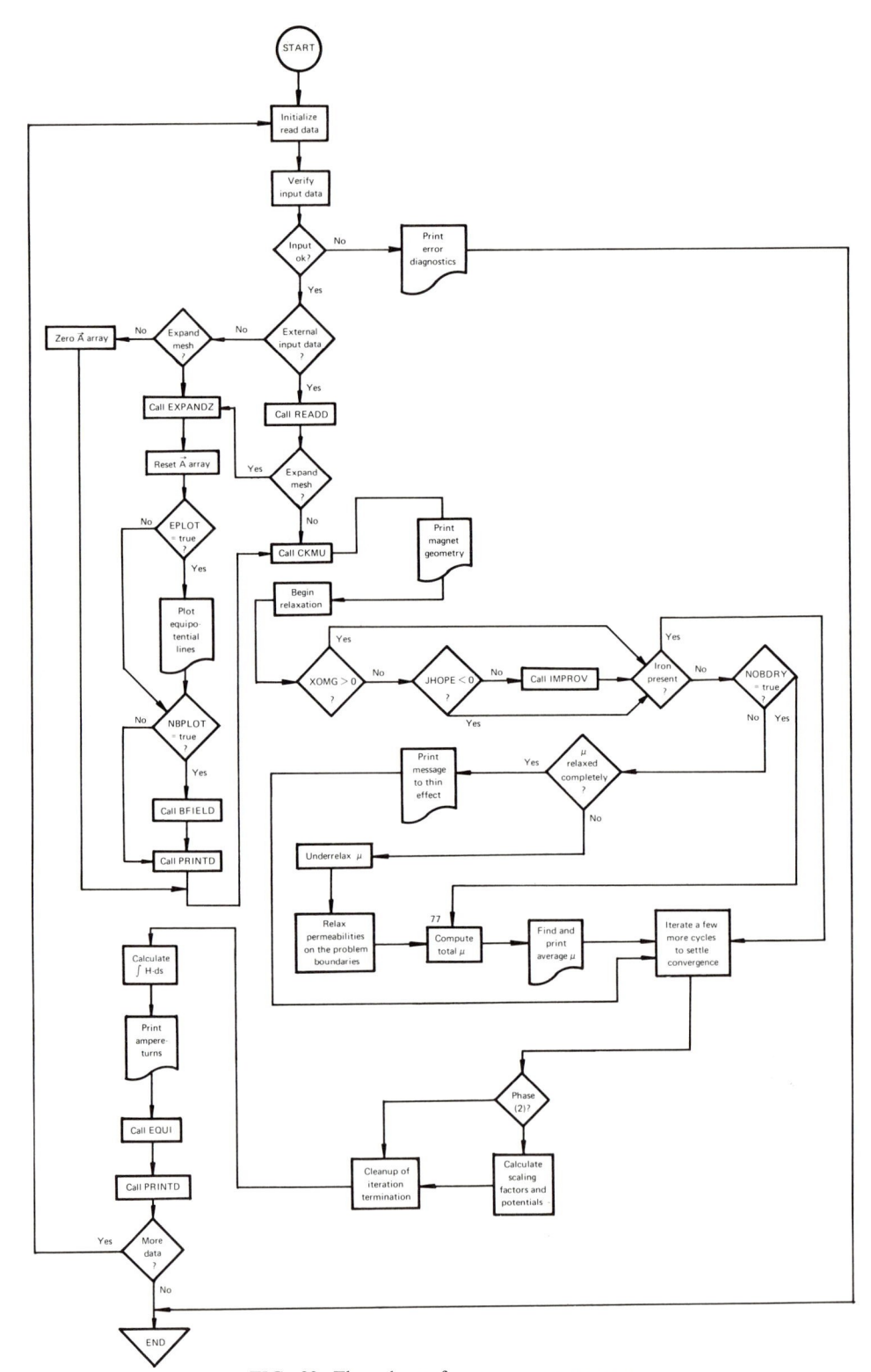

FIG. 29. Flow chart of program NUTCRACKER.

either rectangular or cylindrical symmetry. The main program which is responsible for these functions, also calculates the initial relaxation factors; calls subroutine DRAW, which draws a map of the magnet; and proceeds with the main iterative process by calling the appropriate subroutines that contain the finite difference equations. To increase the rate of convergence, the mesh is swept in columns, and suitable printout is provided at each iteration indicative of the rate of convergence. Once the program has converged, the computed results may be written on magnetic tape for further use before printing via subroutines PRINTD and EQUI.

The program terminates when: 1) the number of requested iterations is exceeded, 2) time limit is exceeded, or 3) when the error is less than a preassigned small number (typically 4×10^{-6}).

c. Preparation of Input Data

The preparation of input for program NUTCRACKER requires considerable effort and attention. Since the program utilizes a square mesh, a very special setup is required to solve the anticipated geometry. The first step involves the

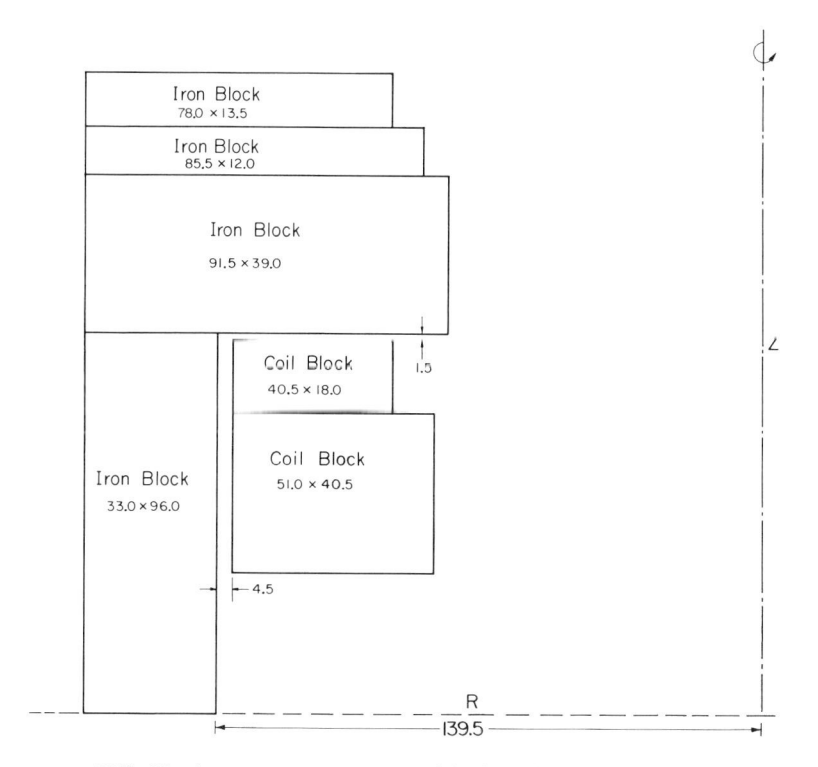

FIG. 30. A test magnet geometry with dimensions in centimeters.

I. Two-Dimensional Magnetostatic Programs

selection of the mesh size to be used; this is a problem in itself, since the square mesh allows only an approximation to the real geometry. Computer programs have been written [23] that find the best possible mesh to fit the geometry of the magnet. Once the mesh size and the magnet layout have been approximated, the recording of the coordinates of air, iron, and coil regions follows, as well as various program variables, which will be outlined in an example.

Figure 30 shows the geometry of a cylindrically symmetrical magnet, and Table 17 shows the set of data required for this particular run. The description of the variables and data included in this run are listed in Table 18. The first card in Table 17 identifies the problem. The second card specifies the cylindrical nature of the problem (F), the mesh size (135×135), convergence criterion, and other quantities described in detail in Table 18.

Table 17

NUTCRACKER Input Data

			Card Column				
1	10	20	30	40	50	60	70
CYLINDRICAL TEST MAGNET.							
F	135.0	135.0	0.00005	4.0	2.0	600.0	30.0
T	F	F	0.0	0.0	2.01	2.01	1.0
	2 100.0	0.0	F	F	F	F	
	5	1	1	134	134		
2.0	2.0	134.0	2.0	134.0	134.0	2.0	134.0
2.0	2.0						
	24	56	51	90 4623304.0			
	51	63	63	90 4623304.0			
	1	93	64	115			
	64	54	90	115			
	90	58	98	115			
	98	63	107	115			
	1	1	134	134	F	T	T
	1	1	134	134	F	F	T
2.0	0.0	0.0	4.0	50.0	0.0	8.0	0.88
		←——— Blank card					
20.0	800.0	4.0	0.0		0 0.00005		

4. Program NUTCRACKER

Table 18
NUTCRACKER Input Parameters and Specifications

Card	Format	Variable	Description
1	80A1	TITLE	Problem title; 80 characters permissible
2	L10	SYSTEM	Coordinate system; T = rectangular, F = cylindrical
	F10.0	XMESH	Number of mesh points in x direction; $0.0 < \text{XMESH} \leq 171.0$
	F10.0	YMESH	Number of mesh points in y direction; $0.0 < \text{YMESH} \leq 151.0$
	F10.0	ERROR	Relative error used to control convergence; typical value = 10^{-5}
	F10.0	NIRON	Number of iron pieces; $0.0 \leq \text{NIRON} < 100.0$, see Card 8
	F10.0	NCOIL	Number of coils; $0.0 \leq \text{NCOIL} < 100.0$, see Card 7
	F10.0	ITLIM	Number of iterations; ITLIM > 0.0, typical value = 500.0
	F10.0	TLIMM	Time limit; iterations cease after (TLIMM - 2.0 minutes), typical value = 15 minutes on IBM 360/91, 25 minutes on Control Data 6600
3	L10	PHASE1	Set PHASE1 = T, for preliminary run to initialize $\vec{A}$, μ arrays; once results have been obtained, Phase II is used exclusively
	L10	DATAIN	Set DATAIN = T for external data input; set = F for nonexternal data input
	L10	DATAO	Set DATAO = T if data output directed to external device (tape); set = F for nonexternal device
	4F10.0	X1,Y1,X2,Y2	Coordinates of a region (meters), see Fig. 27
	F10.0	NCURVE[a]	Number of closed curves for $\int H \, ds$; $0.0 < \text{NCURVE} \leq 20.0$
4	I10	NREGS[b]	Number of regions for field printout; typical value = 2
	F10.0	STARMU	Initial approximation of permeability; typical value for first run with iron is 100.0; on subsequent runs of same problem, the average permeability should be used
	F10.0	CALCOM	Output options for contour plotting: If CALCOM = 0.0, no plotting; if = 1.0, plot vector potential; if = 2.0, plot permeability; if = 3.0, plot B field
	L10	MORDAT	If MORDAT = F, only one problem run; if = T, multiple problems
	L10	XPAND	If XPAND = T, last solution to be used as basis for new problem; if = F, only one solution for one problem run
	L10	NEXP1	If NEXP1 = T, form ratio of new mesh to old mesh; if = F, do not form ratio
	L10	NEXP2	Variables XPAND, NEXP1, NEXP2 are used to run problems with coarse mesh size, calculate potentials, expand mesh to larger size, and use values of potentials interpolated to larger size; not debugged, use default option = F
5	I10	NPOINT	Number of corner points in closed line integral. Count starting point twice (for example, for a rectangular region NPOINT = 5)
5 & 6 Repeat NCURV times	I10 I10 I10 I10	INFLUE(I,1) INFLUE(I,2) INFLUE(I,3) INFLUE(I,4)	Region of influence in mesh points; must enclose paths of all line integrals
6 Use as many cards necessary	F10.0 F10.0 F10.0 F10.0	X1 Y1 X2 Y2	Coordinates of closed line integral followed in counterclockwise direction, specified by minimum of five sets of coordinates with each section parallel to one axis; NPOINT specifies the number of sections

70   I. Two-Dimensional Magnetostatic Programs

Table 18 (cont.)

Card	Format	Variable	Description
7 Repeat NCOIL times	I10 I10 I10 I10 F15.0	I J K L CURRENT	Coordinates x and y of the coil(s), in mesh points (I,J) ───coil─── (K,L) Enter current density (amperes/meter^2)
8 Repeat NIRON times	I10 I10 I10 I10	I J L M	Coordinates x and y of the iron section(s), in mesh points (I,J) ───iron─── (L,M)
9 Repeat NREGS times	I10 I10 I10 I10 L10 L10 L10 F10.3	NARRAY(I,1) NARRAY(I,2) NARRAY(I,3) NARRAY(I,4) MESMET^c PRPLOT BFEQUI^c INCREM^c	Coordinates x and y of the region of the printout, in mesh points NARRAY(I,1),NARRAY(I,2) ───field printout─── NARRAY(I,3),NARRAY(I,4) Enter F (not debugged) Enter T for equipotential plot Enter T (not debugged) Leave blank (not debugged)
10	F10.0	INFO(1)	Boundary conditions determined on left vertical boundary: if INFO(1) = 0.0, Dirichlet; if = 2.0, Neumann. Lower boundary is always Neumann boundary
	F10.0	INFO(2)^d	External unit number to be read; typical value = 0.0
	F10.0	INFO(3)^d	External unit number to be written; typical value = 0.0
	F10.0	INFO(4)	Edit options: if = 0.0, no printout; if = 1.0, $\vec{A}$ and μ only; if = 2.0, $\vec{A}$, μ and B fields; if = 3.0, $\vec{A}$, μ, B and magnetic moments; and 4.0 = printout of B field only. Standard entry = 4.0
	F10.0	INFO(5)^e or ITP1	Number of iterations in Phase I; typical value = 50.0
	F10.0	INFO(6)^f or XONG	Method for approximation to ω. If XMOG = 0.0, Frankel's method is used automatically; any other number becomes starting value for ω
	F10.0	INFO(7)	Frequency of improving ω; if = 0.0, not improved. Uses value of XMONG throughout calculations; typical value = 8.0
	F10.0	INFO(8)	Underrelaxation factor for μ; generally $0.85 \leq \text{INFO}(8) \leq 0.9$
11	F10.0	INFO(9)	If INFO(9) = 1.0, then values of B at 3 mesh points with coordinates specified by INFO(10) to INFO(15) are calculated during iterations. If INFO(9) = 0.0, the above is not done
	F10.0 F10.0 F10.0 F10.0 F10.0 F10.0	INFO(10)x_1 INFO(11)y_1 INFO(12)x_2 INFO(13)y_2 INFO(14)x_3 INFO(15)y_3	Coordinates for 3 mesh points used in INFO(9)
	F10.0	INFO(16)	If INFO(16) = 1.0, printout of field is scaled relative to value of $\|B(x_1,y_1)\|$; if INFO(16) = 0.0, the field is not scaled

Table 18 (*cont.*)

Card	Format	Variable	Description
12	F10.0	INFO(17)	Number of equipotential regions to be drawn; typical values: 15 to 25 for small problems, 25 to 40 for large problems
	F10.0	INFO(18)[g]	If INFO(18) < ITLIM, the permeability is not relaxed after this many iterations
	F10.0	INFO(19)	Number of additional relaxations after relaxing permeability, typically 2 or 3
	F10.0	INFO(20)	Always zero
	I10	IHOPE	Parameter to control "brute force" convergence of the problem. Normally IHOPE = 0, which results in automatic adjustment
	F10.0	GAFE	Permeability no longer relaxed when relative error < GAFE; typical value = 2.0×10^{-5}

[a] Denotes multiple card usage; card 5 will be read "NCURVE" times.

[b] Denotes multiple card usage; card 9 will be read "NREGS" times.

[c] Despite limited print control options, the user will receive sufficient printout for magnet analysis.

[d] User should declare his own external devices (tapes, disks) peculiar to his own computer environment.

[e] This number should be large enough (for example, $\frac{1}{2}$ of ITLIM) for the automatic cutoff of Phase I to be operable. If this automatic cutoff is not desired, ITP1 equals 40 or 50, typically. This is not recommended, but may be done at loss of speed of convergence.

[f] Use 0.0 only for the first-time run of a problem. Then the last ω calculated in the problem can be used by setting INFO(6) = ω.

[g] If ITLIM > INFO(18), results may not converge as rapidly as possible.

To arrive at the desired values of XMESH and YMESH, Fig. 31 and the following procedure will help. Pick a mesh spacing, for instance, $h = 0.01$ meters, then

$$\text{XMESH} = (X_2 - X_1)/h + 1 = (1.50/0.01) + 1 = 151 \text{ Nodes}, \quad (4.6)$$

$$\text{YMESH} = (Y_2 - Y_1)/h + 1 = (1.20/0.01) + 1 = 121 \text{ Nodes}, \quad (4.7)$$

and

$$h = (X_2 - X_1)/(\text{XMESH} - 1). \quad (4.8)$$

The third card specifies the beginning of the problem with PHASE I activated, without external I/O devices. The universe of the problem is next specified (0.0, 0.0, 2.01, 2.01) with NCURVE = 1.0.

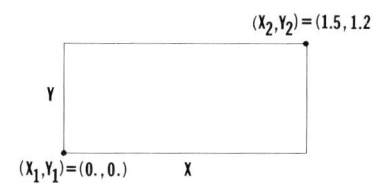

FIG. 31. An example of mesh scaling with program NUTCRACKER.

72   I. Two-Dimensional Magnetostatic Programs

The fourth card specifies one region for field printout, sets initial permeability to 100.0, does not request CalComps, specifies that this is the last case (MORDAT = F), and finally indicates that the parameters XPAND, NEXP1, NEXP2, have been omitted (see Table 18).

The fifth card specifies the region of influence which is in essence the universe of the problem, while the sixth and seventh cards specify the coordinates for the path of the line integral: namely, from (2.0, 2.0), (134.0, 2.0), (134.0, 134.0), (2.0, 134.0) and (2.0, 2.0).

The next 3 cards (8th through 10th) specify the mesh coordinates of the coils and the current density, followed by 4 cards specifying the four iron regions. The 15th card specifies the coordinates of the region where printout is desired (in this case, the whole universe).

The 16th card specifies various quantities described in Table 18.

The 17th card is blank. It must be physically present unless special printout is requested as described in Table 18 (Card No. 11).

```
                           NUTCRACKER VERSION 5.4


     CYLINDRICAL TEST MAGNET.



     ------------------- ELAPSED TIME      0 SECONDS
     COORDINATE SYSTEM IS CYLINDRICAL.
        I
       RI
        I
        +----
            Z
            NODES IN THE X-DIRECTION        135
     NODES IN THE Y-DIRECTION         135
     ERROR    5.0000E-05
     NUMBER OF IRON BLOCKS       4
     NUMBER OF COILS IS       2
     ITERATION LIMIT      600.00
     TIME LIMIT 30.00 MINUTES
     PHASE 1 REQUESTED                T
     AREA
     (    0.  ,    0.   ) (    2.0100,    2.0100) IN METERS
     NUMBER OF CURVES        1.00
        5 POINTS
     AREA OF INFLUENCE . . .         1          1         134         134
     CURVE (X,Y)       2.00       2.00     134.00       2.00     134.00   134.00       2.00    134.00       2.00       2.00
     COIL COORDINATES          24         56         51         90       CURRENT DENSITY          46.2330E+05IN A/SQ.M.
     COIL COORDINATES          51         63         63         90       CURRENT DENSITY          46.2330E+05IN A/SQ.M.
     IRON COORDINATES           1         93         64        115
     IRON COORDINATES          64         54         90        115
     IRON COORDINATES          90         58         98        115
     IRON COORDINATES          98         63        107        115
     TOTAL NO. OF IRON NODES =        4198
     MESH SIZE     .015000 METERS
     CALCULATED AMPERE TURNS       1.368960E+06
        RESULTS FOR FIELDS WILL BE PRINTED OR PLOTTED FOR THE FOLLOWING,  2REGIONS
          1       1      134      134              F              T          T    -0.
          1       1      134      134              F              F          T    -0.
     LEFT VERTICAL BOUNDARY TYPE IS         2
     PRINT TYPE = 4.00    NUMBER OF ITER PHASE 1 IS   50
      METHOD FOR OMEGA=  8.000     UNDERRELAXATION FACTOR FOR MU=   .880
     OMEGA RE-CALCULATETED EVERY        8 ITERATIONS
     OMEGA CALC. BY FRANKELS EQN =   1.953465
       TYPE OF OPTIMIZATION PROCESS =      0
     INITIAL PERMEABILITY =         100
     MU NO LONGER RELAXED         WHEN ERROR (   5.0000E-05
     THERE WILL BE     20 EQUIPOTENTIAL CURVES DRAWN      5ADDITIONAL ITERATIONS DONE TO SEETLE CONVERGENCE
     ALL CURRENT DENSITIES ARE POSITIVE


     ------------------- ELAPSED TIME      2 SECONDS
```

FIG. 32. Printout of input specification for program NUTCRACKER.

d. Output Description

NUTCRACKER provides for a variety of output option as specified by INFO(4). A typical printout begins with a listing of the input parameters used in the program, as well as diagnostics of errors that might have occurred in the input data. Figure 32 shows the input printout.

Next NUTCRACKER draws the geometry of the magnet as shown in Fig. 33. This map is followed by an iterative account of the convergence rate with occasional calls to subroutine IMPROV which modifies the SOR factor ω to improve convergence; Fig. 34 shows a typical printout.

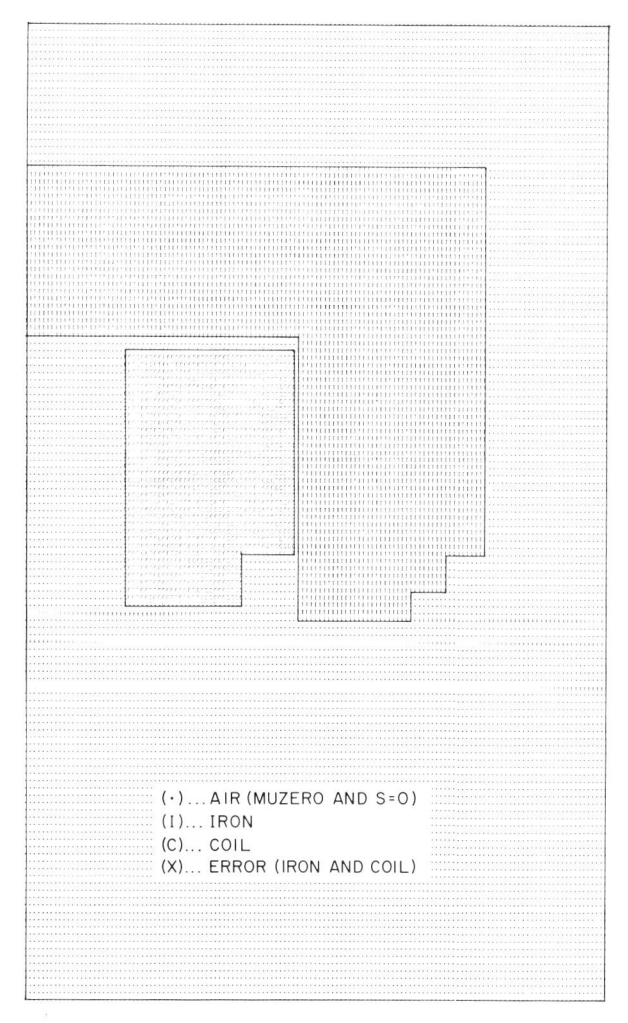

FIG. 33. Printer-plot of the magnet geometry, the border and outline of the geometry was traced for legibility.

74 I. Two-Dimensional Magnetostatic Programs

```
------------------ ELAPSED TIME       7 SECONDS
          CALLING IMPROV( 3), EOLD=    1.000000 OMEGA=  1.953465E+00

AITKEN EXTRAPOLATION AND CARRES METHOD
LAST 15 VALUES OF LAMBDA =  1.413128E+00  7.626025E-01  1.053137E+00  9.611830E-01  9.804661E-01
                            9.755551E-01  9.743589E-01  9.748491E-01  9.734034E-01  9.724258E-01
                            9.724641E-01  9.727515E-01  9.725378E-01  9.726986E-01  9.725684E-01
 AITKEN EXTRAPOLATED VALUE OF LAMBDA =    9.726266E-01
 ESTIMATED SQ.E.V. OF JACKOBI MATRIX =    9.995314E-01
 OMNEW =   1.957625E+00    OMEGA RETURNED (OMEG2) =  1.947031E+00
 ITER =   1                     CONVERGENCE RATE   59.7592593E-03       RELATIVE ERROR   47.6172597E-03
 ITER =   2                     CONVERGENCE RATE   20.8662735E-04       RELATIVE ERROR   45.3192384E-03
 ITER =   3                     CONVERGENCE RATE   47.3491169F-04       RELATIVE ERROR   42.2764457E-03
 ITER =   4                     CONVERGENCE RATE   27.8660742E-04       RELATIVE ERROR   39.8085135E-03
 ITER =   5                     CONVERGENCE RATE   23.9719012E-04       RELATIVE ERROR   37.4939448E-03
             CURVE         1        SCALING FACTOR  13.447879E+01
 MU RELAXED, AVERAGE MU =     1.4299E+02
       .                                   .                                     .
       .                                   .                                     .
       .                                   .                                     .
 ITER =   308                   CONVERGENCE RATE   49.5341612E-06       RELATIVE ERROR   67.2616141E-06
 ITER =   309                   CONVERGENCE RATE   59.6403525E-06       RELATIVE ERROR   64.4701581E-06
 ITER =   310                   CONVERGENCE RATE   50.3192835E-06       RELATIVE ERROR   62.1985024E-06
 MU RELAXED, AVERAGE MU =     7.1541E+02
 ITER =   311                   CONVERGENCE RATE  -44.4575946E-05       RELATIVE ERROR   85.5195564E-06
 ITER =   312                   CONVERGENCE RATE   29.7448733E-05       RELATIVE ERROR   69.0682992E-06
          CALLING IMPROV( 1), EOLD=    .000014 OMEGA=  1.931363E+00

AITKEN EXTRAPOLATION AND CARRES METHOD
 LAST THREE   APPROX TO LAMBDA =    9.709457E-01  9.439143E-01  9.626819E-01
 AITKEN EXTRAPOLATED VALUE OF LAMBDA =    9.549913E-01
 ESTIMATED SQ.E.V. OF JACKOBI MATRIX =    9.988938E-01
 OMNEW =   1.935621E+00    OMEGA RETURNED (OMEG2) =  1.919526E+00
 ITER =   313                   CONVERGENCE RATE   21.7211736E-05       RELATIVE ERROR   52.1164243E-06
 ITER =   314                   CONVERGENCE RATE   47.7690151E-06       RELATIVE ERROR   50.3489485E-06
 ITER =   315                   CONVERGENCE RATE   66.8920472E-06       RELATIVE ERROR   47.9660788E-06

 CONVERGENCE ACHIEVED.
 AVERAGE TIME PER ITERATION=  3.02 SECONDS.    TOTAL ELAPSED TIME=  957 SECONDS.

------------------ ELAPSED TIME       957 SECONDS

------------------ ELAPSED TIME       957 SECONDS
 AMPERE TURNS =    1.379300331E+06
```

FIG. 34. A partial printout of the iteration cycles.

Upon convergence, NUTCRACKER produces an equipotential plot on output paper as shown in the upper half of Fig. 35 while the lower half shows the equivalent equipotential CalComp produced by program TRIM. Following this plot NUTCRACKER prints a map of the magnetic field (or vector potential) components as specified by the parameter, INFO(4); Figure 36 shows part of such printout.

5. Program MAREC

a. Introduction

This program computes two-dimensional static magnetic fields and includes the effects of finite permeability of the iron. MAREC was developed at CERN (see Perin [18], Van der Meer [19], and Perin and Van der Meer [20]), and has been extensively used in the magnet design studies of the Intersecting

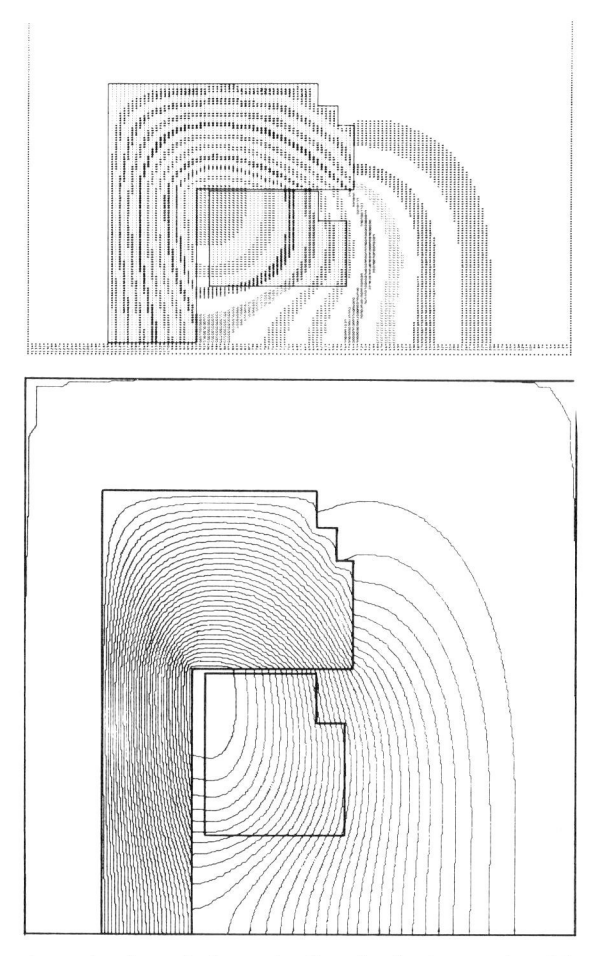

FIG. 35. A printer-plot (upper) shows the flux distribution produced by program NUT-CRACKER, and the flux distribution produced by program TRIM (lower).

Storage Ring project and in the present studies for the new SPS facility at CERN.

Historically, program MAREC was divided into two separately executable programs: MARE-A (Magnet Relaxation in Air), which computes the magnetic field distribution with constant permeability (see Perin [18]); MARE-I (Magnetic Relaxation in Iron), which calculates the flux distribution and the scalar potential on the boundary of two-dimensional saturated magnet cores (see Van der Meer [19]); MARE-C (Magnet Relaxation Combined) combines these two programs to solve completely the magnetic field problem in both air-coil and iron regions (see Perin and Van der Meer [20]).

These codes have been written in Fortran language for the CDC 6600/7600

FLUX DENSITY IN KILOGAUSS - B, BX, BY -
1 TO 5

33	13.83029	13.830	0.	13.83171	13.832	-.030	13.83913	13.839	-.077	13.85251	13.852	-.123	13.87180	13.871	-.169
32	13.87914	13.879	0.	13.88096	13.881	-.027	13.88747	13.887	-.070	13.89961	13.899	-.112	13.91725	13.916	-.153
31	13.92321	13.923	0.	13.92527	13.925	-.024	13.93188	13.932	-.063	13.94298	13.943	-.102	13.95897	13.958	-.139
30	13.96445	13.964	0.	13.96597	13.966	-.020	13.97247	13.972	-.057	13.98317	13.983	-.092	13.99800	13.997	-.126
29	14.00232	14.002	0.	14.00374	14.004	-.017	14.00950	14.009	-.050	14.01964	14.019	-.083	14.03390	14.033	-.114
28	14.03678	14.037	0.	14.03814	14.038	-.015	14.04341	14.043	-.045	14.05280	14.053	-.074	14.06604	14.066	-.103
27	14.06866	14.069	0.	14.06971	14.070	-.012	14.07457	14.075	-.039	14.08324	14.083	-.066	14.09545	14.095	-.092
26	14.09889	14.099	0.	14.09933	14.099	-.010	14.10363	14.104	-.034	14.11137	14.111	-.059	14.12277	14.123	-.082
25	14.12658	14.127	0.	14.12696	14.127	-.008	14.13050	14.130	-.030	14.13746	14.137	-.052	14.14768	14.147	-.073
24	14.15092	14.151	0.	14.15158	14.152	-.006	14.15484	14.155	-.026	14.16112	14.161	-.046	14.17022	14.170	-.065
23	14.17341	14.173	0.	14.17383	14.174	-.005	14.17717	14.177	-.023	14.18288	14.183	-.041	14.19126	14.191	-.057
22	14.19383	14.194	0.	14.19427	14.194	-.003	14.19719	14.197	-.019	14.20262	14.203	-.036	14.21041	14.210	-.051
21	14.21227	14.212	0.	14.21265	14.213	-.002	14.21530	14.215	-.016	14.22025	14.220	-.031	14.22746	14.227	-.045
20	14.22983	14.230	0.	14.22960	14.230	-.000	14.23201	14.232	-.014	14.23659	14.237	-.027	14.24312	14.243	-.039
19	14.24633	14.246	0.	14.24589	14.246	.000	14.24748	14.247	-.011	14.25147	14.251	-.023	14.25764	14.258	-.034
18	14.26157	14.262	0.	14.26107	14.261	.001	14.26227	14.262	-.009	14.26532	14.265	-.019	14.27062	14.271	-.029
17	14.27477	14.275	0.	14.27440	14.274	.001	14.27555	14.276	-.008	14.27830	14.278	-.016	14.28245	14.282	-.025
16	14.28580	14.286	0.	14.28579	14.286	.002	14.28707	14.287	-.006	14.28962	14.290	-.014	14.29364	14.294	-.021
15	14.29686	14.297	0.	14.29635	14.296	.003	14.29752	14.298	-.005	14.29999	14.300	-.012	14.30366	14.304	-.018
14	14.30735	14.307	0.	14.30651	14.307	.003	14.30709	14.307	-.003	14.30930	14.309	-.010	14.31257	14.313	-.016
13	14.31656	14.317	0.	14.31574	14.316	.003	14.31616	14.316	-.003	14.31777	14.318	-.008	14.32085	14.321	-.013
12	14.32488	14.325	0.	14.32422	14.324	.003	14.32446	14.324	-.002	14.32586	14.326	-.006	14.32833	14.328	-.011
11	14.33226	14.332	0.	14.33151	14.332	.003	14.33176	14.332	-.001	14.33287	14.333	-.005	14.33485	14.335	-.009
10	14.33861	14.339	0.	14.33780	14.338	.003	14.33818	14.338	-.001	14.33922	14.339	-.004	14.34112	14.341	-.007
9	14.34440	14.344	0.	14.34365	14.344	.003	14.34382	14.344	-.000	14.34503	14.345	-.003	14.34687	14.347	-.006
8	14.35065	14.351	0.	14.34929	14.349	.003	14.34911	14.349	.001	14.34989	14.350	-.002	14.35162	14.352	-.005
7	14.35624	14.356	0.	14.35483	14.355	.002	14.35423	14.354	.001	14.35450	14.355	-.001	14.35594	14.356	-.003
6	14.36097	14.361	0.	14.36003	14.360	.002	14.35935	14.359	.000	14.35944	14.359	-.001	14.36029	14.360	-.002
5	14.36521	14.365	0.	14.36418	14.364	.002	14.36382	14.364	.000	14.36383	14.364	-.001	14.36430	14.364	-.002
4	14.36855	14.369	0.	14.36739	14.367	.001	14.36691	14.367	.000	14.36721	14.367	-.001	14.36783	14.368	-.001
3	14.37153	14.372	0.	14.37058	14.371	.001	14.36994	14.370	.000	14.37021	14.370	-.000	14.37102	14.371	-.001
2	14.37504	14.375	0.	14.37334	14.373	.001	14.37322	14.373	.000	14.37291	14.373	.000	14.37337	14.373	-.000
1	14.37855	14.379	0.	14.37567	14.376	0.	14.37525	14.375	0.	14.37508	14.375	0.	14.37500	14.375	0.

FLUX DENSITY IN KILOGAUSS - B, BX, BY -
6 TO 10

33	13.89620	13.895	-.213	13.92570	13.923	-.255	13.96032	13.957	-.294	14.00021	13.996	-.332	14.04467	14.040	-.367
32	13.94025	13.939	-.194	13.96808	13.966	-.232	14.00045	13.998	-.268	14.03722	14.034	-.302	14.07877	14.075	-.333
31	13.98025	13.979	-.176	14.00616	14.005	-.211	14.03664	14.035	-.244	14.07106	14.068	-.275	14.10945	14.106	-.303
30	14.01721	14.016	-.159	14.04111	14.040	-.191	14.06924	14.068	-.221	14.10163	14.099	-.249	14.13734	14.135	-.274
29	14.05169	14.051	-.144	14.07363	14.073	-.173	14.09946	14.098	-.200	14.12919	14.127	-.225	14.16266	14.160	-.248
•															
•															
•															
4	14.36891	14.369	-.002	14.37044	14.370	-.002	14.37258	14.373	-.003	14.37560	14.376	-.004	14.37881	14.379	-.004
3	14.37225	14.372	-.001	14.37354	14.374	-.001	14.37475	14.375	-.002	14.37718	14.377	-.002	14.37987	14.380	-.003
2	14.37482	14.375	-.001	14.37623	14.376	-.001	14.37770	14.378	-.001	14.37924	14.379	-.001	14.38107	14.381	-.001
1	14.37611	14.376	0.	14.37767	14.378	0.	14.38119	14.381	0.	14.38098	14.381	0.	14.38388	14.384	0.

FIG. 36. Partial printout from NUTCRACKER showing a flux density edit.

computer systems and occupy approximately $51K_8$ memory locations for a 4000-point mesh. A typical run takes approximately seven to thirteen minutes on the CDC 6600, depending on the field level and the size of the mesh used. The program is general enough to accept any reasonable pole configuration, including concavities, several minima and a large number of coils of any shape.

b. Program Description

MAREC makes use of the modified scalar-potential method applied by several experimenters to magnet computations [24, 24a]. The universe of the problem is divided into two parts: the air-coil and iron regions, which are computed separately thus realizing a considerable saving in computer time and memory space.

We begin with the fundamental relations $\nabla \cdot \mathbf{B} = 0$ and $\nabla \times \mathbf{H} = \mathbf{J}$ and first consider that μ is constant. In areas where $\mathbf{J} = 0$, the magnetic scalar potential is used to define the field on $\mathbf{H} = -\nabla\phi$. However, in current carrying regions, since $\mathbf{H}$ is not irrotational, it is necessary to define a vector $\mathbf{M}$ such that **H** − **M** is irrotational. This leads to the solution of a Poisson-type equation of the form:

$$\nabla^2 \phi^* = -\nabla \cdot \mathbf{M}. \tag{5.1}$$

The solution of this equation by finite difference methods necessitates the construction of a grid on which the magnet geometry is superimposed. The finite difference equations for each node that lie inside the domain are approximated by

$$\partial^2 \phi^*/\partial x^2 \approx 2[(\phi_1^* - \phi_0^*)/d_1 - (\phi_0^* - \phi_3^*)/d_3]/(d_1 + d_3), \tag{5.2}$$

$$\partial^2 \phi^*/\partial y^2 \approx 2[(\phi_2^* - \phi_0^*)/d_2 - (\phi_0^* - \phi_4^*)/d_4]/(d_2 + d_4). \tag{5.3}$$

In Fig. 37, the finite difference equation for node 0 is obtained by substituting

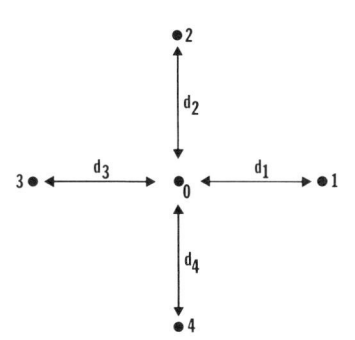

FIG. 37. A mesh cell for program MAREC showing the notation of the localized system used.

the right side of Eqs. (5.2) and (5.3) into Eq. (5.1), resulting in

$$\phi_0^* = (\phi_1^* A + \phi_2^* B + \phi_3^* C + \phi_4^* D + E)/K, \tag{5.4}$$

where

$$A = (d_2 + d_4)/d_1, \quad B = (d_1 + d_3)/d_2,$$
$$C = (d_2 + d_4)/d_3, \quad D = (d_1 + d_3)/d_4, \tag{5.5}$$

$$E = A \int_0^1 M \, dx - C \int_3^0 M \, dx, \tag{5.6}$$

and

$$K = A + B + C + D. \tag{5.7}$$

Refer to Fig. 37 for the localized numbering system and notations. Equation (5.7) is valid for any interior mesh point, as well as for points close to the boundary.

By an appropriate use of a successive overrelaxation scheme, the potentials ϕ^* are successively corrected so that at the $n + 1$st iteration, the correction applied is

$$\phi^{*(n+1)} = \phi_1^{*(n)} + \varrho[A\phi_1^{*(n)} + B\phi_2^{*(n)} + C\phi_3^{*(n)} \\ + D\phi_4^{*(n)} + E - K\phi_0^{*(n-1)}]/K \quad (5.8)$$

where ϱ is the accelerating factor whose optimal value can be introduced in the data or computed by the program.

So far, we have considered μ a constant. In reality the magnetic field intensity **H** satisfies

$$\nabla \times \mathbf{H} = -4\pi \mathbf{J}. \quad (5.9)$$

The magnetic induction **B** may be determined from the magnetic vector potential **A** from

$$\mathbf{B} = \mu \mathbf{H} = \nabla \times \mathbf{A}, \quad (5.10)$$

where μ is the relative permeability. Substituting Eq. (5.10) into Eq. (5.9) we obtain

$$\nabla \times (1/\mu \, \nabla \times \mathbf{A}) = -4\pi \mathbf{J}. \quad (5.11)$$

Since **A** in two dimensions has one nonzero component,

$$\mathbf{A} = (0, 0, A_Z). \quad (5.12)$$

Eq. (5.11) can be written as

$$\nabla \cdot (\gamma \nabla A) = -4\pi J. \quad (5.13)$$

where $\gamma = 1/\mu$.

Performing the indicated operation we obtain

$$\frac{\partial^2 A}{\partial x^2} + \frac{\partial^2 A}{\partial y^2} - \frac{1}{\mu}\left[\frac{\partial \mu}{\partial x}\frac{\partial A}{\partial x} - \frac{\partial \mu}{\partial y}\frac{\partial A}{\partial y}\right] = 0. \quad (5.14)$$

The permeability μ is defined as a function of the magnetic induction and is given as a table of values. The vector potential A is calculated at the nodes of the superimposed grid, and a solution is found by approximating Eq. (5.14) with a set of finite difference equations.

From the calculated vector potentials the value of the field components of B are calculated by

$$B_x = \partial A/\partial y, \quad (5.15)$$

$$B_y = -\partial A/\partial x. \quad (5.16)$$

Thus, the absolute value of B is determined, from which the permeability is found by suitable interpolating schemes. This process is repeated with adjusted permeability values until the change in permeability per cycle is less than a specified percentage for all mesh points.

This brief description of the method does not include the special techniques employed by the program to ensure or accelerate convergence; interested readers may refer to the literature [24, 24a, 24b, 25] for a more thorough description of the various schemes used.

Program Organization

The program is organized as shown in the simplified block diagram (see Fig. 38). The code begins by reading the data that describe the magnet geometry and setting up the mesh necessary for the problem. The computation begins in the air-coil region, with the initial boundary conditions specified in the data. By assuming an equipotential pole (in the absence of a better approximation), the scalar potentials in air are computed, thus establishing the initial boundary condition for the vector potential inside the iron region. The iron computation begins by determining the vector potentials and permeabilities at each mesh point in the iron universe, calculating the mmf drops, and establishing the criteria necessary to revise the values of the current and the scalar potentials on the air-iron interface for the next cycling process. The iterative process continues until a specified convergence criterion is achieved. At the end of the computation loop, magnetic tapes are reserved for separate solutions to the air-coil and the iron regions, which in turn may be used as initial conditions for subsequent problems, with great savings in computer time.

c. Preparation of Input Data

The preparation of input data for MAREC follows the same general pattern as that described in the previous magnet programs, that is, the user must provide information pertaining to the geometry, coil configuration, type of material, and various other constants necessary for the successful operation of the program. The necessary parameters and specifications as well as the sequence and card formats are shown in tabular form in Table 19. The first card of the data deck must have four parameters, MCYCLE, PESO, STACK, MREAD, punched on one card with (I5,2F6.3,2I5) format. Similarly, the third and subsequent NPTS-1 cards will consist of the air-boundary specification given in the $\bar{y}$, $\bar{x}$ coordinate system. These quantities consist of the x and y coordinates of the specified point; the initial guess of the scalar potential for point (x, y); and a code digit specifying the type of boundary point. In a similar fashion all the quantities appearing in Table 19, are punched sequentially on cards. It should be noted that some of the points require specification in either of two

Table 19

MAREC Input Parameters and Specifications

Parameter name	Format	Cards	Description
MCYCLE,PESO,STACK, MREAD,MSMOOT	I5,2F6.3,2I5	1	MCYCLE is the required number of complete cycles. PESO is a weighting factor to control the change of air boundary conditions from cycle to cycle. Normal convergence, set = 1.0. STACK is the packing factor for laminated cores. MREAD is a switch for reading in a previous solution, set MREAD = 0 for No, and MREAD ≠ 0 for Yes, read in approximation. MSMOOT is not debugged, set = 0.
DXA,DYA	2F10.5	1	The horizontal (DXA) and vertical (DYA) spacing in air, with input units in millimeters.
NPTS	45X,I5	1	The number of air boundary points to be specified, NPTS ≦ 499.
$X_i,Y_i,V_i,KODE_i$ [There must be NPTS boundary-point cards arranged in a sequential manner, such that any two points are geometrically neighbors.]	3E15.7,I5	NPTS	X_i and Y_i are the x, y coordinates of a specified boundary point in millimeters. V_i is the value of the modified scalar potential in ampere-turns. Leave KODE blank for normal points; KODE = 2 for a point on horizontal Neumann boundary; KODE = 3 for intersections of Neumann boundaries; and KODE = 4 for a point on the vertical Neumann boundary.
NUMCO,MPRO	2I5	1	NUMCO is the total number of rectangular conductors (≦100). MPRO influences the boundary condition for the scalar potential and informs the code on the choice of the vector $\vec{M}$. If MPRO = 0, the code assumes $\vec{M}$ = 0, is on the upper part of the air boundary. If MPRO = 1, the assumption made is that $\vec{M}$ = 0, is on the lower part of the air boundary.
COABL,COABR,COORL, COORU,CUDE	4F9.2,F7.3	NUMCO	The first four parameters are the coordinates (millimeters) of a rectangular conductor. The order shown should be maintained. x_1,y_1 (COABR,COORU) x,y (COABL,COORL) CUDE is the current density of the conductor in amperes/millimeter^2.
EWANT	E14.8	1	The desired accuracy (%) for the scalar potential, air-coil region.
XFE,YFE	2F10.3	1	The coordinates (millimeters) of the last point at which the flux crossing the air-iron interface is computed. This point must be given in the $\bar{x}, \bar{y}$ coordinate system (see Fig. 39) and they define a point which must lie at the intersection of a mesh line of the air region with the air-iron interface.
NINS	I5	1	To facilitate the code in reaching the corner points of the air-iron interface, e.g., coil-window, it is advisable (not compulsory) to enter in data cards, such corner points. NINS specifies the number of such points entered (NINS≦10).
$XINS_i,YINS_i$	2F10.3	NINS	The corner-point coordinates, given in the $\bar{x}, \bar{y}$ system (see Fig. 39), due to rounding error it is recommended to displace these corner-points by the product of, 0.0001 × DXA, toward the concave side of the corner. Units are in millimeters.
NADD	I5	1	The number of points needed to close the iron boundary (NADD≦20).

Table 19 (*cont.*)

Parameter name	Format	Cards	Description
$XADD_i, YADD_i, NCADD_i,$ XFL_i, YFL_i	2F10.3,I5,2F10.3	NADD	$XADD_i$ and $YADD_i$ are the coordinates of the point (millimeters) in the $\bar{x}', \bar{y}'$ coordinate system (see Fig. 39). $NCADD_i = 2$ for all points on a horizontal Neumann boundary of the iron region. $NCADD_i = 3$ for points lying on a vertical Neumann boundary, and $NCADD_i = 0$ for all the remaining points. The flux value in a point $[XADD(I), YADD(I)]$, will always be taken to equal to the flux value at some point $[XFL(I), YFL(I)]$ of the air-iron interface (see Fig. 39). The coordinates XFL(I) and YFL(I) must be given in the $\bar{x}', \bar{y}'$ coordinate system, and define a point which must not necessarily lie on a mesh line of the air region (millimeters).
DXI,DYI	2F10.5	1	The horizontal and vertical mesh spacing (millimeters) in iron.
W	F10.3	1	The width (millimeters) of the return yoke (see Fig. 39).
TIME	F10.5	1	The time limit (minutes) for the iteration in iron.
EMU,EFINAL,NTABLE	2F10.0,I5	1	EMU is the requested precision (percent) for the permeability. EFINAL is the requested precision (percent) for the vector potentials in iron. NTABLE is the number of points in the permeability versus induction table (NTABLE $\leq$ 100).
B_i, AMU_i	6F10.0 (three pairs per card)	NTABLE	B_i is the induction (webers/meter). B_i must be zero, $B_i + 1 > B_i$. The highest B value must be larger than the maximum induction expected. AMU_i is the corresponding permeability value.
NPR	I5	1	A number of points for checking purposes. After each approximation, the permeability in each of these points is printed (NPR $\leq$ 8).
KPR	8I5	1	The point number of each of the check points (see NPR).
XPE,YPE	2F10.3	1	The coordinates (millimeters) of the last point at which the mmf drops are computed at each intersection of a mesh line of the iron region with the iron boundary (see Fig. 39). The coordinates of XPE and YPE are given in the $\bar{x}', \bar{y}'$ coordinate system, and they define a point which must lie at the intersection of a mesh line of the iron region with the iron boundary, beyond the air-iron interface.
XEL,YEL	2F10.3	1	The mmf drop along the iron boundary is used to modify the scalar potential at the intersection points of the mesh lines of the air region with the air-iron interface. The coordinates of XEL, YEL (millimeters) must be given in the $\bar{x}, \bar{y}$ coordinate system and they define a point coincident with the intersection of the artificial closing-boundary and the air-iron interface (see Fig. 39).
XE,YE	2F10.3	1	The coordinates of a point (millimeters) needed to calculate the ampere-turns necessary to achieve the required field strength discarding saturation effects. This is accomplished by computing $\int \bar{H} \cdot ds$, between some point in the median plane and the chosen point (XE,YE) on the interface. The coordinates of (XE,YE) must be given in the $\bar{x}, \bar{y}$ coordinate system (see Fig. 39).

(Any data required by subroutine USER)

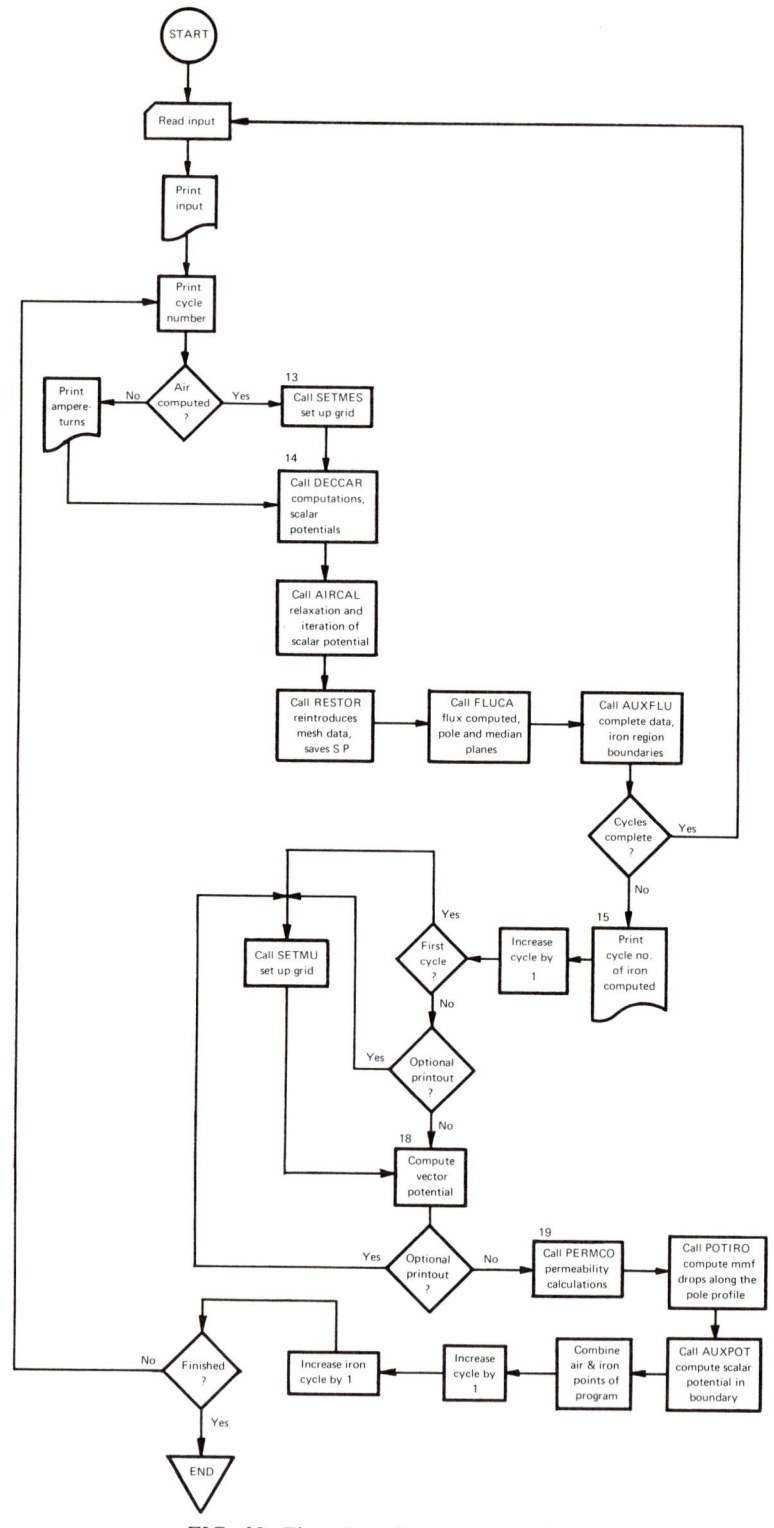

FIG. 38. Flow chart for program MAREC.

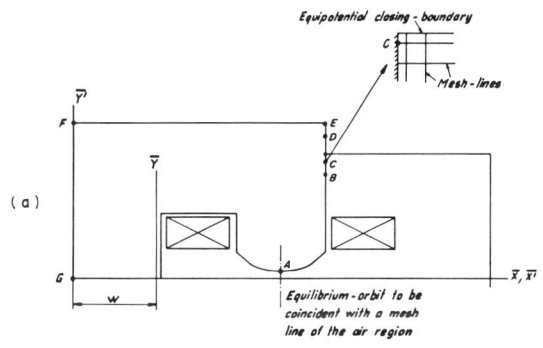

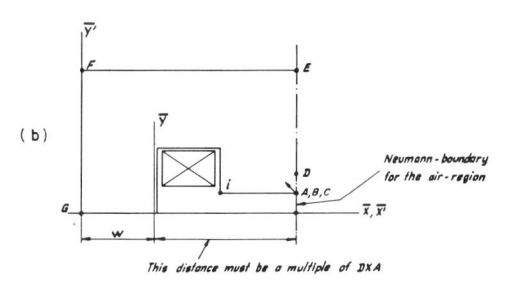

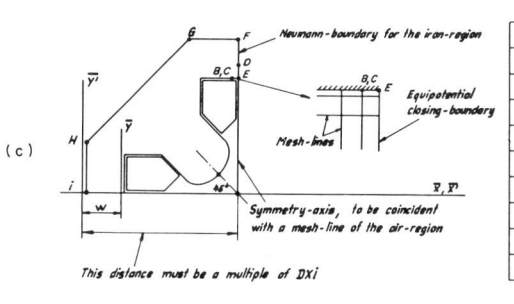

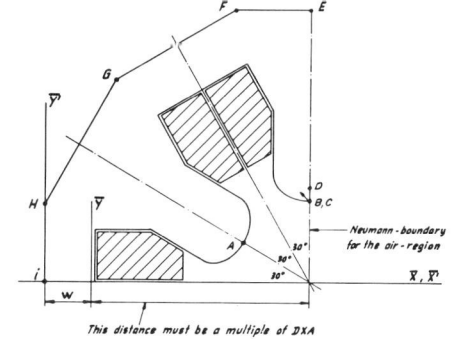

FIG. 39. Definition of various input quantities for various types of magnets. (a) C magnet; (b) symmetrical H magnet; (c) quadrupole; (d) sextupole.

84   I. Two-Dimensional Magnetostatic Programs

distinct coordinate systems pertaining to the air ($\bar{x}, \bar{y}$) and iron ($\bar{x}', \bar{y}'$) portions of the program. Figure 39 shows the location of these points in various sample magnet geometries with their respective coordinate systems.

With the input data complete, the program proceeds with the calculation of the resulting scalar potential in the air region. From this potential, we derive other useful quantities: field components, gradients, etc. The computation of these derived quantities is performed in subroutine USER supplied by the user.

```
      SUBROUTINE USER
C
C     COMPUTES THE FIELD IN THE MEDIAN PLANE FROM X=MIN TO X=MAX
C
      COMMON NCYCLE,DUM1(7),DX,DUM2(23507),X(250),B(250)
      IF(NCYCLE-1) 1,1,2
    1 READ(1,100) XMIN,XMAX
      I1=XMIN/DX+1.0
      I2=XMAX/DX+1.0
    2 DO 3 I=I1,I2
      CALL DERI(I,BX,B(I))
      X(I)=DX*FLOAT(I-1)
    3 CONTINUE
      WRITE(2,200) (X(I),B(I),I=I1,I2)
  100 FORMAT(2F10.0)
  200 FORMAT(4(F9.3,2X,F10.3,5X),F9.3,2X,F10.3)
      RETURN
      END
```

FIG. 40. Sample coding of subroutine USER to produce a field edit.

Figure 40 shows a typical structure of USER. An examination of this routine reveals that the user must specify the location of the field components, or gradients to be printed, in terms of XMIN and XMAX, which specify the minimum and maximum value of the x coordinate on the median plane. XMIN, XMAX (format 2F10.3) is the input for subroutine USER.

d. An Illustrative Example

To further show the process of data preparation and the correct sequence of cards to produce an accurate description of the contemplated magnet geometry, the CERN-PS magnet will be used. Table 20 shows a complete listing of the required data for this magnet prepared in conjunction with the input description shown in Table 19. The pole profile was simulated with 40 points and the mesh size was set to 1 centimeter to conform with programs TRIM and LINDA, so that some comparable statistics could be obtained. The permeability curve used was identical to TRIM and LINDA and the lamination stacking factor was set to 0.95. The data appearing in Table 20 corresponds to a computation at 10.0 kilogauss.

Table 20

MAREC Input Data for the CERN-PS Magnet

```
                              Card Column
  1     5    10    15    20        30        40        50        60       70

        4   1.0   0.95    0      0
      10.0        10.0
                                                                40
                 0.0           0.0       2000.0
                 0.0         220.0       2000.0
               220.0         220.0       2000.0
               220.0          85.26      2000.0
               235.0          62.00      2000.0
               245.0          49.40      2000.0
               255.0          40.60      2000.0
               260.0          37.69      2000.0
               270.0          33.69      2000.0
               280.0          33.22      2000.0
               290.0          34.14      2000.0
               300.0          35.22      2000.0
               310.0          36.39      2000.0
               320.0          37.62      2000.0
               330.0          38.82      2000.0
               340.0          40.10      2000.0
               350.0          41.47      2000.0
               360.0          42.93      2000.0
               370.0          44.51      2000.0
               380.0          46.20      2000.0
               390.0          48.02      2000.0
               400.0          50.00      2000.0
               410.0          52.15      2000.0
               420.0          54.48      2000.0
               430.0          56.51      2000.0
               440.0          59.85      2000.0
               450.0          62.95      2000.0
               460.0          66.39      2000.0
               470.0          70.23      2000.0
               480.0          74.54      2000.0
               490.0          79.41      2000.0
               500.0          84.97      2000.0
               580.0         220.00      2000.0
               580.0         520.00      2000.0
              1500.0         520.00      2000.0
              1500.0           0.00      2000.0
               790.0           0.00      2000.0
               590.0           0.00     42250.0
               210.0           0.00     42250.0
                10.0           0.00      2000.0
         2      1
        10.0         210.0        80.0     210.0     -1.548
       590.0         790.0        80.0     210.0      1.548
         1.0E-2
       580.0         500.0
         2
         0.0         220.0
       220.0         220.0
         3
       940.0         580.0       0       940.0       500.0
         0.0         580.0       0       940.0       500.0
         0.0           0.0       2       940.0       500.0
        20.0          20.0
       360.0
        15.0
        10.0          10.0      40
         0.0        4444.44      1.2     3246.75       1.4      2222.22
         1.5        1499.25      1.55    1069.52       1.6       709.22
         1.65        458.72      1.7      308.64       1.75      230.41
         1.8         176.37      1.85     139.08       1.9       111.11
         1.95         90.09      2.0       75.19       2.05       60.98
         2.1          49.50      2.125     44.25       2.15       39.06
         2.175        33.00      2.2       26.53       2.25       18.0
         2.2797       15.0       2.3069    13.0        2.3443     11.0
         2.3996        9.0       2.4905     7.0        2.5627      6.0
         2.6706        5.0       2.8498     4.0        3.2074      3.0
         3.5644        2.5       4.2782     2.0        4.8134      1.8
         5.7052        1.60      6.4186     1.5        7.4887      1.4
         9.2721        1.3      10.6989     1.25      12.8389      1.2
        23.5390        1.1
         3
        13    256   1267
       940.0         560.0
       580.001       510.001
       400.0          50.0
        10.0        1000.0
```

The output produced by MAREC begins with a listing of the input data and thereafter for each cycle the following tables are printed for the air portion of the code:

1. Output produced by subroutine USER (see Fig. 41)
2. Flux crossing the iron-air interface
3. Flux crossing the median plane.

```
CYCLE NUMBER    5         AIR COMPUTATION

REQUIRED AMPERE-TURNS=    40535.633


CONVERGENCE OBTAINED AFTER      8 ITERATIONS

  60.00    -761.044      70.00    -998.365      80.00   -1237.903      90.00   -1485.656     100.00   -1746.840
 110.00   -2026.702     120.00   -2331.088     130.00   -2666.956     140.00   -3042.978     150.00   -3479.366
 160.00   -3964.168     170.00   -4545.572     180.00   -5246.606     190.00   -6121.320     200.00   -7276.897
 210.00   -8973.196     220.00   -9006.782     230.00   -9619.413     240.00  -10567.675     250.00  -11708.282
 260.00  -12863.628     270.00  -13787.740     280.00  -14272.943     290.00  -14339.089     300.00  -14140.694
 310.00  -13809.896     320.00  -13423.043     330.00  -13016.883     340.00  -12634.979     350.00  -12191.329
 360.00  -11777.098     370.00  -11362.723     380.00  -10948.592     390.00  -10535.032     400.00  -10122.333
 410.00   -9710.717     420.00   -9300.016     430.00   -8889.501     440.00   -8478.806     450.00   -8069.560
 460.00   -7664.312     470.00   -7265.774     480.00   -6876.997     490.00   -6501.819     500.00   -6145.204
 510.00   -5013.276     520.00   -5513.027     530.00   -5251.930     540.00   -5038.215     550.00   -4882.493
 560.00   -4802.139     570.00   -4832.414     580.00   -5557.842     590.00   -5712.675     600.00   -4568.335
 610.00   -3848.230     620.00   -3313.477     630.00   -2875.358     640.00   -2495.131     650.00   -2153.161
 660.00   -1838.069     670.00   -1542.471     680.00   -1261.065     690.00    -989.660     700.00    -724.693
 710.00    -462.342     720.00    -199.005     730.00      70.177     740.00     351.835     750.00     656.366
 760.00    1002.177     770.00    1426.525     780.00    2016.512     790.00    3039.267     800.00    2179.164
 810.00    1752.466     820.00    1493.132     830.00    1314.924     840.00    1181.527     850.00    1075.536
 860.00     987.675     870.00     912.571     880.00     846.905     890.00     788.517     900.00     735.939
 910.00     688.136     920.00     644.353     930.00     604.024     940.00     566.711     950.00     532.068
 960.00     499.816     970.00     469.722     980.00     441.592     990.00     415.260    1000.00     390.583
```

FIG. 41. Printout produced by subroutine USER showing the total field B in gauss from 10.0 mm to 1000.0 mm in steps of 10.0 mm.

The iron printout section begins with the iron cycle number followed by the successive values of the permeability at the check points specified in the input for each iron iteration. Then the following tables are printed:

1. Permeability map
2. Flux map (Fig. 42)
3. Mesh-point numbers corresponding to the two preceding maps.
4. The mmf drops at successive points of the air-iron interface.

6. Evaluations: TRIM, LINDA, NUTCRACKER, and MAREC

It would be advantageous to compare some of the two-dimensional magnetostatic programs (TRIM, LINDA, NUTCRACKER, and MAREC) as to their memory and time requirements, rate of convergence, and other pertinent geometric and mathematical criteria. Other similar programs will be found later in this text; however, in the opinion of the author, the programs described thus far represent the two-dimensional magnetostatic type most commonly used. A cylindrical magnet (see Fig. 30) was selected for comparative tests with program

6. Evaluations: TRIM, LINDA, NUTCRACKER, and MAREC

FLUX MAP   THESE FIGURES, DIVIDED BY 1000, GIVE THE FLUX IN WB/M

```
378  378  378  378  378  378  378  378  378  378  378  378  378  378  378  378  378
378  375  372  369  367  365  363  362  362  363  365  367  369  372  375  377
378  372  367  361  356  352  349  347  347  349  351  356  361  367  372  376
378  369  361  352  345  338  334  331  331  334  338  344  352  361  369  376
378  366  355  343  333  325  319  315  315  318  324  333  343  355  366  376
378  363  349  335  322  311  303  299  299  302  310  321  334  349  363  376
378  360  343  326  310  297  287  282  282  286  295  309  325  343  361  377
378  357  337  317  298  282  271  265  265  270  280  296  316  337  358  378
378  354  331  308  286  267  254  248  247  253  265  283  307  331  356  379
378  351  325  299  274  252  237  229  229  235  249  270  297  326  354  380
378  349  319  290  262  237  219  211  210  217  232  257  287  320  352  382
378  346  313  281  249  221  200  191  191  198  214  242  278  314  350  383
378  343  308  272  237  204  181  171  171  178  196  228  268  309  348  384
378  340  302  263  224  187  160  150  150  157  176  213  258  304  347  386
378  338  297  255  212  170  138  129  129  135  155  198  249  299  346  388
378  335  292  247  201  153  115  107  107  113  132  183  241  295  345  389
378  333  287  240  190  136   89   84   85   90  110  169  233  292  345  392
378  331  283  234  181  121   62   61   63   66   87  158  227  290  346  395
378  329  279  228  172  110   33   38   41   43   69  149  222  288  347  398
378  327  276  222  165  102   30                      65  145  219  288  349  404
378  326  273  217  159   96   27                      62  142  218  288  353  411
378  324  270  213  154   91   23                      60  141  218  290  357  418
378  323  267  210  150   86   19                      60  141  219  293  361
378  322  265  207  146   83   16                      61  143  222  296  366
378  321  263  204  143   80   14                      63  147  227  301  370
378  320  262  202  141   77   11                      65  151  233  308  374
378  320  261  201  139   74    8                      63  156  241  316
378  319  260  200  137   73    5                          164
378  319  260  199  137   71    2
378  319  259  199  136   71    0
```

POINT NUMBERS CORRESPONDING WITH PRECEDING MAPS

```
1422 1425 1428 1431 1434 1437 1440 1443 1446 1449 1452 1455 1458 1461 1464 1467
1373 1376 1379 1382 1385 1388 1391 1394 1397 1400 1403 1406 1409 1412 1415 1418
1324 1327 1330 1333 1336 1339 1342 1345 1348 1351 1354 1357 1360 1363 1366 1369
1275 1278 1281 1284 1287 1290 1293 1296 1299 1302 1305 1308 1311 1314 1317 1320
1226 1229 1232 1235 1238 1241 1244 1247 1250 1253 1256 1259 1262 1265 1268 1271
1177 1180 1183 1186 1189 1192 1195 1198 1201 1204 1207 1210 1213 1216 1219 1222
1128 1131 1134 1137 1140 1143 1146 1149 1152 1155 1158 1161 1164 1167 1170 1173
1079 1082 1085 1088 1091 1094 1097 1100 1103 1106 1109 1112 1115 1118 1121 1124
1030 1033 1036 1039 1042 1045 1048 1051 1054 1057 1060 1063 1066 1069 1072 1075
 981  984  987  990  993  996  999 1002 1005 1008 1011 1014 1017 1020 1023 1026
 932  935  938  941  944  947  950  953  956  959  962  965  968  971  974  977
 883  886  889  892  895  898  901  904  907  910  913  916  919  922  925  928
 834  837  840  843  846  849  852  855  858  861  864  867  870  873  876  879
 785  788  791  794  797  800  803  806  809  812  815  818  821  824  827  830
 736  739  742  745  748  751  754  757  760  763  766  769  772  775  778  781
 687  690  693  696  699  702  705  708  711  714  717  720  723  726  729  732
 638  641  644  647  650  653  656  659  662  665  668  671  674  677  680  683
 589  592  595  598  601  604  607  610  613  616  619  622  625  628  631  634
 540  543  546  549  552  555  558  561  564  567  570  573  576  579  582  585
 491  494  497  500  503  506  509                      521  524  527  530  533  536
 442  445  448  451  454  457  460                      472  475  478  481  484  487
 393  396  399  402  405  408  411                      423  426  429  432  435  438
 344  347  350  353  356  359  362                      374  377  380  383  386
 295  298  301  304  307  310  313                      325  328  331  334  337
 246  249  252  255  258  261  264                      276  279  282  285  288
 197  200  203  206  209  212  215                      227  230  233  236  239
 148  151  154  157  160  163  166                      178  181  184  187
  99  102  105  108  111  114  117                          132
  50   53   56   59   62   65   68
   1    4    7   10   13   16   19
```

FIG. 42. Partial printout from MAREC showing flux maps of the CERN-PS magnet.

88   I. Two-Dimensional Magnetostatic Programs

TRIM, LINDA, and NUTCRACKER. The CERN Proton Synchrotron magnet (CERN-PS) (see Fig. 9) was selected for comparative analysis using TRIM, LINDA, and MAREC.

Operational characteristics of these codes, are given in Table 21. Each program has its own unique capabilities, and the choice of which code to use depends on the requirements of the user. For example, TRIM reliably produces good results for problems of virtually any geometry. This flexibility, however, is offset by slow convergence in problems with large iron regions, such as the CERN-PS magnet. LINDA produces very accurate results, but is restricted to certain types of magnet geometries.

Another important consideration is the amount of memory required by these programs, since this dictates the size of the computer required, and relates in-

Table 21

Magnetostatic Program Comparisons

	TRIM	LINDA	MAREC	NUTCRACKER
Type of potential	Vector	Scalar/vector	Scalar/vector	Vector
Mesh type	Triangular	Rectangular	Rectangular	Square
Difference operator	7 points	5 points	5 points	5 points
Initial conditions:				
Potentials	0.0	0.0	0.0	0.0
Permeabilities (μ)	1.0	10^4	1.0	10^3
Overrelaxation (ω)	1.94	1.50	-	1.93
Optimization process:				
Recalculation of (ω)	Each 20 iterations	Variable (Program controlled)	-	Each 6 iterations
Magnetization data	6 tables/100 points	1 table/50 points	1 table/100 points	1 table/40 points
Convergence criterion	10^{-6}	rms of residual 10^{-7} for infinite μ, program controlled for finite μ	10^{-6}	4×10^{-6}
Iterations to converge CERN-PS Magnet	1 000	318	323	-
Host computer	IBM360/91 CDC 6600	IBM360/91 CDC 6600	CDC 6600	CDC 6600 IBM360/91
Maximum mesh points	4 000	10 000	16 000	22 500
Memory requirements	156 K_8 words	210 K_8 words	125 K_8 words	297 K_8 bytes
Precision	14 digits	14 digits	14 digits	7 digits
Computation time (CDC 6600):				
CERN-PS Magnet	19.66 min	7.70 min	4.32 min	-
Cylindrical test magnet	6.40 min	14.76 min	-	16.40 min

directly to the degree of the attainable accuracy. Although MAREC and NUTCRACKER have only been tested with arrays needing $125K_8$ and $212K_8$ memory locations respectively, TRIM and LINDA have been used extensively at the Lawrence Berkeley Laboratory with various mesh configurations having different memory requirements. All of the programs in this section are capable of reducing the appropriate arrays to effect a decrease in memory requirements. For example, for a 2000-point problem, TRIM can be accomodated in $120K_8$ memory locations; LINDA can solve a 2500-point problem in less than $100K_8$ locations, but with a corresponding reduction in the attainable accuracy.

There are methods by which the user may increase the accuracy of the results obtained from a reduced mesh. The author suggested in 1967 (see Colonias [11]) a strategy by which the contemplated magnet geometry is first solved with a coarse mesh to obtain initial potentials. Next, the region of interest is isolated and magnified, and the initially calculated potentials are inserted as well as those for the interpolated points. Then, potentials on the boundary of the problem are "frozen" and the enlarged region is solved. This scheme has been successfully used with program TRIM by the author and by Lari [26] with varying degrees of success.

The rate of convergence, also an important factor to be considered, mainly depends on: the size of the mesh used, the ratio of the number of points in air to the number of points in iron, the excitation level, and, to some extent, on the type of magnet. In TRIM, the rate of convergence is dependent upon the uniformity of the triangular mesh, and, somewhat, upon the type of triangles used. Therefore, the potential user should exercise particular care in distributing the available mesh.

The ability of NUTCRACKER to calculate magnetic fields is greatly restricted by its square mesh, and by the large number of mesh points required if sufficient accuracy is to be obtained. These two limitations restrict the use of this program to magnets with approximate geometries and without fine structure. LINDA provides internally programmed mechanisms for the optimum rate of convergence. For example, the code applies ratio-limit and absolute-limit conditions on the amount by which reluctivities may change during their recalculation procedure. These conditions sense the manner by which all reluctivities are changing and allow them to move in groups, rather than by individual points, which prevent oscillations.

LINDA also permits the use of a different relaxation factor for the vector potential by recognizing whether the vector potential is in a region of uniform reluctivity or not, and by adjusting the overrelaxation factor ω accordingly. This mechanism relieves the user of the need to decide whether a region of a magnet is a high or low reluctivity area.

Also in LINDA, where the contemplated magnet geometry has more than one pathway for return flux (two separated flux-normal boundaries), the value of the vector potential is changed by a special $\int \mathbf{H} \cdot d\mathbf{l}$ program which allows for the balancing of flux from one leg to the other. In general, LINDA provides for many mechanisms that tend to minimize computational errors and relieve the user from making decisions regarding the potential difficulties that the code might encounter during the calculation process.

Table 22

CERN-PS Magnet Field Computations

x^a (centimeters)	TRIM BY (teslas)			LINDA BY (teslas)			MAREC BY (teslas)		
	1.0	1.201	1.442	1.0	1.201	1.442	1.0	1.201	1.442
-8.0	1.3270	1.5781	1.8509	1.3260	1.5800	1.8497	1.3406	1.6134	1.8649
-7.0	1.2873	1.5330	1.8066	1.2865	1.5350	1.8064	1.3000	1.5664	1.8225
-6.0	1.2468	1.4863	1.7587	1.2462	1.4890	1.7594	1.2589	1.5185	1.7766
-5.0	1.2060	1.4391	1.7086	1.2055	1.4416	1.7100	1.2176	1.4701	1.7277
-4.0	1.1651	1.3916	1.6568	1.1646	1.3941	1.6587	1.1761	1.4214	1.6763
-3.0	1.1241	1.3437	1.6038	1.1236	1.3463	1.6061	1.1348	1.3724	1.6231
-2.0	1.0829	1.2956	1.5497	1.0825	1.2982	1.5523	1.0934	1.3232	1.5686
-1.0	1.0418	1.2472	1.4948	1.0413	1.2500	1.4980	1.0521	1.2738	1.5132
0.0	1.0005	1.1986	1.4392	1.0000	1.2010	1.4420	1.0109	1.2246	1.4571
1.0	0.9593	1.1499	1.3830	0.9585	1.1519	1.3860	0.9698	1.1751	1.4005
2.0	0.9182	1.1010	1.3262	0.9170	1.1025	1.3285	0.9288	1.1257	1.3436
3.0	0.8770	1.0519	1.2689	0.8754	1.0530	1.2705	0.8878	1.0763	1.2862
4.0	0.8357	1.0028	1.2111	0.8338	1.0034	1.2123	0.8468	1.0268	1.2285
5.0	0.7944	0.9536	1.1530	0.7925	0.9541	1.1542	0.8059	0.9774	1.1706
6.0	0.7531	0.9043	1.0945	0.7515	0.9050	1.0960	0.7654	0.9284	1.1130
7.0	0.7119	0.8549	1.0358	0.7107	0.8560	1.0380	0.7256	0.8802	1.0562
8.0	0.6708	0.8058	0.9771	0.6701	0.8073	0.9796	0.6868	0.8332	1.0006

[a] Denotes distance on the median plane in the vacuum chamber area.

A computer solution of the CERN-PS magnet with TRIM, LINDA, and MAREC provides additional basis for comparison. Three cases were calculated for the B_y component of the field at three excitation levels (see Table 22); both the measured [27] and the computed normalized gradients were plotted (see Fig. 43). In Fig. 43, the normalized field index k, defined as $k = 1/B_0 \, (dB_y/dx)$, is shown as a function of radial distance from the equilibrium orbit. For all three programs the computed gradients compare favorably with those measured.

To present a more accurate picture of performance by NUTCRACKER, TRIM, and RINDA (a cylindrical-coordinate version of LINDA), a cylindrical test magnet

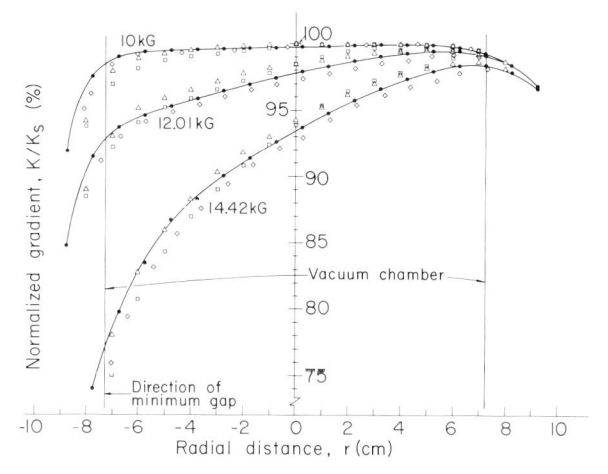

FIG. 43. Normalized gradient plots for the CERN-PS magnet computed by program TRIM ($\triangle$), LINDA ($\square$), and MAREC ($\diamond$); (–●–) measured values.

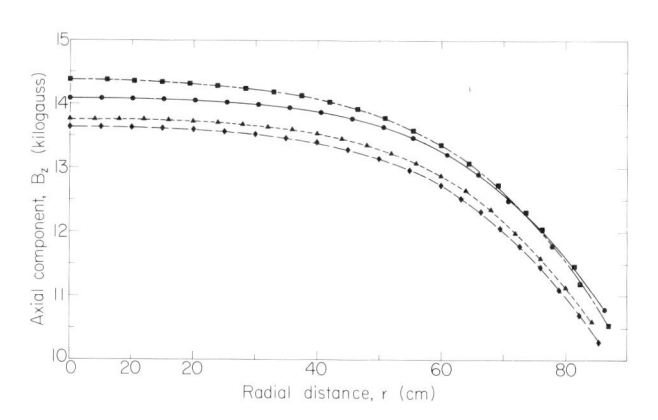

FIG. 44. Comparison curves of the axial component B_z of the test magnet. Program RINDA is a special cylindrical version of LINDA. Cylindrical magnet test case. At = 1368960.0. Edit on median plane. (■) NUTCRACKER, (●) measured, (▲) RINDA, (◆) TRIM.

was run with each program. The results of this test are shown in Fig. 44 where the axial component of the measured field is compared with those calculated by TRIM, RINDA, and NUTCRACKER.

7. Program GRACY

This program was developed by Jellett and Parzen [28] to be used in the study of high-field superconducting magnets. It uses the vector potential method, since it was feared that saturation effects prevailing with such high fields would lead to convergence problems if the two-potential method were used. Program GRACY uses a uniform rectangular mesh of about 10,000 points. The current density at each mesh point is calculated by computing the vector potential from

$$\nabla \cdot \frac{1}{\mu} \nabla A = -4\pi \mathbf{J}, \tag{7.1}$$

using **A** computed at the mesh points. In the air portion of the problem the iteration procedure replaces the differential Eq. (7.1) by its finite difference equivalent

$$\sum_{i=0}^{4} \alpha_i A_i = -4\pi J_0, \tag{7.2}$$

where α_i's are iteration coefficients whose values are derived from the problem geometry and the reluctivity, and A_i's are the vector potentials for the four neighboring points on the mesh. Each interior point in the mesh is iterated by

$$A_0^{n+1} = A_0^n - \varrho R^{(n)}, \tag{7.3}$$

$$R^{(n)} = [\alpha_1 A_1^{(n)} + \alpha_2 A_2^{(n)} + \alpha_3 A_3^{(n+1)}$$
$$+ \alpha_4 A_4^{(n+1)} + \alpha_0 A_0^{(n)} - 4\pi J_0]/\alpha_0, \tag{7.4}$$

where n is the iteration number and ϱ is the overrelaxation parameter. For linear problems (iron is infinitely permeable) the code chooses $\varrho \approx 1.9$.

In the iron part of the problem, the vector potential A obeys the equation

$$\partial/\partial x(\gamma\, \partial A/\partial x) + \partial/\partial y(\gamma\, \partial A/\partial y) = 0, \tag{7.5}$$

where γ = reluctance = $1/\mu$ and μ = permeability of the material. In addition, at the air-iron interface, the boundary condition

$$(\partial \mathbf{A}/\partial n_{\text{air}}) = \gamma(\partial \mathbf{A}/\partial n_{\text{iron}}), \tag{7.6}$$

must be satisfied, where $(\partial \mathbf{A}/\partial n)$ is the derivative of **A** along the normal to the iron surface.

Program GRACY computes γ after each iteration by

$$\gamma^{(n)} = \beta \gamma(B) + (1 - \beta)\gamma^{(n-1)}, \tag{7.7}$$

where n is the iteration number and $\gamma(B)$ represents the dependence of γ on B. β is an underrelaxation parameter whose choice depends on how nonlinear the γ versus B relationship is. A common choice of β is approximately 0.1, which means that γ cannot change by more than ≈ 10 percent in each iteration, gives convergence in most iron magnet problems.

8. University of Colorado Magnet Program

The contributions of the University of Colorado's magnet program have been significant. As early as 1964, Ahamed et al. [29, 29a, 30] published papers on computer programs for calculating magnetic fields associated with rotating electrical machinery. Basically, their approach is the vector potential method. They use the basic partial differential equation

$$\partial/\partial x(\gamma\, \partial A/\partial x) + \partial/\partial y(\gamma\, \partial A/\partial y) = -J, \qquad (8.1)$$

which, in essence, is the nonlinear equivalent of the usual Poisson's equation, in which γ represents the nonlinear reluctivity given by

$$\gamma = a(1 - b\,|\,B\,|)^{-1}, \qquad (8.2)$$

where a and b are constants of the magnetic material. Therefore, Eq. (8.2) is a valid approximation for a limited region; however, use outside this region will produce a negative value of reluctivities.

In the air region of the problem universe, Eq. (8.1) reduces to the familiar Laplace's equation

$$\partial^2 A/\partial x^2 + \partial^2 A/\partial y^2 = 0, \qquad (8.3)$$

while in current-carrying regions Eq. (8.1) reduces to Poisson's equation

$$\partial^2 A/\partial x^2 + \partial^2 A/\partial y^2 = -J/\gamma_0. \qquad (8.4)$$

In the iron regions the reluctivity is a function of the magnetic induction **B**. The current density vector is zero due to the infinite resistivity of the laminations in the axial direction; therefore, Eq. (8.1) becomes

$$\partial/\partial x(\gamma\partial A/\partial x) + \partial/\partial y(\gamma\partial A/\partial y) = 0. \qquad (8.5)$$

Equations (8.1), (8.3), (8.4), (8.5) form a set of partial differential equations describing the magnetic vector potential **A** in the cross section of heteropolar machines, subject to boundary conditions. These equations are transformed into a set of difference equations to obtain numerical solutions. The two-dimensional problem-space is subdivided into a nonuniform rectangular mesh, at the nodes of which the vector potential is evaluated.

The relaxation of the vector potential is carried out in two steps: first, the reluctivities are assumed to be constants and the vector potential is relaxed successively using the formulated algorithm; and second, the reluctivities are recalculated by Eq. (8.2) using the numerical values of the vector potential calculated in the first step. This procedure is repeated until the vector potentials have arrived near enough to their final value. The rate of convergence is accelerated by the following scheme.

94   I. Two-Dimensional Magnetostatic Programs

After every iteration of the vector potential the line integral $\int \mathbf{H} \cdot d\mathbf{l}$ is computed on a closed path, one mesh away from the boundary of the universe and is compared with the surface integral $\iint \mathbf{J} \cdot d\mathbf{s}$. The ratio of these two integrals, c_n, is used to multiply all vector potentials

$$c_n = \left(\iint \mathbf{J} \cdot d\mathbf{s}\right) \Big/ \left(\int \mathbf{H}_n \cdot d\mathbf{l}\right) \neq 1. \tag{8.6}$$

The relaxation procedure and the adjustment of the vector potentials continue until the scalar c_n is equal, or nearly equal, to unity.

The authors of this program have reported significant saving of computer time with this method.

9. Program COILS

a. Introduction

Program COILS is a general purpose computer program used to calculate vector potentials, axial and radial components of magnetic fields, forces and mutual inductances between coaxial conductors, and to plot lines of force, lines of constant B, B_z, etc., of any system composed of current-carrying elements. These elements may be any combination of circular filaments, thin solenoids, cylindrical or plane current sheets, or thick cylindrical coils with rectangular cross sections. The computed fields are accurate even within the windings of the coils, and in most cases, the accuracy is to a minimum of seven significant figures.

COILS was written at Oak Ridge National Laboratory by Garrett [31] and has been modified by Mumaugh [32]. The version described herein is that modified at Lawrence Berkeley Laboratory by members of the author's group. It is written in Fortran and is operational on the CDC 6600/7600 computer systems. The program occupies 60K$_8$ memory locations and a typical run is completed in less than 30 seconds on the CDC 6600 computer.

b. Program Organization

COILS is composed of a main program and eight subroutines. The flow of information from the main program to the subroutines is shown in Fig. 45. The function of each subroutine is outlined below:

1. Subroutine FAP is a Fortran translation of the original routine written in Fap language (for the IBM 7090 series). The original routine has four entry points: GAUSS, LOOP, KANDE, and SELF. Their functions are:

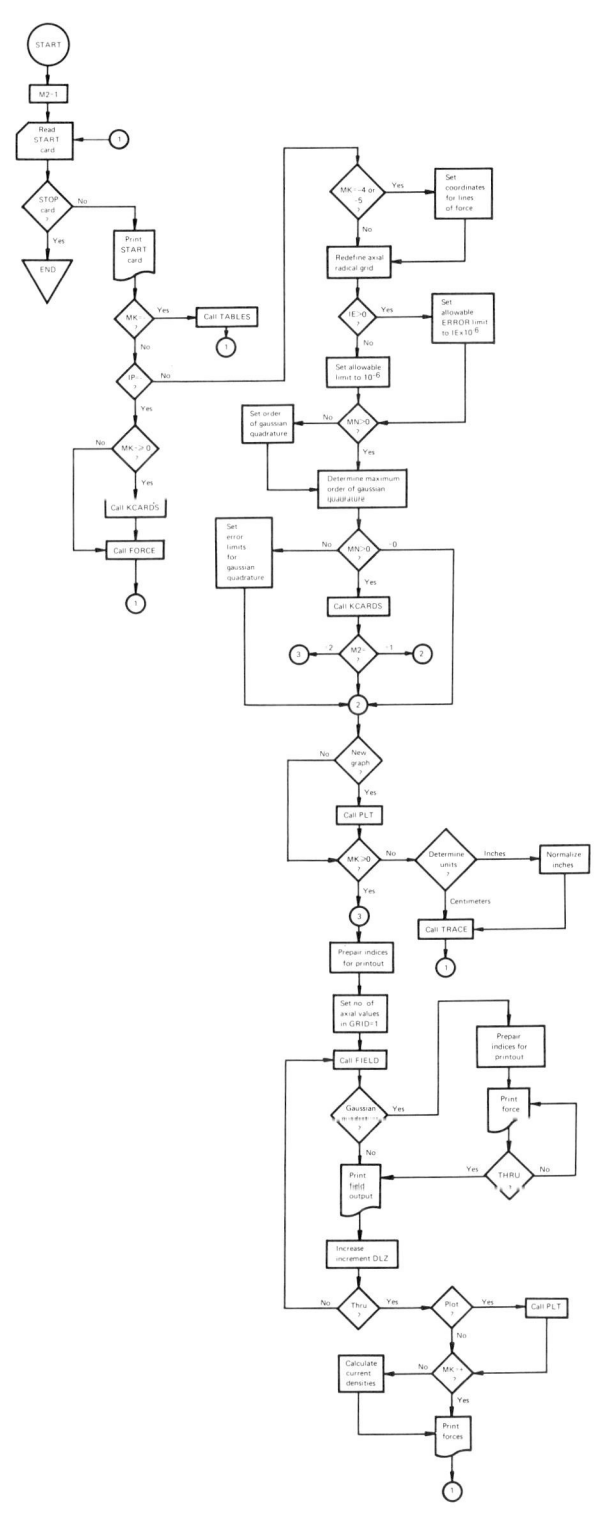

FIG. 45. Flow chart of program COILS.

Entry GAUSS: Transmits complete sets of abscissas and weight coefficients for gaussian numerical integration into COMMON as one-dimensional array.

Entry LOOP: Computes field components B_z, B_r, and vector potential **A**, for a field point at $(0, r)$ and a circular loop at (z, a); also computes elliptic integrals.

Entry KANDE: Computes the elliptic integrals K, $K - E$, and E. These integrals are computed from the Hastings' series approximations.

Entry SELF: Computes self-inductance of ideal solenoids or of thick cylindrical coils.

2. Subroutine KCARDS reads in the geometry and currents for a system of coaxial circular loops, ideal solenoids, thick cylindrical coils, etc.

3. Subroutine FIELD computes B_z, B_r, B, A, rA, and $\partial B_r/\partial z$ quantities at a single field point, for a system of coaxial loops and/or thick coils.

4. Subroutine JONES performs the iterative procedure and computes and stores the final values of fields (with or without the gaussian integration in radial depth), forces, and mutual inductances between two solenoids.

5. Subroutine TABLES tabulates elliptic integrals, magnetic fields, and vector potentials of a single loop or solenoid source at all points of a Cartesian grid.

6. Subroutine FORCE computes axial force, and mutual and self-inductance between designated pairs of coaxial ideal solenoids and/or thick rectangular coils.

7. Subroutines TRACE and PLOT trace lines of constant magnetic flux (rA), constant total B, constant B_z, etc., for any axially symmetric current system.

c. Preparation of Input Data

There are two sets of data cards necessary to successfully execute this program: the first set, consisting of one card called START card, contains the information pertinent to the number of coils, number of points for which field values are desired, the order of gaussian quadrature, error allowances, etc. The second set consists of K-cards, each specifying the individual coil geometry, current density, radius, thickness and other dimensions.

1. START Card. A START card is read in at the beginning, and again whenever the calculations are completed for all points of the particular field grid. Any number of systems and field grids can be computed in succession, since the program ends only when a STOP card is read. Table 23, Card 1, describes the parameters needed for the START card in the order of their placement on the card. Figure 46 shows the geometrical interpretation of some of the pertinent quantities.

9. Program COILS

Table 23

COILS Input Parameters and Specifications

Card	Format	Parameter	Description
1	A6	NAME	Problem identification, for example, START1
	I3	MK	Number of coils or coil pairs in coil system
	I3	IN	Input dimensions: if IN = 1, inches; IN = 0, centimeters
	I3	IP	Matrix printout option: if IP = 0, print matrices; IP = 1, do not print matrices
	I3	ID	Not used
	4I3	IE,MN,NDL,NN	Used only when comparisons desired between two or more orders of gaussian quadrature; standard value = 0
	I3	NZ	Number of axial values in field grid
	I3	NA	Number of radial values in field grid
	F9.4	DIZ	Axial grid spacing
	F9.4	Z_1	Axial distance of grid from origin
	F9.4	DIA	Radial grid spacing
	F9.4	A_1	Radial distance of grid from origin
2	A4	NAME	Coil identification, for example, COL1
	I2	IS	Symmetry flag: if IS = 0, mirror pair; if = 1, single coil; if = 2, opposed pair
	2F6.4	ZREF,AREF	Axial and radial reference points: if = 0.0, they identify origin
	2F9.4	Z_0,A_0	The Z and R coordinates of the central filament
	F9.4	B	Axial breadth
	F9.4	D	Radial breadth: the combination of B and D = 0.0 identifies a circular filament or loop; B ≠ 0.0 and D = 0.0 identifies an ideal solenoid; B = − and D = 0.0 calculates self-inductance of an ideal solenoid
	F9.4	CD	Volume current density of coil, surface current density of solenoid, or total current of loop; units are ampere/inches^2 or ampere/centimeters^2 depending on whether IN = 1 or IN = 0
	F9.0	BREF	Normalization flag: if BREF = 0.0, fields are computed without applying a normalizing factor; a nonzero value entered defines the magnitude of axial field component B_z at any arbitrarily assigned axial point (ZREF,0.0); note that BREF is without effect except on the last K card
3	A4	NAME	Program termination: enter STOP

2. K-cards. The parameters necessary to describe a K-card are shown in Table 23 as Card 2.

3. STOP Card. This card terminates all runs.

d. Output Description

The output produced by COILS is simple. Figure 47 shows a partial output for the coil arrangement of the Chalk River 100-cm spectrometer [33]. The output begins by listing the input data, then, tables of values of the coordinates

98   I. Two-Dimensional Magnetostatic Programs

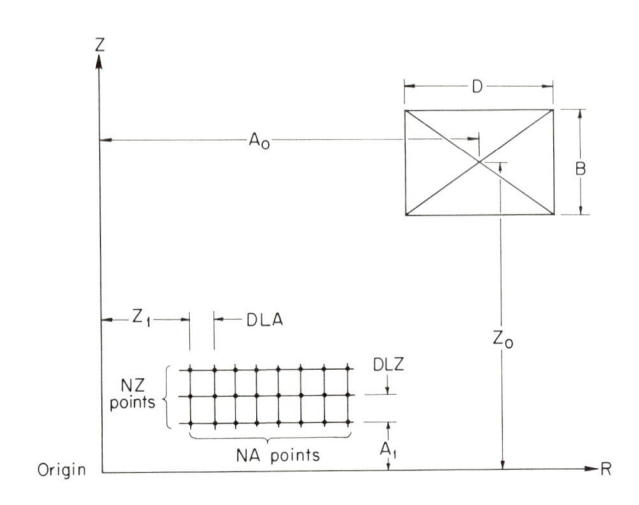

FIG. 46. Notation and interpretation of input quantities used in program COILS.

R=	0.	Z	B AXIAL	B RADIAL	B TOTAL	VECTOR A	RA	DBR/DZ	M	I	C)E6
6		0.	-.007139	0.	.007139	0.	0.	0.	0	0.	1000000
6		1.00	-.007120	0.	.007120	0.	0.	0.	0	0.	1000000
6		2.00	-.006784	0.	.006784	0.	0.	0.	0	0.	1000000
6		3.00	-.005326	0.	.005326	0.	0.	0.	0	0.	1000000
6		4.00	-.001484	0.	.001484	0.	0.	0.	0	0.	1000000
6		5.00	.006333	0.	.006333	0.	0.	0.	0	0.	1000000
R=	1.00	Z	B AXIAL	B RADIAL	B TOTAL	VECTOR A	RA	DBR/DZ	M	I	C)E6
6		0.	-.007129	-.000000	.007129	-35.676999E-04	-35.676999E-04	.000039	6	4.0	865637
6		1.00	-.007179	-.000008	.007179	-35.753574E-04	-35.753574E-04	-.000102	6	4.0	865637
6		2.00	-.007047	-.000294	.007054	-34.584254E-04	-34.584254E-04	-.000513	6	4.0	865637
6		3.00	-.005909	-.001112	.006013	-28.092896E-04	-28.092896E-04	-.001156	6	4.0	865637
6		4.00	-.002472	-.002664	.003635	-98.945768E-05	-98.945768E-05	-.001971	6	4.0	865637
6		5.00	.004899	-.005081	.007058	28.081477E-04	28.081477E-04	-.002869	6	4.0	865637
R=	2.00	Z	B AXIAL	B RADIAL	B TOTAL	VECTOR A	RA	DBR/DZ	M	I	C)E6
6		0.	-.006990	.000000	.006990	-70.884417E-04	-14.176883E-03	.000292	6	4.0	748191
6		1.00	-.007255	.000195	.007258	-72.101830E-04	-14.420366E-03	.000051	6	4.0	748191
6		2.00	-.007750	-.000184	.007752	-72.866094E-04	-14.573219E-03	-.000847	6	4.0	748191
6		3.00	-.007595	-.001662	.007775	-64.750106E-04	-12.950021E-03	-.002182	6	4.0	748191
6		4.00	-.005404	-.004668	.007141	-34.515746E-04	-69.031493E-04	-.003876	6	4.0	748191
6		5.00	.000590	-.009476	.009494	34.633796E-04	69.267591E-04	-.005753	6	4.0	748191
R=	3.00	Z	B AXIAL	B RADIAL	B TOTAL	VECTOR A	RA	DBR/DZ	M	I	C)E6
6		0.	-.006397	.000000	.006397	-10.335424E-03	-31.006271E-03	.000989	6	4.5	644656
6		1.00	-.007035	.000836	.007084	-10.791502E-03	-32.374507E-03	.000531	6	4.6	644656
6		2.00	-.008618	.000767	.008652	-11.704606E-03	-35.113817E-03	-.000810	6	4.7	644656
6		3.00	-.010180	-.001044	.010233	-11.742947E-03	-35.228840E-03	-.002928	6	4.8	644656
6		4.00	-.010177	-.005288	.011469	-88.035324E-04	-26.410597E-03	-.005638	6	4.9	644656
6		5.00	-.006615	-.012425	.014077	-19.963412E-05	-59.890236E-05	-.008663	6	5.0	644656
R=	4.00	Z	B AXIAL	B RADIAL	B TOTAL	VECTOR A	RA	DBR/DZ	M	I	C)E6
6		0.	-.004780	.000000	.004780	-12.709391E-03	-50.837563E-03	-.002386	6	5.0	552699
6		1.00	-.005968	.002169	.006350	-13.848129E-03	-55.392517E-03	.001737	6	5.0	552699
6		2.00	-.009170	.003053	.009665	-16.618616E-03	-66.474463E-03	-.000171	6	5.0	552699
6		3.00	-.013298	.001448	.013377	-19.123005E-03	-76.492021E-03	-.003213	6	5.0	552699
6		4.00	-.016596	-.003671	.016998	-18.339726E-03	-73.358904E-03	-.007147	6	5.0	552699
6		5.00	-.016743	-.013019	.021209	-10.366165E-03	-41.464661E-03	-.011594	6	5.0	552699

FIG. 47. Partial printout from program COILS showing computed field values of the 100-cm Chalk River spectrometer.

9. Program COILS   99

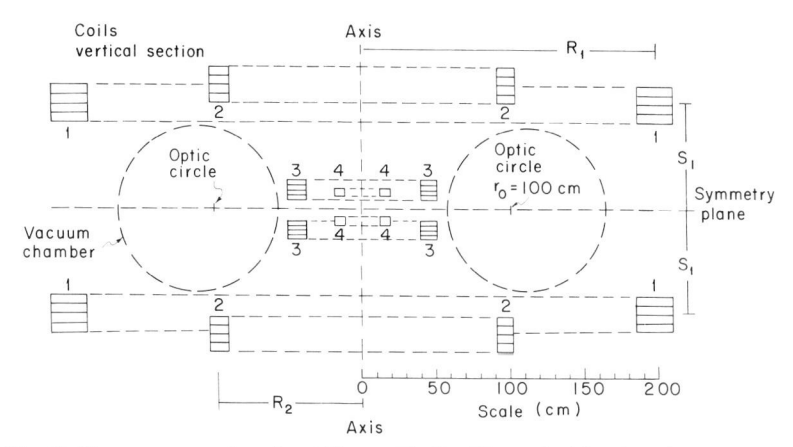

FIG. 48. The geometry for the 100-cm Chalk River double focusing spectrometer (from Ewan *et al.* [33]).

Table 24

COILS Input Data for the 100-cm Chalk River Spectrometer

				Card Column			
1	10		30				70
START1 13	0 0 0	0 0 0 0	1140	1.0	0.0	2.0	0.0
COL1 0	0.0	0.0	75.388	197.552	3.048	19.729 1.963754	0.0
COL2 0	0.0	0.0	69.688	197.552	3.048	19.729 1.963754	0.0
COL3 0	0.0	0.0	63.988	197.552	3.048	19.729 1.963754	0.0
COL4 0	0.0	0.0	58.288	197.552	3.048	19.729 1.963754	0.0
COL5 0	0.0	0.0	88.321	96.670	3.048	8.334 1.965384	0.0
COL6 0	0.0	0.0	82.621	96.670	3.048	8.334 1.965384	0.0
COL7 0	0.0	0.0	76.921	96.670	3.048	8.334 1.965384	0.0
COL8 0	0.0	0.0	71.221	96.670	3.048	8.334 1.965384	0.0
COL9 0	0.0	0.0	17.153	45.061	1.524	8.545 -2.079297	0.0
CO10 0	0.0	0.0	14.247	45.061	1.524	8.545 -2.079297	0.0
CO11 0	0.0	0.0	11.341	45.061	1.524	8.545 -2.079297	0.0
CO12 0	0.0	0.0	8.435	45.061	1.524	8.545 -2.079297	0.0
CO13 0	0.0	0.0	8.887	15.467	3.048	3.393 -1.963705	0.0
STOP							

100 I. Two-Dimensional Magnetostatic Programs

of the points and their associated field values resulting from the computations are printed—most are self-explanatory; the remaining three columns are described as follows:

The monitor count M represents the number of entries to subroutine JONES —that is, the product of $\Sigma \eta$ and the number of independent coils. The summation is taken over all the orders used and represents the total number of elementary gaussian approximations. The monitor count I is the average number of iterations. The value C is a measure of the radial separation between the field point and the nearest gaussian solenoid used in the calculation. Note that it is multiplied by 10^6 before printout.

To illustrate the usefulness of the code in the design of coils with arbitrary geometry, the coil arrangement of the Chalk River, $\pi\sqrt{2}$, double-focusing spectrometer is analyzed and the computed results compared with those measured. Figure 48 shows the coil arrangement with $r_0 = 100$ cm. The radial and axial tolerances for the coils are ≈ 0.50 millimeters for the external coils, and ≈ 0.25 millimeters for the other coils. The corresponding data for this coil arrangement is shown in Table 24. In brief, the first card (START1) specifies the problem parameters: namely, the number of coils (13), the axial and radial grid size (11, 40), the axial grid spacing (DLZ = 1), etc. The next thirteen cards specify the quantities necessary to describe the coil geometry and their associated current density.

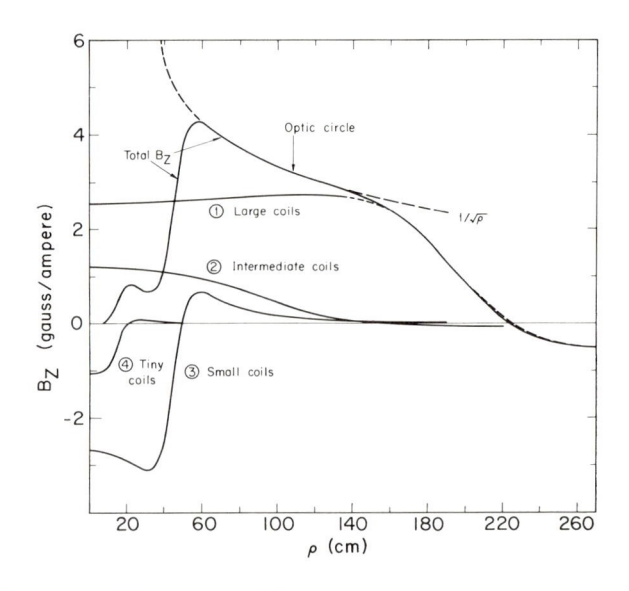

FIG. 49. The computed magnetic field form showing contributions due to individual pairs of coils. The final field B_z is shown to vary as $1/\sqrt{\varrho}$ in the region of the optic circle (from Ewan et al. [33]).

9. Program COILS

When plotted, the calculated fields show the contribution of each pair of coils to the total field strength on the plane of symmetry as a function of the radial distance from the symmetry axis (see Fig. 49). The field has the desired shape $(r^{-\frac{1}{2}})$ from approximately $r = 60$ centimeters to $r = 140$ centimeters.

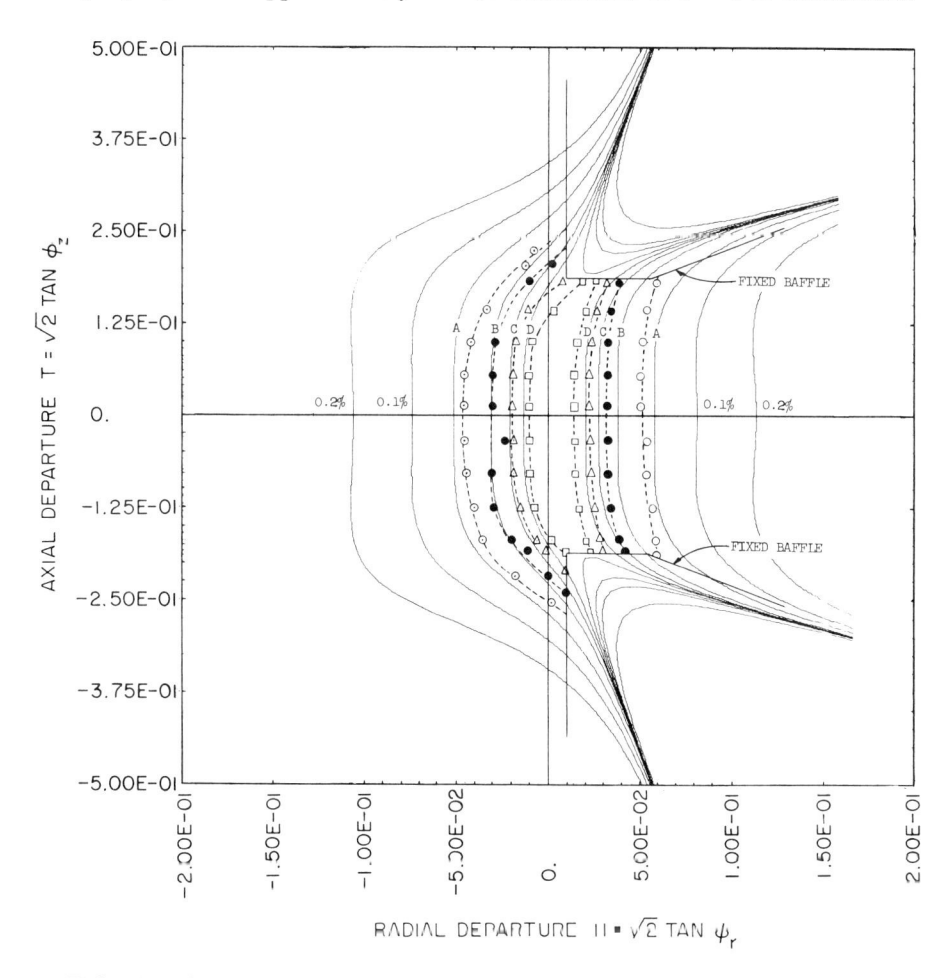

FIG. 50. CalComp plots of isoaberration curves giving the permissible range of source-departure angles (ϕ_r, ϕ_z) for various radial aberration values in the image plane. Curves plotted for $-.200$, $-.100$, $-.050$, $-.0,02$ $-.010$, $-.005$, $0.$, $.005$, $.010$, $.030\%$ resolution. (⊙) A = 0.05%, (●) B = 0.02%, (△) C = $0,01\%$, (□) D = 0.005%.

The calculated field values were used to determine the focusing properties of this spectrometer in terms of the isoaberration contours (Fig. 50). The contours were obtained by a special version of program GOC3D described elsewhere in this text. The agreement of these contours, with those determined experimentally is excellent.

II | Calculation of Electrostatic Fields

10. Program JASON

a. Introduction

JASON, written by Sackett and Healey [34] is a two-dimensional electrostatic program capable of solving a variety of problems encountered in the design of electrostatic lenses, deflecting systems, spectrometers and in the calculation of electrostatic fields. Outstanding characteristics of the program include: 1) usage of both cylindrically symmetrical and two-dimensional Cartesian systems 2) generalized boundary conditions (Neumann, Dirichlet), 3) generalized quadrilateral mesh, 4) use of block iterative methods for the solution of the equations, 5) ease and simplicity of input, 6) consideration of nonhomogeneous, anisotropic media, 7) flexible output and CalComp plots of field and equipotential lines.

Program JASON has been written in Fortran and is operational on the CDC 6600/7600 computer systems. The program occupies $110K_8$ words of central memory for a 1300-point mesh configuration and a typical run takes less than one minute on the CDC 6600 computer system.

b. Program Description

JASON solves the linear Poisson equation

$$\nabla \cdot (\varkappa \nabla \phi) + \varrho = 0, \qquad (10.1)$$

where ϕ and ϱ are scalar functions of position and $\varkappa$ is a second-rank tensor whose components are functions of position. Appropriate use of the tensor $\varkappa$

allows the consideration of nonhomogeneous anisotropic media. The mathematical model that solves Eq. (10.1) has been elaborately described (see Sackett and Healey [34]) and it is sufficient to say that the method employed uses the finite-element approach to generate the difference equations approximating Eq. (10.1). The solution algorithm uses normalized successive block overrelaxation.

The problem universe consists of a nonuniform quadrilateral mesh upon which the problem geometry is outlined. The method of generating this mesh is similar to that utilized in program TRIM, and use is made of the equipotential zoning method. The nonuniform mesh provides the user with the advantage of specifying finer or coarser mesh size in accordance with the degree of accuracy desired. The specification of a geometry with completely arbitrary boundaries is also an advantage inherent in nonuniform meshes.

It may be argued that a mesh generation with boundaries is probably unstable for some types of geometries and the problem might diverge. Our experience with generators of this type, used in JASON and TRIM, has indicated that there is always convergence to a well-defined nonuniform mesh regardless of geometry. However, when this geometry is solved for the potentials or fields, in some cases the convergence is very slow, occasionally oscillations occur and the problem takes an abnormally long time to converge. To correct conditions of this type requires thorough familiarity with the program in order to make the appropriate change of parameters relating to the overrelaxation scheme.

Once the geometry has been generated, the program proceeds with the main iterative portion of the code responsible for the determination of the potentials that result from the geometry specified, subject to the boundary conditions imposed. During this phase convergence is monitored and an optimum overrelaxation factor is calculated, to increase the rate of convergence. The problem is considered converged when the difference between two successive iterates is less than a preassigned value (10^{-7}). Then the program is directed to a set of subroutines which produce an edit of the negative gradient of the potential throughout the problem geometry. The edit-routine fits a harmonic polynomial, in the least-squared sense, to a specified set of mesh points surrounding the point of evaluation.

The flow chart shown in Fig. 51 shows the flow of information, while Table 25 outlines the functions of the subroutines utilized.

c. Preparation of Input Data

The sequence of data cards necessary to successfully operate program JASON is similar to the data structure required for program MESH (see the generator portion of program TRIM). Specifically, JASON requires data specifying

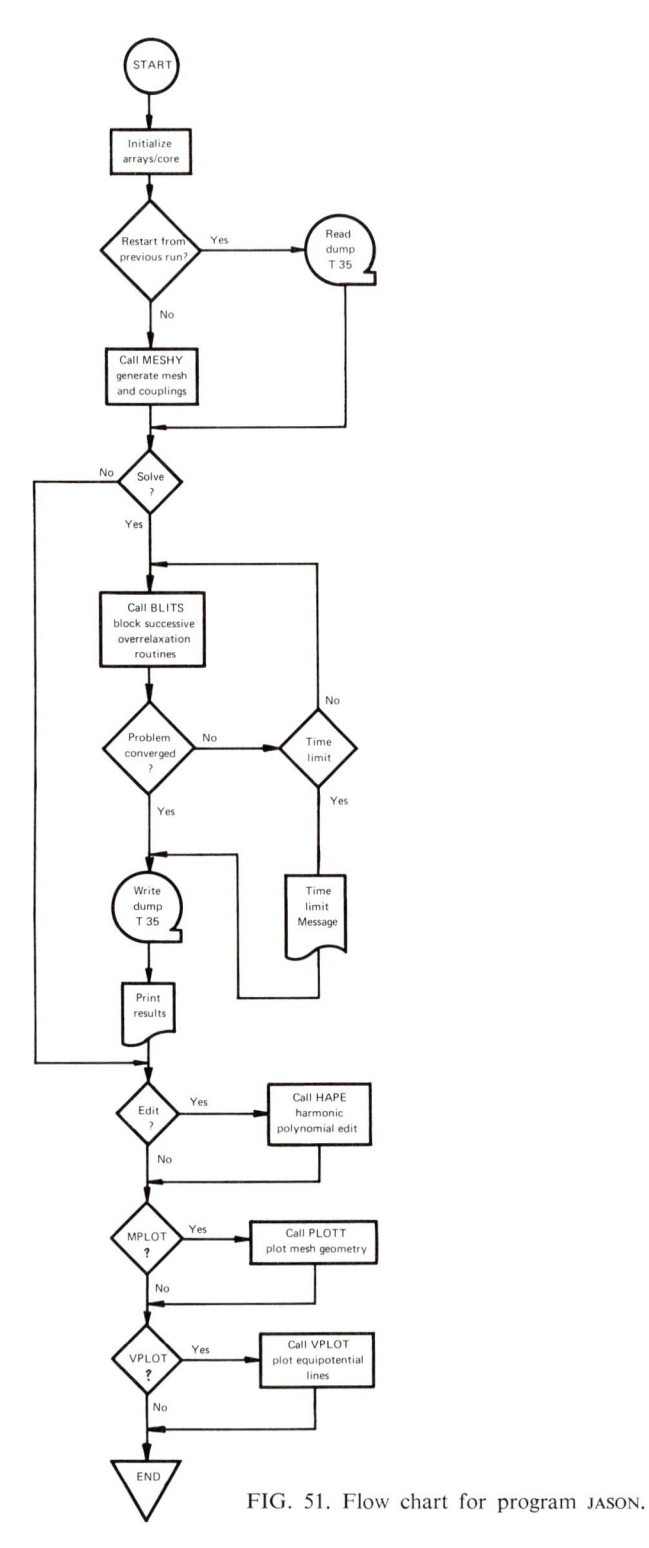

FIG. 51. Flow chart for program JASON.

10. Program JASON

Table 25

JASON Subroutines

Subroutine	Description
JASON	The main control section
MESHY FILUP SMOOTH SCOUP	Perform calculations for mesh generator
BLITS	Blocks successive overrelaxation
CAM CYDER HAPE ICASE RAM	Edit functions of the program that differentiates between Cartesian edit (RAM) and cylindrical edit (CYDER)
CLINE CONPLO INSIDE ONEDGE QUAD2 RECKON VPLOT	Perform necessary calculations as well as preparation of appropriate arrays for plotting generated mesh (PLOTT) and equipotential lines resulting from solution of program (VPLOT)
DATA DATAI IRDBYTE RDNUM CCNARG MINNY MOCK	Free-format data-reading routines peculiar to the CDC 6600/7600 computer installation at Lawrence Berkeley Laboratory. Potential user may substitute his own reading routines peculiar to his installation

the geometry of the problem, the boundary, and data necessary for plotting the generated mesh or the resulting equipotential lines.

JASON requires six sets of data for proper execution. The presence or absence of any of these sets of data is dictated by six execution control cards, whose function is outlined as follows:

II. Calculation of Electrostatic Fields

1. GENERATE. This control card instructs the program to generate a non-uniform quadrilateral mesh based on the data presented and to calculate all pertinent quantities. If this card is missing, the program assumes that the generator output is to be read from the dump file.

2. SOLVE. This card instructs the program to solve for the resulting potentials at the mesh nodes (no other input cards are required).

3. EDIT. This card instructs the program to calculate the negative gradient of the potential at the quadrilateral centroids (no other input cards are required).

4. MPLOT. The presence of this card instructs the program to plot the generated mesh showing all regions and boundaries (no other input data are required).

5. VPLOT. This card makes it possible to plot equipotential lines.

6. Execution control cards must be terminated by a blank card.

From the previous description it is evident that the only data required are for the GENERATE and VPLOT "execution control" cards, which are described as follows:

1st card   *Title card.* This card contains any title or identification data; all 80 columns may be used.

2nd card   *Universe card.* Parameters KMAX, LMAX, NR, LIN

    KMAX = maximum mesh coordinate in the x direction

    LMAX = maximum mesh coordinate in the y direction

    NR = number of regions including universe

    LIN = sentinel with the following options:
If LIN = 0, problem has cylindrical symmetry and the lower universe boundary is automatically set to $\partial\phi/\partial n = 0$. If LIN = 1, problem is given in two-dimensional Cartesian coordinates and all the universe boundaries are set to Dirichlet type of boundary with $\phi = 0$.

3rd card   *Region card.* Parameter NP, KR, KZ, RHO

    NP = number of points in region (an integer quantity).

    KR, KZ = dielectric constant (KR and KZ = 1 in vacuum). Since K is a tensor quantity, KR and KZ define the components for oriented material

    RHO = source density

    L,K,X,Y   L, K are the mesh coordinates (integers) and X, Y are the dimensional coordinates (floating point quantities). NP cards are required. The region *must* be closed. That is, the first and last points, must coincide:

$$(L,K,X,Y)_1, =,(L,K,X,Y)_{NP}$$

In a similar fashion all regions included in the problem universe must be specified. Once all region information has been defined the program expects data pertaining to the boundary conditions. Data for these boundaries are as follows:

1. *Neumann boundaries* (one card per boundary)

K1,L1 = mesh coordinates of first point
K2,L2 = mesh coordinates of last point
  IQ = direction of the normal (see Fig. 52)
  Q = value of the normal component of $(\varkappa\nabla\phi)$ at that boundary or interface. For example, if there is a surface charge then

$$(\varkappa_1\nabla\phi)_{n+} - (\varkappa_2\nabla\phi)_{n-} = D_{n+} - D_{n-} = \sigma = Q \tag{10.2}$$

where $\varkappa$ = dielectric tensor
  $-\nabla\phi$ = electric field
  $n+, n-$ = normal on either side of boundary
  D = displacement field
  σ = surface charge density

Once the Neumann boundary cards have been specified, the program expects a card on which 0 s is punched, format free. This card must be present even if no Neumann boundaries are specified since it indicates the end of Neumann boundary cards.

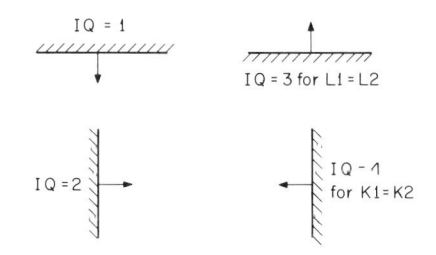

FIG. 52. Definition of Neumann boundaries.

2. *Dirichlet boundaries* are specified by two sets of cards: The first set specifies the number of points (NP) used to describe the boundary and the potential at each of the inclusive points; the second set consists of NP cards giving the coordinates K,L,X,Y for each point, in the same manner as specified in the region cards. Again we have the restriction that no diagonal lines are allowed.

The last card of the GENERATE execution control contains only two zeros: 0 0. This card must be present even if no Dirichlet boundaries are present, and signals the end of data input for the GENERATE phase.

3. *Input for* VPLOT *execution control card* consists of a set of cards, one for each plot, specifying the number of equipotential lines to be plotted (≤ 100) and limits defining the plotting region; i.e., NPLOT, XMIN, XMAX, YMIN, YMAX. The last card must have NPLOT $= 0$, to terminate the data set.

d. Output Description

Program JASON provides output for the mesh generator portion as well as for the main algorithmic solver. The mesh generator output lists the complete input data set, followed by an iteration by iteration printout of the progress of mesh convergence. The quantities CONX and CONY refer to the value of the convergence criterion at the cycle indicated, while RHOX and RHOY refer to the overrelaxation parameters.

The main algorithmic solver provides a printout for monitoring the progress of convergence. The quantity RESIDUAL/LENGTH printed in the output (Fig. 53) corresponds to the convergence criterion defined by

$$\varepsilon^\eta = \| \mathbf{x}^\eta - \mathbf{x}^{\eta-1} \| / \| \mathbf{x}^\eta \|, \tag{10.3}$$

where η denotes the iteration number and $\mathbf{x}^\eta$ denotes an element of the solution sequence. For convergence, it is normally required that ε be less than 10^{-7}.

The quantities ETA and OMEGA refer to parameters needed in the calculation of the optimum overrelaxation factor and are defined by the following relations

$$\omega_{\text{opt}} = 2/[1 + (1 - \lambda^2)^{1/2}]. \tag{10.4}$$

For $\omega < \omega_{\text{opt}}$, λ can be estimated from ω and the convergence rate ETA by

$$\lambda = \frac{[\omega + (\text{ETA}) - 1]}{[\omega(\text{ETA})^{1/2}]}, \tag{10.5}$$

where ETA is defined as

$$\text{ETA} = \frac{\| \mathbf{x}^{\eta+1} - \mathbf{x}^\eta \|}{\| \mathbf{x}^\eta - \mathbf{x}^{\eta-1} \|}, \tag{10.6}$$

η being the iteration number. Therefore, we may update our estimate for ω_{opt} at any stage in the iterative procedure.

Once convergence has been achieved, the program proceeds to list in tabular form the quantities calculated. There are two maps of potentials that may be obtained by JASON, depending on the specification of the execution control cards. If the SOLVE card has been used, then the listing of the potentials is in the form shown in Fig. 53. The quantities listed correspond to the coordinates (K,L,X,Y) of the point at which the potential (V) is calculated. In case of axially symmetric problems the dimensional coordinates will be Z, R.

10. Program JASON

ELECTROSTATIC QUADRUPOLE DEFOCUSSING MAGNET

CYCLE	RESIDUAL/LENGTH	ETA	OMEGA
5	9.528758E-02	7.814411E-01	1.564545
10	3.723297E-02	8.591992E-01	1.647656
15	2.218826E-02	8.904087E-01	1.716925
20	1.457501E-02	8.876943E-01	1.756868
25	8.212078E-03	8.648153E-01	1.772844
30	3.888414E-03	8.492795E-01	1.779603
35	1.734675E-03	8.486919E-01	1.784688
40	7.970410E-04	8.548512E-01	1.790249
45	3.737051E-04	8.536720E-01	1.794298
50	1.729673E-04	8.497844E-01	1.796634
55	7.715611E-05	8.407944E-01	1.796792
60	3.180918E-05	8.274247E-01	1.794938
65	1.240734E-05	8.193233E-01	1.792386
70	4.749455E-06	8.181389E-01	1.789955
75	1.821986E-06	8.207003E-01	1.788061
80	7.034965E-07	8.214774E-01	1.786401
85	2.739440E-07	8.232092E-01	1.785330
90	1.083994E-07	8.296540E-01	1.785332
95	4.432766E-08	8.384880E-01	1.786958
100	1.893977E-08	8.456764E-01	1.789803
109	1.084329E-09	9.495664E-01	1.471165

PROBLEM CONVERGED IN   23.844 SECONDS   ***   ***   ***

L	K	X	Y	V	L	K	X	Y	V
1	1	-.290	0.	0.	2	1	-.290	.025	0.
1	2	-.225	0.	7.43740E+00	2	2	-.225	.025	7.43669E+00
1	3	-.160	0.	1.48842E+01	2	3	-.160	.025	1.48829E+01
1	4	-.095	0.	2.23484E+01	2	4	-.095	.025	2.23467E+01
1	5	-.030	0.	2.98356E+01	2	5	-.030	.025	2.98337E+01
1	6	.035	0.	3.73478E+01	2	6	.035	.025	3.73460E+01
1	7	.100	0.	4.48833E+01	2	7	.100	.025	4.48819E+01
1	8	.165	0.	5.24368E+01	2	8	.165	.025	5.24360E+01
1	9	.230	0.	6.00000E+01	2	9	.230	.025	6.00000E+01
1	10	.262	0.	6.00000E+01	2	10	.263	.025	6.00000E+01
1	11	.295	0.	6.00000E+01	2	11	.295	.025	6.00000E+01
1	12	.327	0.	6.00000E+01	2	12	.328	.025	6.00000E+01
1	13	.360	0.	6.00000E+01	2	13	.361	.025	6.00000E+01
1	14	.385	0.	5.85806E+01	2	14	.386	.025	5.85973E+01
1	15	.410	0.	5.70654E+01	2	15	.410	.025	5.71021E+01
1	16	.435	0.	5.54552E+01	2	16	.435	.025	5.55023E+01
1	17	.460	0.	5.37503E+01	2	17	.460	.025	5.37976E+01
1	18	.485	0.	5.19510E+01	2	18	.485	.025	5.19982E+01
1	19	.510	0.	5.00572E+01	2	19	.510	.025	5.01043E+01
1	20	.535	0.	4.80691E+01	2	20	.535	.025	4.81161E+01
1	21	.560	0.	4.59870E+01	2	21	.560	.025	4.60339E+01
1	22	.585	0.	4.38113E+01	2	22	.585	.025	4.38580E+01
1	23	.610	0.	4.15421E+01	2	23	.610	.025	4.15887E+01
1	24	.635	0.	3.91797E+01	2	24	.635	.025	3.92262E+01
1	25	.660	0.	3.67244E+01	2	25	.660	.025	3.67708E+01
1	26	.685	0.	3.41763E+01	2	26	.685	.025	3.42227E+01
1	27	.710	0.	3.15356E+01	2	27	.710	.025	3.15819E+01
1	28	.735	0.	2.88021E+01	2	28	.735	.025	2.88485E+01
1	29	.760	0.	2.59759E+01	2	29	.760	.025	2.60224E+01
1	30	.785	0.	2.30568E+01	2	30	.785	.025	2.31034E+01
1	31	.810	0.	2.00445E+01	2	31	.810	.025	2.00913E+01
1	32	.835	0.	1.69390E+01	2	32	.835	.025	1.69857E+01
1	33	.860	0.	1.37394E+01	2	33	.860	.025	1.37864E+01
1	34	.885	0.	1.04459E+01	2	34	.885	.025	1.04930E+01
1	35	.910	0.	7.05824E+00	2	35	.910	.025	7.10542E+00
1	36	.935	0.	3.57628E+00	2	36	.935	.025	3.62339E+00
1	37	.960	0.	0.	2	37	.960	.025	0.
1	38	.985	0.	0.	2	38	.985	.025	0.
1	39	1.010	0.	0.	2	39	1.010	.025	0.
1	40	1.035	0.	0.	2	40	1.035	.025	0.
1	41	1.060	0.	0.	2	41	1.060	.025	0.
1	42	1.085	0.	0.	2	42	1.085	.025	0.

FIG. 53. Partial output of program JASON.

II. Calculation of Electrostatic Fields

If the EDIT execution control card was also specified, then a special edit follows which fits an harmonic polynomial, in the least-squared sense, to a specified set of mesh points surrounding the point of evaluation. The derivatives of the polynomial are then taken as approximations to the derivatives of the potential. To produce better averaging or less error, the centroids of the mesh elements, rather than the nodes, are taken as the points of evaluation. If the problem has $\partial \phi/\partial n = 0$ on the lower universe boundary, as in the case of cylindrically symmetric problems, an edit is taken at the centers of the element sides composing this boundary.

Figure 54 shows a partial listing of the harmonic polynomial fit produced by the sample case described below.

ELECTROSTATIC QUADRUPOLE DEFOCUSSING MAGNET

HARMONIC POLYNOMIAL EDIT *** CARTESIAN COORDINATES

DERIVATIVES ARE FROM THE NEGATIVE GRADIENT OF THE POTENTIAL

LLL	KLL	X	Y	V	VX	VY	VXX	VXY	VYY
1	1	-.257	0.	3.71744E+00	-1.14404E+02	0.	-5.23438E-02	0.	5.23438E-0
1	2	-.192	0.	1.11591E+01	-1.14595E+02	0.	1.21208E-03	0.	-1.21208E-0
1	3	-.127	0.	1.86137E+01	-1.14854E+02	0.	1.85785E-03	0.	-1.85785E-0
1	4	-.062	0.	2.60889E+01	-1.15195E+02	0.	2.26040E-03	0.	-2.26040E-0
1	5	.003	0.	3.35886E+01	-1.15566E+02	0.	2.33692E-03	0.	-2.33692E-0
1	6	.067	0.	4.11129E+01	-1.15911E+02	0.	2.04252E-03	0.	-2.04252E-0
1	7	.132	0.	4.86583E+01	-1.16177E+02	0.	1.39667E-03	0.	-1.39667E-0
1	8	.197	0.	5.62171E+01	-1.16375E+02	0.	-5.43698E-02	0.	5.43698E-0
1	9	.246	0.	6.00000E+01	1.60286E-09	0.	-1.58188E-09	0.	1.58188E-0
1	10	.279	0.	6.00000E+01	6.63706E-10	0.	-2.68954E-09	0.	2.68954E-0
1	11	.312	0.	6.00000E+01	-6.25949E-10	0.	-2.74595E-09	0.	2.74595E-0
1	12	.344	0.	6.00000E+01	-1.62474E-09	0.	-1.52191E-09	0.	1.52191E-0
1	13	.373	0.	5.92825E+01	5.68526E+01	0.	1.37754E-02	0.	-1.37754E-0
1	14	.398	0.	5.78243E+01	6.06316E+01	0.	3.26878E-02	0.	-3.26878E-0
1	15	.423	0.	5.62693E+01	6.44099E+01	0.	1.14696E-02	0.	-1.14696E-0
1	16	.447	0.	5.46146E+01	6.81907E+01	0.	5.32533E-03	0.	-5.32533E-0
1	17	.472	0.	5.28625E+01	7.19744E+01	0.	2.33618E-03	0.	-2.33618E-0
1	18	.498	0.	5.10159E+01	7.57531E+01	0.	6.88635E-04	0.	-6.88635E-0
1	19	.523	0.	4.90749E+01	7.95234E+01	0.	1.57804E-04	0.	-1.57804E-0
1	20	.548	0.	4.70398E+01	8.32829E+01	0.	1.47987E-05	0.	-1.47987E-0
1	21	.573	0.	4.49109E+01	8.70311E+01	0.	-8.05664E-05	0.	8.05664E-0
1	22	.598	0.	4.26883E+01	9.07682E+01	0.	-1.92429E-04	0.	1.92429E-0
1	23	.623	0.	4.03725E+01	9.44948E+01	0.	-3.13298E-04	0.	3.13298E-0
1	24	.647	0.	3.79637E+01	9.82126E+01	0.	-4.30687E-04	0.	4.30687E-0
1	25	.672	0.	3.54620E+01	1.01924E+02	0.	-5.29049E-04	0.	5.29049E-0
1	26	.697	0.	3.28675E+01	1.05631E+02	0.	-5.90952E-04	0.	5.90952E-0
1	27	.722	0.	3.01304E+01	1.09338E+02	0.	-5.99234E-04	0.	5.99234E-0
1	28	.747	0.	2.74006E+01	1.13048E+02	0.	-5.40141E-04	0.	5.40141E-0
1	29	.772	0.	2.45280E+01	1.16764E+02	0.	-4.07624E-04	0.	4.07624E-0
1	30	.797	0.	2.15623E+01	1.20490E+02	0.	-2.05305E-04	0.	2.05305E-0
1	31	.822	0.	1.85034E+01	1.24228E+02	0.	4.79543E-05	0.	-4.79543E-0
1	32	.847	0.	1.53508E+01	1.27977E+02	0.	3.32343E-04	0.	-3.32343E-0
1	33	.872	0.	1.21044E+01	1.31738E+02	0.	4.04062E-04	0.	-4.04062E-0
1	34	.898	0.	8.76389E+00	1.35502E+02	0.	1.75506E-03	0.	-1.75506E-0
1	35	.923	0.	5.32895E+00	1.39275E+02	0.	-2.05111E-02	0.	2.05111E-0
1	36	.948	0.	1.78915E+00	1.43055E+02	0.	-2.13082E-02	0.	2.13082E-0
1	37	.973	0.	0.	-0.	0.	-0.	0.	0.
1	38	.998	0.	0.	-0.	0.	-0.	0.	0.
1	39	1.023	0.	0.	-0.	0.	-0.	0.	0.
1	40	1.048	0.	0.	-0.	0.	-0.	0.	0.
1	41	1.073	0.	0.	-0.	0.	-0.	0.	0.
1	42	1.098	0.	0.	-0.	0.	-0.	0.	0.
⋮									
1	71	2.193	0.	0.	-0.	0.	-0.	0.	0.
1	72	2.238	0.	0.	-0.	0.	-0.	0.	0.

LLL	KLL	X	Y	V	VX	VY	VXX	VXY	VYY
1	1	-.257	.012	3.71802E+00	-1.14413E+02	1.52989E-02	-1.11223E+00	4.51293E-01	1.11223E+0
1	2	-.192	.012	1.11588E+01	-1.14559E+02	4.27542E-02	-3.19526E+00	3.66180E-01	3.19526E+0
1	3	-.127	.012	1.86134E+01	-1.14829E+02	6.44091E-02	-4.80799E+00	2.50305E-01	4.80799E+0
1	4	-.062	.012	2.60885E+01	-1.15185E+02	7.66339E-02	-5.71168E+00	9.49267E-02	5.71168E+0
1	5	.003	.012	3.35882E+01	-1.15574E+02	7.73544E-02	-5.75563E+00	-7.84315E-02	5.75563E+0
1	6	.067	.012	4.11126E+01	-1.15936E+02	6.61487E-02	-4.91410E+00	-2.43473E-01	4.91410E+0
1	7	.132	.012	4.86580E+01	-1.16215E+02	4.44938E-02	-3.30138E+00	-3.73012E-01	3.30138E+0
1	8	.197	.012	5.62177E+01	-1.16367E+02	1.60784E-02	-1.15506E+00	-4.71897E-01	1.15506E+0
1	9	.246	.012	6.00000E+01	1.79056E-09	-4.89739E-10	-1.80884E-08	-2.65620E-01	1.80884E-0
1	10	.279	.012	6.00000E+01	7.14610E-10	-1.17379E-09	-3.74810E-08	-6.93348E-09	3.74810E-0
1	11	.312	.012	6.00000E+01	-6.80023E-10	-1.20212E-09	-4.01741E-08	7.51626E-09	4.01741E-0
⋮									

FIG. 54. Partial output obtained by a special fitting routine called by JASON when an EDIT control card is used.

```
NERATE
LVE
IT
LOT
LOT
                                    $ BLANK CARD REQUIRED $
ECTROSTATIC QUADRUPOLE DEFOCUSSING MAGNET
   45    12      1                  $ KMAX,LMAX,NR,LIN $
   1.    1.      0.                 $ BEGIN REGION 1 $
   1     -0.29   0.
   1     0.23    0.
   1     0.36    0.
   1     0.435   0.
   1     0.935   0.
   1     0.96    0.
   1     1.36    0.
   1     2.26    0.
   33    2.26    0.8
   45    2.26    1.28
   45    0.36    1.28
   45    -0.29   1.28
   33    -0.29   0.8
   1     -0.29   0.                 $ END REGION 1 $
   1.    1.      0.                 $ BEGIN REGION 2 $
   33    -0.29   0.8
   33    0.23    0.8
   45    0.23    1.28
   45    -0.29   1.28
   33    -0.29   0.8                $ END REGION 2 $
   1.    1.      0.                 $ BEGIN REGION 3 $
   33    0.23    0.8
   33    0.36    0.8
   45    0.36    1.28
   45    0.23    1.28
   33    0.23    0.8
   1.    1.      0.                 $ BEGIN REGION 4 $
   33    0.36    0.8
   33    0.710   0.80
   33    1.11    0.8
   33    2.26    0.8
   45    2.26    1.28
   45    0.36    1.28
   33    0.36    0.8                $ END REGION 4 $
   1.    1.      0.                 $ BEGIN REGION 5 $
   1     0.435   0.
   1     0.935   0.
   8     0.935   0.175
   8     0.485   0.175
   8     0.435   0.175
   1     0.435   0.                 $ END REGION 5 $
   1.    1.      0.                 $ BEGIN REGION 6 $
   8     0.485   0.175
   8     0.935   0.175
   12    0.935   0.275
   12    0.535   0.275
   12    0.485   0.275
   8     0.485   0.175              $ END REGION 6 $
   1.    1.      0.                 $ BEGIN REGION 7 $
   12    0.535   0.275
   12    0.935   0.275
   12    0.960   0.275
   15    0.960   0.350
   15    0.585   0.350
   15    0.535   0.350
   12    0.535   0.275              $ END REGION 7 $
   1.    1.      0.                 $ BEGIN REGION 8 $
   15    0.585   0.350
   15    0.960   0.350
   15    0.985   0.350
   17    0.985   0.400
   17    0.610   0.4000
   17    0.585   0.400
   15    0.585   0.350              $ END REGION 8 $
   1.    1.      0.                 $ BEGIN REGION 9 $
   17    0.610   0.400
   17    0.985   0.400
   19    0.985   0.450
   19    0.710   0.450
   19    0.610   0.450
   17    0.610   0.400              $ END REGION 9 $
   1.    1.      0.                 $ BEGIN REGION 10 $
   19    0.710   0.450
   19    0.985   0.450
   19    1.010   0.450
   23    1.010   0.550
   23    0.760   0.550
   23    0.710   0.550
   19    0.710   0.450              $ END REGION 10 $
   1.    1.      0.                 $ BEGIN REGION 11 $
   23    0.760   0.550
   23    1.010   0.550
   23    1.060   0.550
   29    1.060   0.700
   29    0.760   0.700
   23    0.760   0.550              $ END REGION 11 $
   1.    1.      0.                 $ BEGIN REGION 12 $
   29    0.710   0.700
   29    0.760   0.700
   29    1.060   0.700

43   29   1.110     0.700
43   33   1.110     0.800
27   33   0.710     0.800
27   29   0.710     0.700              $ END REGION 12 $
1    1    73 1      1 0.               $ NEUMANN BOUNDARY $
0    5                                 $ THIS CARD REQUIRED $
21   0.                                $ BEGIN DIRICHLET BOUNDARY 1 (GRID B) $
37   1    0.96      0.
37   2    0.96033   0.025
37   3    0.96130   0.05
37   4    0.96293   0.075
37   5    0.96519   0.1
37   6    0.96810   0.125
37   7    0.97165   0.150
38   7    0.97366   0.1625
38   8    0.97582   0.175
38   9    0.98061   0.2
38   10   0.98601   0.225
38   11   0.99202   0.25
38   12   0.99524   0.2625
39   12   0.99861   0.275
39   13   1.00578   0.3
39   14   1.01352   0.325
39   15   1.01843   0.34
40   15   1.0225    0.34
53   15   1.36      0.34
73   15   2.26      0.34
73   1    2.26      0.
62   60.                               $ DIRICHLET BOUNDARY GRID A $
13   1    0.36      0.
13   2    0.36087   0.025
13   3    0.36346   0.050
13   4    0.36773   0.075
13   5    0.37048   0.0875
14   5    0.37363   0.100
14   6    0.38108   0.125
14   7    0.39      0.150
14   8    0.39498   0.1625
15   8    0.40028   0.175
15   9    0.41183   0.20
15   10   0.41804   0.2125
16   10   0.42453   0.225
16   11   0.43829   0.250
16   12   0.44554   0.2625
17   12   0.45302   0.275
17   13   0.46861   0.300
18   13   0.47671   0.3125
18   14   0.485     0.325
18   15   0.49346   0.3375
19   15   0.5021    0.350
19   16   0.51983   0.375
20   16   0.5283    0.3875
20   17   0.5381    0.400
21   17   0.5475    0.4125
21   18   0.55698   0.425
21   19   0.56657   0.4375
22   19   0.57628   0.450
22   20   0.59601   0.475
23   20   0.60602   0.4875
23   21   0.61612   0.500
24   21   0.625304  0.5125
24   22   0.63657   0.525
24   23   0.64692   0.5375
25   23   0.65734   0.550
25   24   0.66784   0.5625
26   24   0.6784    0.575
26   25   0.6862    0.6
26   26   0.69      0.63
26   27   0.6883    0.650
26   28   0.6812    0.675
26   29   0.67533   0.6875
25   29   0.66747   0.700
25   30   0.65714   0.7125
24   30   0.64331   0.725
24   31   0.62332   0.7375
23   31   0.61796   0.740
22   31   0.59182   0.748
21   31   0.57      0.75
20   31   0.5225    0.75
19   31   0.51      0.75
18   31   0.485     0.75
17   31   0.465     0.75
16   31   0.435     0.75
15   31   0.41      0.75
14   31   0.385     0.75
13   31   0.36      0.75
12   31   0.3275    0.75
11   31   0.295     0.75
10   31   0.2625    0.75
9    31   0.23      0.75
9    1    0.23      0.
0    0.                                $THIS CARD REQUIRED$ $END GENERATE DATE$
25                                     $ PLOT N=25 EQUIPOTENTIAL LINES $
                                       $ END DATA $
```

FIG. 55. Listing of input data for the electrostatic quadrupole magnet. Note that data are entered format free.

112   II. Calculation of Electrostatic Fields

e. Illustrative Example

The example used is a study made at Lawrence Berkeley Laboratory, Berkeley, California in conjunction with the modification of the downstream deflector electrodes of the 88-inch cyclotron to produce radial electrostatic quadrupole focusing to the deflected beam [35]. The deflected beam, as it crosses the main magnet pole edge, is defocused radially by the fringing magnetic field of the cyclotron. The distortion of the phase space by the nonlinear field increases the effective phase-space area and reduces the transmission to the experimental area. This study uses ray-tracing through the electric and magnetic fields of the deflector region to determine the amount of focusing; the shape required for the downstream deflector electrodes is determined by the program JASON. The problem universe for the electrode arrangement has been separated into twelve regions for maintaining uniformity of mesh lines and two regions with Dirichlet boundaries that constitute the electrodes at the potentials indicated. This geometry evolved from repeated JASON runs by comparing the amount of focusing that was obtained (through ray-tracing) until the desirable properties were achieved. Figure 55 shows the input data that conforms with the previously established rules and some explanatory remarks that follow.

The first five cards specify the functions to be performed. The next card gives the title of the problem, followed by the universe specification quantities (KMAX, LMAX), the number of regions into which the problem is divided (NR), and the geometry sentinel (LIN) indicating the Cartesian-coordinate system. The next twelve sets of cards specify the region card with the appropriate predesignated parameters followed by the logical and dimensional coordinates for the

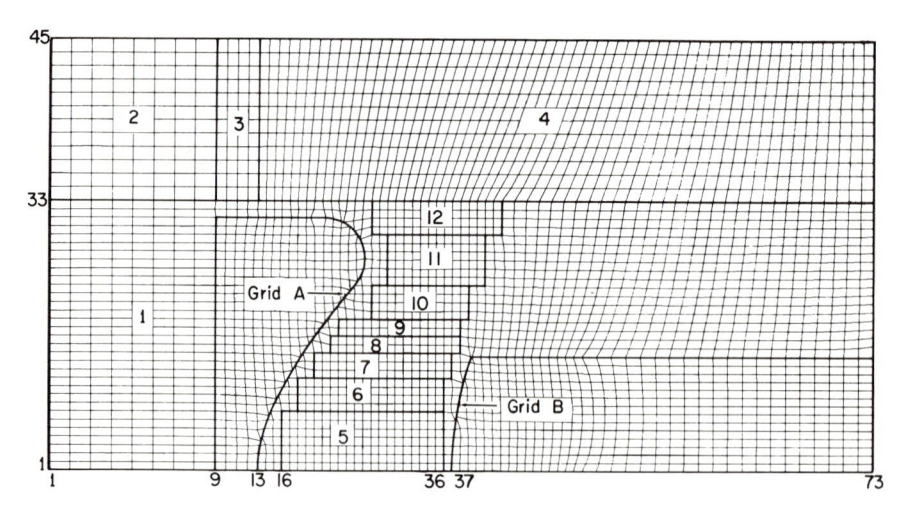

FIG. 56. Quadrilateral mesh generated by program JASON. The intensified lines are boundary region lines.

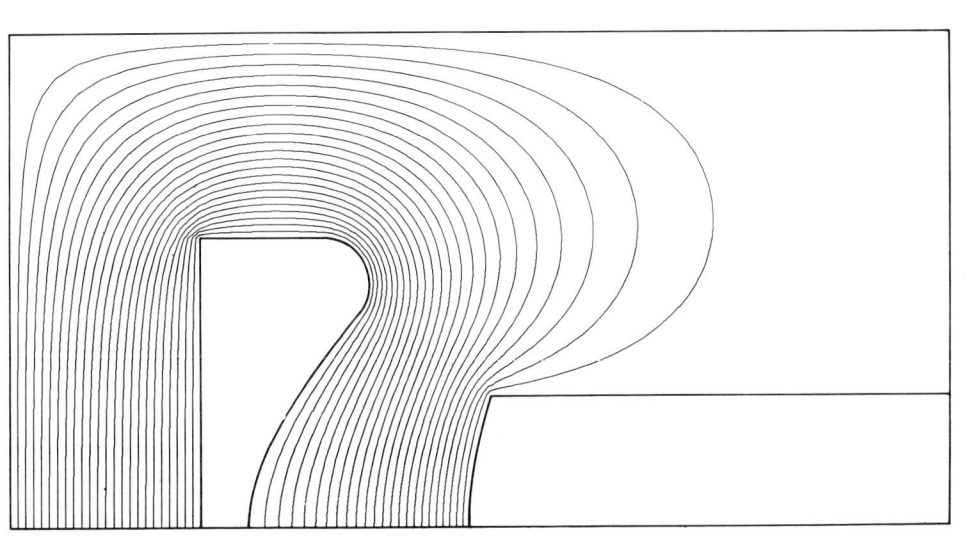

FIG. 57. CalComp plot of equipotential lines for the electrostatic quadrupole shown in Fig. 56.

points that define the region. The logical coordinates refer to the mesh lines allocated for each dimensional interval.

For curved boundaries (e.g., GRID A, GRID B in Fig. 56), enough points are specified to accurately describe the curved geometry. Once all regions have been specified, a Neumann boundary condition is inserted that specifies the mesh coordinates of the boundary, the direction of the normal component of the electric field and its value at that boundary. The next card, 0 s, is required for historical reasons.

Now the definition of the Dirichlet boundaries begins for both grids by specifying the number of points describing the grid (e.g., GRID B = 21 points) and the potential at that grid, followed by the logical and dimensional coordinates for each point in the same manner as for the region points.

The last card of the data set instructs the program to prepare an equipotential plot consisting of 25 lines (see Fig. 57). The output obtained from this example has been explained in Section d.

III Three-Dimensional Magnetostatic Programs

11. Introduction

In the previous chapter we discussed programs capable of solving problems relating to the calculation of magnetic fields in two-dimensional geometries. In this section we review the progress made in calculations of three-dimensional linear and nonlinear problems and outline some of the current effort attempting to formulate some working computer programs.

The transition to three-dimensional problems has developed logically after the many years of experience with their two-dimensional forerunners although the implementation of three-dimensional nonlinear programs with general boundaries has been slow, painstaking, and incomplete. This can be attributed to the user's indecision as to the best method of approach, and inadequate capability of computing machinery that is neither large enough nor fast enough to accommodate the enormous storage space required to solve such problems within a reasonable computer time. Perhaps the most important reason has been the decrease in research funding for the scientific world. Nevertheless, various experimenters considering this problem have offered suggestions as to the appropriate method of solution: the one-potential method (vector or scalar potential), the two-potential method (combination of scalar in air and vector potential in iron), or the superposition of the effects of currents and induced dipoles.

For the implementation of linear, three-dimensional problems with arbitrary current elements there has been considerable progress. Fields can be calculated directly by the Biot-Savart law with considerable ease and very little computational effort.

12. Lawrence Berkeley Laboratory Magnet Program

a. Introduction

Some preliminary work using a one-potential method throughout (scalar-potential) resulted in program SIMONE which handled only cases of infinite μ. The currents were handled by straight-line filaments and the interface had to lie on mesh lines. The program was discontinued because it did not show sufficient merit to justify its continuance in the face of limited funding. Next, program TAMI [36] was developed, which uses the dipole approximation to calculate the vector potential due to the magnetic material. Essentially, the program uses the relation

$$\mathbf{A}_M = \frac{\mu}{4\pi} \int_v \frac{\mathbf{M} \times \mathbf{r}}{r^3} \, dv, \qquad (12.1)$$

where **M** is the magnetization of the material,

  r is the vector from each dipole to the point at which the vector potential is being computed, and

  μ is the permeability of the material.

In Cartesian coordinates

$r = |\mathbf{r}|$

$\mathbf{r} = (x' - x)\mathbf{i} + (y' - y)\mathbf{j} + (z' - z)\mathbf{k}$

where x, y, z = coordinates of integration,

  x', y', z' = coordinates point of observation.

Note that **M** depends only on the variables of integration and not on the location of the point at which the vector potential is being calculated. TAMI divides the volume integral into small volume elements, integrates over each volume element and sums the results. The magnetization **M** is assumed to be constant in each volume element, (a sufficiently correct assumption if a large number of volume elements is being considered). Under these conditions the x components of the vector potential **A** is derived to be

$$A_x = \frac{\mu_0}{4\pi} \left[M_{y1} \int_{V_1} \frac{r_{z1}}{r_1^3} dV_1 - M_{z1} \int_{V_1} \frac{r_{y1}}{r_1^3} dV_1 \right.$$
$$\left. + \cdots + M_{yn} \int_{V_n} \frac{r_{zn}}{r_n^3} dV_n - M_{zn} \int_{V_n} \frac{r_{yn}}{r_n^3} dV_n \right]. \qquad (12.2)$$

Similar expressions are used for the y and z component and

$$\mathbf{A}_m = A_x \mathbf{i} + A_y \mathbf{j} + A_z \mathbf{k}. \tag{12.3}$$

The magnetization vector $\mathbf{M}$ is determined from

$$\mathbf{B} = \mathbf{H} + 4\pi \mathbf{M}. \tag{12.4}$$

Since $\mathbf{B} = \mu \mathbf{H}$

$$\mathbf{M} = \mathbf{B}(1 - \gamma)/4\pi, \tag{12.5}$$

where $\gamma = 1/\mu$.

Eqs. (12.4) and (12.5) are determined by simultaneous equations in n unknowns of the form

$$(1 - \gamma_j) \sum_{n=1}^{i} \mathbf{B}_{nj} - 4\pi \mathbf{M}_j = 0, \tag{12.6}$$

where i is the number of volume elements and $\mathbf{B}_{nj}$ is the B field of volume element n at the centroid of volume element j. An initial approximation of $\mathbf{M}$ is made and the corresponding γ is determined. The system is solved for the $\mathbf{M}_j$'s by successive overrelaxation. We tested this program in small rectangular iron bars of finite permeability in the uniform field of a Helmholtz pair of coils, with the iron bar divided into several different arrangements of cubes (5 cubes, 125 cubes etc.). In all these cases the code converged to a reasonable solution. However, when the coil was placed close to the iron the problem became unstable and diverged. TAMI was doing something incorrectly. It took a relatively long time to solve and occupied a large computer memory; further efforts in this direction were abandoned.

b. Program THOR

Early in 1970 work began again in the formulation and development of a new three-dimensional magnetostatic program based on the vector potential method.

The algorithm developed by Sackett [37] utilizes a variational principle for the nonlinear magnetostatic field problem. A finite-element approximation for an arbitrary mesh is constructed from this principle, and the resulting equations are solved by nonlinear successive overrelaxation. Briefly the method is as follows:

In a three-dimensional Cartesian space, the magnetostatics equation for the vector potential is

$$\nabla \times (\gamma \nabla \times \mathbf{A}) - \mathbf{J} = 0, \tag{12.7}$$

where $\gamma = 1/\mu$ (reluctance),

$\mathbf{A}$ = vector source potential, and

$\mathbf{J}$ = source current.

If we let V denote a closed, simply-connected region in Cartesian three-dimensional space, then we define the residual magnetic energy in V to be

$$T = \tfrac{1}{2} \int_V [U(\nabla \times \mathbf{A} \cdot \nabla \times \mathbf{A}) - 2\mathbf{J} \cdot \mathbf{A}]\, dV, \qquad (12.8)$$

where

$$\gamma(\nabla \times \mathbf{A} \cdot \nabla \times \mathbf{A}) = dU/d(\nabla \times \mathbf{A} \cdot \nabla \times \mathbf{A}). \qquad (12.9)$$

Equation (12.7) is the Euler equation associated with the above functional. The variational problem of rendering T stationary in V is therefore equivalent to finding the solution to Eq. (12.17) in V. Moreover, the natural boundary condition for the variation problem,

$$(\gamma \nabla \times \mathbf{A}) \mathsf{x} \hat{\mathbf{n}} = 0 \quad \text{on} \quad S \qquad (12.10)$$

where $\hat{\mathbf{n}}$ is the outward unit normal to the surface S bounding V, is consistent with the continuity requirements of the solution to Eq. (12.7). The more general boundary condition,

$$(\gamma \nabla \times \mathbf{A}) \mathsf{x} \hat{\mathbf{n}} + \mathbf{Q} = 0 \quad \text{on} \quad S, \qquad (12.11)$$

may be obtained by writing

$$T = \tfrac{1}{2} \int_V [U(\nabla \times \mathbf{A} \cdot \nabla \times \mathbf{A}) - 2\mathbf{J} \cdot \mathbf{A}]\, dV + \int_S \mathbf{Q} \cdot \mathbf{A}\, ds. \qquad (12.12)$$

The mesh used is constructed from triangular prisms, one element of which appears in Figure 58. The mesh is generated in a manner similar to that used in program TRIM. In this manner the simplicity and generality of input present in TRIM is carried over in THOR. The nonlinear algebraic equation resulting from the finite element approximation are solved by nonlinear PSOR. Concus [38] has applied this method to the solution of two-dimensional problems and has found that convergence is achieved in fewer iterations, although not always in less computational time than by linearized successive overrelaxation. In addition, he found that the nonlinear method would converge on problems that the linearized method would not.

THOR has been programmed but not thoroughly tested; however, preliminary observations indicate that the program is potentially powerful provided further experimentation and testing is carried out.

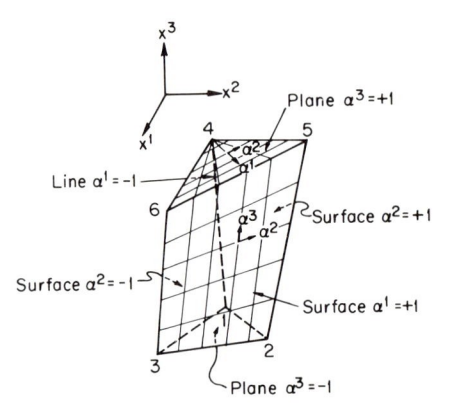

FIG. 58. An element of a triangular prism mesh.

13. Carnegie Mellon Magnet Program

Foss [39, 40] has reported at various conferences his contributions on the development of a three-dimensional program. Specifically, he uses a one potential method—scalar potential. The program is capable of calculating two and three-dimensional, Cartesian or cylindrical problems by using subroutines that modify the program. The available mesh could contain about 10,000 points, and special routines have been written to provide for complex boundary calculations and simple interior points.

The execution of the program is dictated by a set of instruction cards. Each instruction card contains information regarding the action that the program should take. This action may be to return to the calculation, to direct the program to print a brief explanation of the action taken, or to produce informative diagnostics in case the instruction was not executed.

The problem is solved when the divergence of the flux is zero, which is achieved by repeated adjustments of the potential of each mesh. Although the method works well with small problems, it is very slow for large mesh sizes. Foss has described (see Foss [39]) alternative ways of solution.

This program is written in machine language for the CDC-G20 computer and handles problems with infinitely permeable iron only.

14. University of Nevada Magnet Program

Program RENO, developed by Halacsy [41] and his associates produces solutions for three-dimensional magnetostatic problems and includes the effects of finite permeability of the magnetic material. The program is still in the developmental stage so only a summary of the progress is given here.

RENO [42] consists of three separate programs: RENO1 which computes the geometrical coordinates and indices of points within and on the surface of a parallelepiped, and calculates the field induced by electric currents in empty space (air); RENO2 calculates the scalar magnetic potential ϕ and the relative permeability; RENO3 calculates the H-field in the space outside the magnetic materia. The backbone of the whole program is subroutine PHICAL, which calculates the scalar magnetic potential in specified points in the magnetic material where dipole moments were induced by the electric currents considered in RENO1. The status of this program as of the writing of this manuscript is still experimental.

15. CERN Magnet Program

The efforts at CERN to arrive at a three-dimensional magnetostatic program have been reported by Caeymaex [43] and are initially restricted to solving only the linear problem (infinite μ in iron).

This preliminary program, called MIMI, follows the same pattern as its two-dimensional counterpart program MAREC described previously. The problem volume is subdivided into two parts; the region solved by program MIMI-A, and an iron region solved by MIMI-I as yet unfinished. MIMI-A has been written to solve geometries of the C or H-type magnet with mirror symmetries and flat coils. The program uses the scalar potential method outlined in program MAREC. The magnet geometry is outlined in a three-dimensional lattice of 126K points.

Preliminary results reported by Caeymaex show the program converging to a solution after 25 minutes of central processor time on a CDC 6600 computer. Excellent agreement between measured and calculated field results has also been reported. Future plans include further testing of this program, incorporation of the second portion (MIMI-I), and in addition, investigation of alternative methods of increasing the speed of convergence.

16. Program MAFCO

a. Introduction

Program MAFCO by Brown and Perkins [44] is capable of calculating the magnetic fields resulting from a given set of current-carrying conductors of arbitrary two- or three-dimensional geometry in which no permeable material is present.

The elements which comprise the generalized coil geometry are:

1. *Circular loops* with designated position and orientation in space.
2. *Circular arcs* with designated position and orientation in space.
3. *Helices* along the z axis (in the cylindrical coordinate system) with any designated pitch, starting point, and ending point.
4. *Straight lines* with any arbitrary orientation.
5. *General elements* specified by a list of points which the program connects with straight lines.

All of these elements are assumed to be infinitely thin.

Program MAFCO was written in Fortran and is operational on the CDC 6600/7600 computer systems. It requires approximately $60K_8$ memory locations and a typical run takes approximately less than one minute.

The flexibility of this program to calculate magnetic fields for a variety of current-carrying conductor configurations will be further enhanced by some desired features currently being implemented. These features include plots of the field lines and isobars of constant $|\mathbf{B}|$ and input generator routines to eliminate the large number of entries necessary to simulate an acceptable coil configuration. Additionally, an interacting version of this program is being prepared to allow the user to display and modify input.

b. Program Organization

Program MAFCO consists of a main program and eight subroutines:

1. *The Main Program* (MAFCO) is responsible for reading and printing the input data, calling the appropriate subroutines to perform the desired calculations for the elements specified by input (circular loops, arcs, helices, etc.) and calculating the magnetic field lines (or lines of force) passing through any desired point in space.

2. *Subroutine* LFDLP uses the equations appearing in the work of Brown and Perkins [44] to calculate the field of a circular loop when given coordinates of the center (x, y, z), and the radius of the loop, the Euler angles (α, β) specifying the orientation of the loop, the current (I) passing through the loop, and the coordinates of the point $P(R, \theta, Z)$ where the field is to be calculated.

3. *Subroutine* ARC calculates the field produced by a circular arc at the desired point $P(R, \theta, Z)$.

4. *Subroutine* HELIX calculates the field produced by a helix.

5. *Subroutine* LNCALC calculates the field at a point $P(R, \theta\ Z)$ produced by a straight line element.

6. *Subroutine* FLDSUM. This routine calculates the total field as the sum of the contributions of the fields calculated for each current element specified by input (circular loop, arc, etc.) It also calculates the magnitude of the field at the point as the square-root of the sum of the squares of the three field components.

7. *Subroutine* ILIPTK solves the incomplete elliptic integrals of the first and second kind

$$F(\omega, k) = \int_0^\omega (1 - k^2 \sin^2 \phi)^{-\frac{1}{2}} d\phi, \tag{16.1}$$

$$E(\omega, k) = \int_0^\omega (1 - k^2 \sin^2 \phi)^{\frac{1}{2}} d\phi. \tag{16.2}$$

ILIPTK calls subroutine IEI which performs the calculations specified by Eqs. (16.1) and (16.2).

8. *Subroutine* ELIPTK calculates the complete elliptic integrals of the first and second kind:

$$K(k) = \int_0^{\pi/2} (1 - k^2 \sin^2 \phi)^{-\frac{1}{2}} d\phi, \tag{16.3}$$

$$E(k) = \int_0^{\pi/2} (1 - k^2 \sin^2 \phi)^{\frac{1}{2}} d\phi. \tag{16.4}$$

Figure 59 shows a simplified flow chart of this program.

c. Preparation of Input Data

Preparation of input data for MAFCO is straightforward, and follows well-defined rules; the data structure is flexible, allowing for many jobs to be executed in sequence and presents little difficulty to the user. The data cards, as they appear in the deck, are as follows:

1st card	NOP = number of problems to be processed. (Format I5)
2nd card	NLOOP, NARC NHELIX, NLINE, NGCL, MPNTS, MFIELD, TITLE (Format 7I5, 6A6)
	NLOOP = number of circular loops
	NARC = number of circular arcs

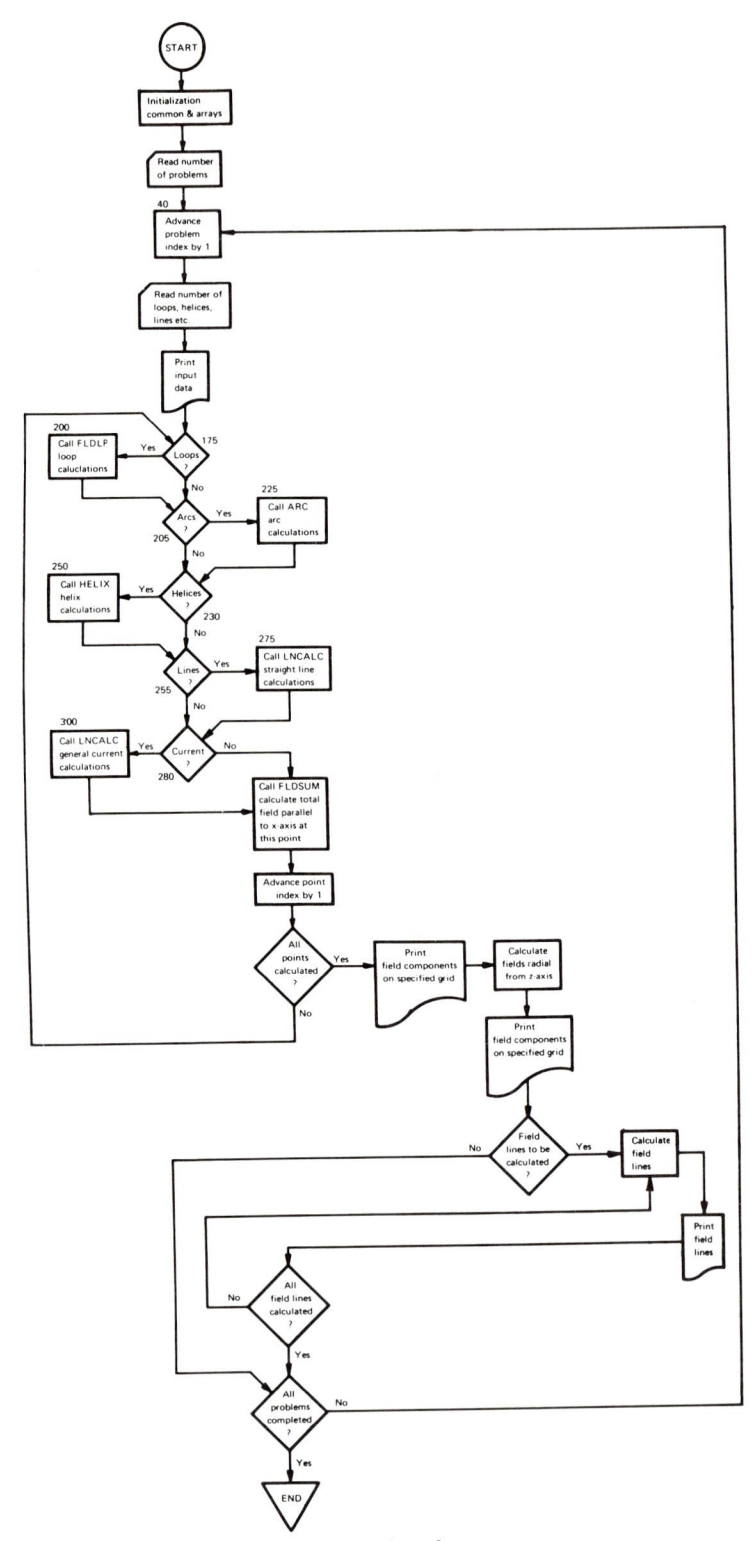

FIG. 59. Flow chart for program MAFCO.

16. Program MAFCO

NHELIX = number of helices

NLINE = number of straight line elements

NGCL = number of general coil elements

MPNTS = number of grid points for calculating fields

MFIELD = number of flux lines to be calculated

TITLE = any descriptive problem title

As many cards as required ⎤ Configuration of current-carrying conductors: one card per conductor, with parameters described in Table 26.

Table 26

MAFCO Specifications for Current-Carrying Conductors

Element	Parameters	Format	Description[a]
Circular loop	X,Y,Z,A,α,β,I (Maximum number of loops = 200)	7F10.1	X, Y, Z = coordinates of center of loop; A = radius; α, β = Euler angles specifying orientation of loop; I = current passing through loop
Circular arc	X,Y,Z,A,α,β,I ϕ_1, ϕ_2	7F10.1 2F10.1	See parameter description for circular loop ϕ_1 = angle of starting point of arc, and ϕ_2 = angle at which arc ends ($\phi_2 > \phi_1$). Two cards required for each arc; maximum number of arcs = 200
Helix	$A, D, \phi_1, \phi_2, I, Z_0$ (Maximum number of helices = 50)	6F10.1	A = radius, D = half distance between turns in Z direction, ϕ_1 = angle of starting point of helix, and ϕ_2 = angle at which helix ends; ($\phi_2 > \phi_1$) ϕ_2 may be greater than 360 degrees I = current in helix; Z_0 = Z coordinate of starting point
Straight line	$X_1,Y_1,Z_1,X_2,Y_2,Z_2,I$	7F10.1	X_1,Y_1,Z_1 = coordinates of one end of line; X,Y,Z = X_2,Y_2,Z_2 = coordinates of the other end; maximum number of straight lines = 300
General	X_i,Y_i,Z_i,I,N (On first card only) X_{i+1},Y_{i+1},Z_{i+1} (On subsequent cards)	4F10.1,I5	X_i,Y_i,Z_i = coordinates of points along general current with i running from 1 to N total number of points. I = current in general element (plus direction toward increasing I); N = maximum number of general current elements; maximum number of general current elements = 100 or 2400 points

[a] Units used are the following: all distances are in centimeters, angles in degrees, currents in amperes, and magnetic field B in gauss.

As many cards as specified by MPNTS (Format 7F10.1)	RMIN, DR, RMAX, THETA, ZMIN, DZ, ZMAX. These parameters determine the grid at which fields are to be calculated. RMIN and RMAX are the minimum and maximum value of the grid in the R direction, DR is the step size, THETA is the angle at which fields will be computed. Similarly ZMIN and ZMAX corresponds with the grid size in the z direction and DZ is the step size.
As many cards as specified by MFIELD (Format 5F10.1, I5)	RBAR, THEBAR, ZBAR, DELMAX, STOTAL, NSKIP. These parameters describe the quantities necessary for the calculation and printout of the field lines (one card per line). RBAR, THEBAR, ZBAR, R, Z and are the coordinates of the origin of the field line to be printed, DELMAX is the arc length increment, STOTAL is the total length of the line, and NSKIP is the printout spacing integer; this flag allows the printout of every NSKIP point on the line.

d. Illustrative Examples

Example 1. To illustrate the manner in which the data are prepared for a successful run and to compare the computed fields with previously known fields, the spectrometer coil arrangement computed by program COILS was chosen.

Because of the large number of input cards necessary to properly simulate the area of the coils with circular loops, only the large coil is processed in the manner shown in Fig. 60. Here, the area of each conductor is simulated by 30 circular loops (in contrast to program COILS in which the area of the conductor is specified) resulting in 240 circular loops for the configuration of the coil. Although data describing the coordinates of these 240 loops are not listed, the input format is similar to data in Table 27. A partial output listing for this example in shown in Fig. 61. The field components computed with this program and with program COILS indicated that both beam programs produced almost identical results.

Example 2. In this example, the coil configuration consists of two four-turn coils and four longitudinal bars with arcs connecting the ends as shown in Fig. 62[†]. The input data are shown in Table 27, in which the mirror coils are approximated by eight circular loops, the three arcs by circular arcs and the four bars by four straight line segments. The calculated isobars of $|B|$ are field lines shown in Fig. 63[†].

[†] This figure was reproduced from Brown and Perkins [44] with the kind permission of the author.

16. Program MAFCO

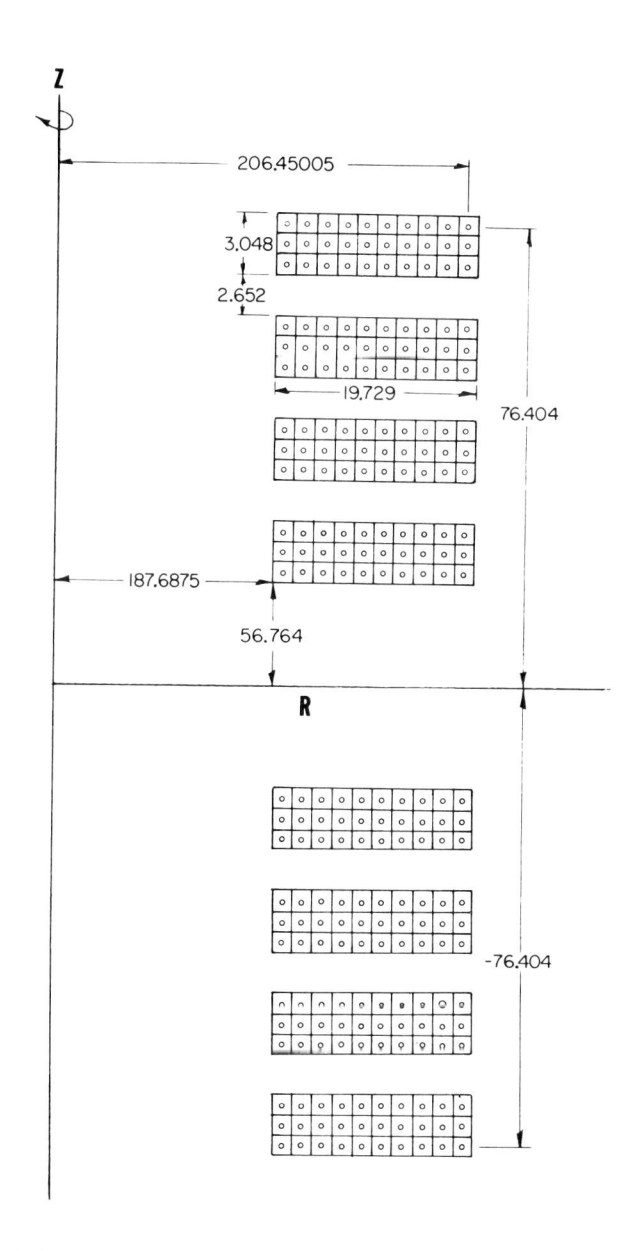

FIG. 60. Simulation of large coil (coil no. 1) of the Chalk River spectrometer shown in Fig. 48.

FIELDS IN A RADIAL DIRECTION

Z	THETA	RMIN	DR	RMAX	R	B(Z)	B(R)	B(0)	1B1
0.	0.	0.	5.00	260.00	0.	4.24481859E-02	0.	0.	4.24481859E-02
					5.00	4.24571074E-02	6.50250253E-17	0.	4.24571074E-02
					10.00	4.24838054E-02	1.42667453E-16	0.	4.24838054E-02
					15.00	4.25282166E-02	2.29254549E-16	0.	4.25282166E-02
					20.00	4.25902060E-02	2.85717265E-16	0.	4.25902060E-02
					25.00	4.26695725E-02	3.84241250E-16	0.	4.26695725E-02
					30.00	4.27660370E-02	4.24248310E-16	0.	4.27660370E-02
					35.00	4.28792257E-02	5.03449279E-16	0.	4.28792257E-02
					40.00	4.30086482E-02	6.22440467E-16	0.	4.30086482E-02
						⋮		⋮	
					165.00	3.76383369E-02	5.85295701E-15	0.	3.76383369E-02
					170.00	3.52990054E-02	6.13918638E-15	0.	3.52990054E-02
					175.00	3.25887285E-02	6.45490605E-15	0.	3.25887285E-02
					180.00	2.95297217E-02	6.53098049E-15	0.	2.95297217E-02
					185.00	2.61708267E-02	7.01955855E-15	0.	2.61708267E-02
					190.00	2.25878643E-02	6.82787160E-15	0.	2.25878643E-02
					195.00	1.88795403E-02	6.63184785E-15	0.	1.88795403E-02
					200.00	1.51588559E-02	6.35342473E-15	0.	1.51588559E-02
					205.00	1.15413456E-02	6.06632800E-15	0.	1.15413456E-02
					210.00	8.13257353E-03	5.69076036E-15	0.	8.13257353E-03
					215.00	5.01762676E-03	6.15306417E-15	0.	5.01762676E-03
					220.00	2.25468159E-03	5.61356517E-15	0.	2.25468159E-03
					225.00	-1.26602351E-04	5.11743425E-15	0.	1.26602351E-04
					230.00	-2.12246406E-03	4.52849563E-15	0.	2.12246406E-03
					235.00	-3.74968026E-03	4.40619763E-15	0.	3.74968026E-03
					240.00	-5.03914447E-03	4.26221558E-15	0.	5.03914447E-03
					245.00	-6.02974395E-03	4.20063290E-15	0.	6.02974395E-03
					250.00	-6.76334125E-03	3.53926957E-15	0.	6.76334125E-03
					255.00	-7.28114781E-03	3.20229954E-15	0.	7.28114781E-03
					260.00	-7.62141990E-03	2.91085599E-15	0.	7.62141990E-03

FIG. 61. Partial printout of the calculated field components.

Table 27

Mafco Input Data for Example 2

				Card Column			
5	10	15	20	30	40	50	60	70

5	10	15	20	30	40	50	60	70
1								
8	3	0	4	0	2	16	MIRROR AND QUADRUPOLE COILS	
	0.0		0.0	-23.0	20.0	0.0	0.0	1.0
	0.0		0.0	-24.0	20.0	0.0	0.0	1.0
	0.0		0.0	-25.0	20.0	0.0	0.0	1.0
	0.0		0.0	-26.0	20.0	0.0	0.0	1.0
	0.0		0.0	23.0	20.0	0.0	0.0	1.0
	0.0		0.0	24.0	20.0	0.0	0.0	1.0
	0.0		0.0	25.0	20.0	0.0	0.0	1.0
	0.0		0.0	26.0	20.0	0.0	0.0	1.0
	0.0		0.0	40.0	16.0	0.0	0.0	4.0
	45.0		135.0					
	0.0		0.0	40.0	16.0	0.0	0.0	4.0
	225.0		315.0					
	0.0		0.0	-40.0	16.0	0.0	0.0	4.0
	135.0		225.0					
	11.3		11.3	-40.0	11.3	11.3	40.0	4.0
	-11.3		11.3	-40.0	-11.3	11.3	40.0	-4.0
	-11.3		-11.3	-40.0	-11.3	-11.3	40.0	4.0
	11.3		-11.3	-40.0	11.3	-11.3	40.0	-4.0
	0.0		0.5	11.0	0.0	-24.0	0.5	24.0
	0.0		0.5	11.0	180.0	-24.0	0.5	24.0
	1.0		0.0	0.0	0.2	20.0		
	1.0		0.0	0.0	-0.2	20.0		
	1.0		180.0	0.0	0.2	20.0		
	1.0		180.0	0.0	-0.2	20.0		
	2.0		0.0	0.0	0.2	20.0		
	2.0		0.0	0.0	-0.2	20.0		
	2.0		180.0	0.0	0.2	20.0		
	2.0		180.0	0.0	-0.2	20.0		
	4.0		0.0	0.0	0.2	12.0		
	4.0		0.0	0.0	-0.2	20.0		
	4.0		180.0	0.0	0.2	12.0		
	4.0		180.0	0.0	-0.2	20.0		
	8.0		0.0	0.0	0.2	6.0		
	8.0		0.0	0.0	-0.2	20.0		
	8.0		180.0	0.0	-0.2	20.0		
	8.0		180.0	0.0	0.2	6.0		

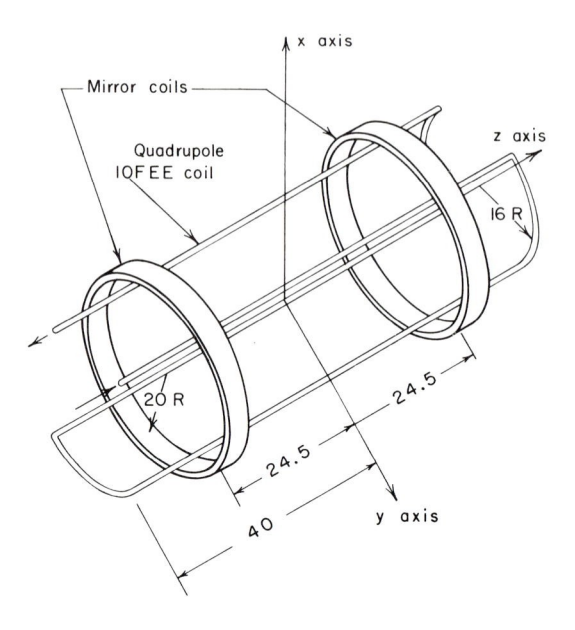

FIG. 62. Two-four-turn solenoidal (mirror) coils and quadrupole Ioffe coil. The arrows show the direction of the current in the quadrupole coil (from Brown and Perkins [44]).

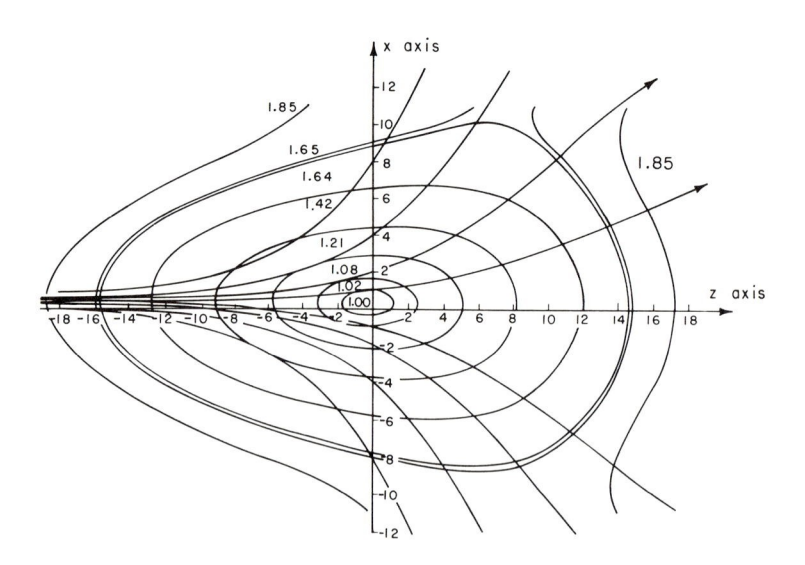

FIG. 63. Isobars of constant $|B|$ and field lines in x, z plane for the current configuration of Fig. 62. The relative values of $|B|$ are given near each isobar (from Brown and Perkins [44]).

17. Program FORCE

a. Introduction

This program by Henning [45] and others [46] is an extension of MAFCO, and is used to calculate the magnitude and direction of the local magnetic force. Once the magnetic field resulting from the specified coil configuration has been calculated, FORCE (using Faraday's law) calculates the magnetic force at the center of each straight line section in the order that these appear in the input data.

Since program FORCE calculates forces for only straight line sections, one must approximate curvilinear conductors by a series of straight line segments. Obviously, the length of each line segment will determine the degree of refinement of the resulting force distribution, thus providing the means for fine or coarse calculations at the expense or saving of computer time.

The program provides for a maximum of 2400 points or 1000 different current elements or any combination thereof.

b. Preparation of Input Data

Data are prepared in the same fashion as for program MAFCO, however, the following changes should be noted:

1. Columns 26–30 in the second card of the FORCE input, calls for the number of straight line segments for which the magnetic forces are to be calculated.
2. The format of this card was changed to (7I5, 7A6) instead of (7I5, 6A6) of MAFCO.

c. Illustrative Example

To illustrate the use of this program, the superconductive saddle coil appearing in Fig. 64 is utilized as an example. This coil consists of two straight sections with compound curves at the ends (turn-around areas). These curves are approximated by using an ellipse that was mapped onto the surface of the cylinder; straight lines approximating the ellipse are generated by an auxiliary computer program. Data necessary for this example are shown in Table 28. The first card contains the number of problems to be solved (integer, right adjusted in columns 1–5). The second card (6I5, 7A6) indicates that there are 28 straight-line element cards following, and that the magnetic forces are to be calculated in four of those straight-line segments. The reason for this being that this

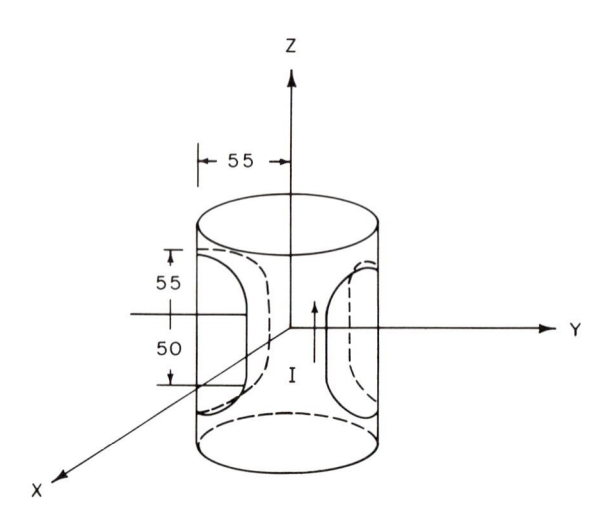

130   III. Three-Dimensional Magnetostatic Programs

FIG. 64. Saddle coil geometry. Distances are in centimeters (from Henning [45]).

configuration has three planes of symmetry and eight symmetrical sections, therefore it is sufficient to determine the magnetic forces on only one section that is composed of four straight line segments.

The rest of the data cards (28 cards) include the quantities necessary to describe this configuration, in the same manner as elaborated in Program MAFCO.

d. Output Description

Upon completion of the execution cycle, the program produces the output appearing in Figure 63, where the quantities printed have the following meaning

ST L NO	Number of straight-line segment
R-AXIS	R coordinate of the midpoint of straight line (cm)
THETA	Theta (θ) coordinate of the midpoint of straight line (degrees)
Z-AXIS	Z coordinate of the midpoint straight line (cm)
B(R)	Radial component of magnetic field (gauss)
B(θ)	Azimuthal component of magnetic field (gauss)
B(Z)	Axial component of the magnetic field (gauss)
BMAG	Magnitude of magnetic field (gauss)
LENGTH	Length of straight line
FMAG	Magnitude of force (pounds/cm)
F(R)	Radial component of force (pound/cm)
F(θ)	Azimuthal component of the force (pounds/cm)

Table 28

Program FORCE Input Data for Illustrative Example[a]

5	10	15	20	25	30	40	50	60	70
1									
0	0	0	28	0	4	SADDLE COIL			
	-0.0		55.0		71.2	47.8	27.3	55.8	100000.0
	47.8		27.3		55.8	53.4	13.1	40.4	100000.0
	53.4		13.1		40.4	54.3	8.7	25.0	100000.0
	54.3		8.7		25.0	54.3	8.7	-25.0	100000.0
	54.3		8.7		-25.0	53.4	13.1	-40.4	100000.0
	53.4		13.1		-40.4	47.8	27.3	-55.8	100000.0
	47.8		27.3		-55.8	0.0	55.0	-71.2	100000.0
	0.0		55.0		-71.2	-47.8	27.3	-55.8	100000.0
	-47.8		27.3		-55.8	-53.4	13.1	-40.4	100000.0
	-53.4		13.1		-40.4	-54.3	8.7	-25.0	100000.0
	-54.3		8.7		-25.0	-54.3	8.7	25.0	100000.0
	-54.3		8.7		25.0	-53.4	13.1	40.4	100000.0
	-53.4		13.1		40.4	-47.8	27.3	55.8	100000.0
	-47.8		27.3		55.8	0.0	55.0	71.2	100000.0
	-0.0		-55.0		71.2	47.8	-27.3	55.8	100000.0
	47.8		-27.3		55.8	53.4	-13.1	40.4	100000.0
	53.4		-13.1		40.4	54.3	-8.7	25.0	100000.0
	54.3		-8.7		25.0	54.3	-8.7	-25.0	100000.0
	54.3		-8.7		-25.0	53.4	-13.1	-40.4	100000.0
	53.4		-13.1		-40.4	47.8	-27.3	-55.8	100000.0
	47.8		-27.3		55.8	0.0	-55.0	-71.2	100000.0
	0.0		-55.0		-71.2	-47.8	27.3	55.0	100000.0
	-47.8		-27.3		-55.8	-53.4	-13.1	-40.4	100000.0
	-53.4		-13.1		-40.4	-54.3	-8.7	-25.0	100000.0
	-54.3		-8.7		-25.0	-54.3	-8.7	25.0	100000.0
	-54.3		-8.7		25.0	-53.4	-13.1	40.4	100000.0
	-53.4		-13.1		40.4	-47.8	-27.3	55.8	100000.0
	-47.8		-27.3		55.8	0.0	-55.0	71.2	100000.0

[a] The first card contains the number of problems; the second card indicates 28 cards (one for each straight-line element in the problem) including 4 straight-line cards for which the field and force are to be calculated.

ST	LNC	R-AXIS	F(R)	THETA	F(Θ)	Z-AXIS	F(Z)				
		SUMF(X)	SUMF(Y)	SUMF(Z)	B(R)	B(Θ)	B(Z)	F(X)	F(Y)	F(Z)	
			BMAG			LENGTH	FMAG				
1		47.587		59.852		63.500					
		3.3356E+00	-3.3372E+00	1.1983E+01	5.5325E+02	1.4722E+02	-1.1300E+02	5.8355E+02		57.352	1.2879E+01

ST	LNC	R-AXIS	F(R)	THETA	F(Θ)	Z-AXIS	F(Z)				
		SUMF(X)	SUMF(Y)	SUMF(Z)	B(R)	B(Θ)	B(Z)	F(X)	F(Y)	F(Z)	
			BMAG			LENGTH	FMAG				
2		54.484		21.762		48.100					
		9.0462E+00	-1.4496E+01	1.4331E+01	2.6158E+01	2.6930E+01	6.8728E+02	4.5610E+00	1.2083E+00	21.683	2.2301E+01
					9.0737E+02	4.3229E+02	-1.3550E+02	1.0142E+03			

ST	LNC	R-AXIS	F(R)	THETA	F(Θ)	Z-AXIS	F(Z)				
		SUMF(X)	SUMF(Y)	SUMF(Z)	B(R)	B(Θ)	B(Z)	F(X)	F(Y)	F(Z)	
			BMAG			LENGTH	FMAG				
3		54.942		11.443		32.700					
		9.5785E+00	-2.7546E+01	8.0394E+00	5.6029E+01	-1.4989E+02	9.9801E+02	1.3776E+01	-1.6109E+01	1.4331E+01	
					1.2765E+03	4.4910E+02	1.8027E+01	1.3533E+03		16.042	3.0253E+01

ST	LNC	R-AXIS	F(R)	THETA	F(Θ)	Z-AXIS	F(Z)				
		SUMF(X)	SUMF(Y)	SUMF(Z)	B(R)	B(Θ)	B(Z)	F(X)	F(Y)	F(Z)	
			BMAG			LENGTH	FMAG				
4		54.993		9.103		0.					
		9.2556E+00	-2.7454E+01	0.	7.9856E+02	-5.5253E+02	1.1272E+03	1.4853E+01	-2.5100E+01	8.0394E+00	
					1.2213E+03	4.1173E+02	-3.4106E-13	1.2888E+03		50.000	2.8972E+01
					1.4727E+03	-1.8347E+03	1.1273E+03	1.3482E+01	-2.5644E+01	0.	

FIG. 65. Partial listing of the output obtained by program FORCE. The quantities listed are defined in the text.

SUMF(X)	Sum of all forces in x direction (pounds)
SUMF(Y)	Sum of all forces in y direction (pounds)
SUMF(Z)	Sum of all forces in z direction (pounds)
F(X)	Force component in x direction, (pounds/cm)
F(Y)	Force component in y direction, (pounds/cm)
F(Z)	Force component in z direction (pounds/cm).

Part B | **ORBIT CALCULATION PROGRAMS**

18. Introduction

In this part of the book we deal with programs capable of computing particle orbits without detailed descriptions of how orbit programs carry out their calculations since that is far beyond the scope of this work. However, to make effective use of these codes it is necessary to have some general idea of the manner in which they solve the problem. For this purpose, it would be desirable to separate the programs into those that solve the equations of motion by numerical integration and those that use a matrix formalism. We might think of these programs respectively as the detailed, or empirical approach, and the ideal or "black box" approach. Numerical integration allows the use of actual measured fields **E** and **B** and also allows us to define fields in any suitable analytic fashion, if we are using analytically derived fields. By numerically integrating through these fields, we obtain results that contain all the higher order (nonlinear) effects of the field on the particles, provided the fields themselves are correctly specified. On the other hand, given certain rather ideal assumptions as to the behavior of the electromagnetic field and the range of travel of the particles with respect to a reference orbit, it is possible to solve the equations of motion exactly. As a result we define the transformation that maps the initial values of the particle onto the final values and thus arrive at a matrix formalism. The distinction drawn here is not clear-cut since programs such as SOTRM use numerical integration to generate the desired matrix transformation. However, as a rule of thumb, the matrix formalism tends to represent the electromagnetic fields as known black boxes represented by matrix transformations. Accordingly, programs GOC, TRAJ, SOTRM, CYDE, and PINWHEEL are based upon numerical integration of the equations of motion; programs TRANSPORT, OPTIK, TRAMP, 4P, and SYNCH are basically of the matrix-formalism type.

In this chapter we will deal with programs that relate in some manner to the motion of charged particles in an electromagnetic field. These programs in some sense solve the equation

$$d\mathbf{P}/dt = q(\mathbf{E} + \mathbf{V} \times \mathbf{B}), \qquad (18.1)$$

where $\mathbf{P}$ and q are, respectively the particle's momentum and charge, and $\mathbf{B}$ and $\mathbf{E}$ are the magnetic induction and electric field intensity.

The statement of the problem in this form, however, is too general for practical applications in design problems. A quick glance at the program descriptions will show that the programs seem to do just about everything but solve Eq. (18.1). The reason for this discrepancy, which is more apparent than real, can be easily stated: These programs have been developed to aid in the design of systems that control the motion of charged particles, i.e., they help to design particle accelerators and beam transport systems. Actually this division of systems is somewhat arbitrary, since a beam must be transported while being accelerated and can be accelerated while being transported. However, questions may be asked as to the type of information needed to design a system and this will depend, to a certain extent, on the application being considered.

Thus, programs have evolved that have different applications and that have been built to answer questions pertaining to their particular application. For example, program CYDE was developed to help design cyclotrons—in particular, the variable energy, 88-inch cyclotron located at the Lawrence Berkeley Laboratory (LBL), in California. Program TRANSPORT was developed originally at the Stanford Linear Accelerator Center (SLAC) in California to assist in designing their beam transport system. Both of these programs model the system under consideration and do much more than just solve Eq. (18.1).

While it is true that simple design problems can be solved without the use of high-speed computers, it is only the modern large-memory high-speed digital computer that makes it possible to set up the entire system and solve for the appropriate design parameters. It is now possible to effectively model a system—whether it be a cyclotron, a linear accelerator or a general beam-transport system—and then study its behavior as a function of the appropriate design parameters. With the availability of computer graphic display equipment, it has become possible to actually picture the model on a cathode-ray tube. For example, the whole beam transport system can be displayed; interaction is possible; and by changing parameters the user can actually see what is happening to the model.

The properties and parameters that are selected for study are dependent upon the system under consideration. Such a profusion of specialized programs exists, based upon different features and combinations of similar features, that we cannot give a detailed discussion of each parameter for every system of interest. However, we can give an overview of what some of these programs can accomplish, and will comment on the types of quantities that are calculated by the various codes. More detail can be found in the introduction to the specific programs.

In summary, the programs presented here under the broad classification of orbit programs divide themselves rather naturally into those that use numerical integration to calculate the orbits and those which, in some sense, idealize the problem and obtain their results using matrices to represent the system elements. In practice, both methods lead to solutions from assumptions and considerations based upon what type of data is available, what answers are being sought, how much time is available for computation purposes, how well the system can be, and need be, modeled, etc. For the user to make a final intelligent choice, the problem to be solved must be well-defined, and the program used must be adequately understood. We can do nothing here to help with the problem definition; however, the program descriptions and sample problems that follow should contribute to the intelligent application of one of the presented programs to solving a well-formulated problem. Whether the program used is suitable for the particular application can only be determined by the experience gained by the user.

IV | Programs Employing Matrix Formalism

19. Introduction

A general beam transport program such as TRANSPORT describes the beam with respect to a reference trajectory (paraxial ray). Evaluations are usually based upon the transformations of the beam envelope (six-dimensional ellipsoid) or upon single particle transformations.

Programs of this type can tell us, for example, what happens to the beam along the system if we change its size, angular divergence, or momentum. The effects of element misalignments can be estimated. Beam envelopes, focus points, energy resolution, element locations, etc. can be found. Special electromagnetic devices can be represented by the program and their separate effects studied. In short, a rather detailed description of the beam can be obtained for any particular system configuration.

The matrix formalism used in these codes follows logically if we think of the solution $(\mathbf{P}, \mathbf{r})$—$\mathbf{P}$ being the momentum and $\mathbf{r}$ the radius vector—as a function of the initial conditions. That is, if we write $\mathbf{z}$ as the phase-space position vector, then

$$\mathbf{z} = (z_1, z_2, z_3, z_4, z_5, z_6) = (P_1, P_2, P_3, r_1, r_2, r_3), \tag{19.1}$$

and with the ith component

$$z_i = z_i^0 + \sum_j \frac{\partial z_i}{\partial z_j^0} \delta z_j^0 + \tfrac{1}{2} \sum_{j,k} \frac{\partial z_i}{\partial z_j^0 \partial z_k^0} \delta z_j^0 \delta z_k^0 + \cdots, \tag{19.2}$$

where $\mathbf{z}^0$ is the reference-particle coordinate vector and

$$\delta z_j^0 = z_j(t_0) - z_j^0(t_0), \tag{19.3}$$

represents the difference between the starting values of the reference particle and the nearby particle. Defining

$$F_{ij} = \partial z_i/\partial z_j^0 \quad \text{and} \quad S_{ijk} = \frac{\partial z_i}{\partial z_j^0 \, \partial z_k^0}, \tag{19.4}$$

we can write

$$z_i^0 = z_i^0 + \sum_j F_{ij} \, \delta z_j^0 + \sum_{j,k} S_{ijk} \, \delta z_j^0 \, \delta z_k^0. \tag{19.5}$$

The first-order terms are represented by the matrix F, and with suitable re-indexing the second-order terms can also be represented by a matrix S.

This is not the usual way by which matrix transformations are derived, although one program, SOTRM, does indeed obtain them in this fashion; it is used here as a convenient way to directly illustrate how a matrix formalism can be used to solve the problem. The more conventional way to handle such problems is to change coordinates to arc length along some reference curve, usually the paraxial trajectory, and then derive the equations of motion of a nearby particle with respect to the reference trajectory. For first-order terms this approach leads to standard linear equations the solution of which can be represented by a matrix that multiplies the initial conditions. Higher-order terms can be obtained, although in practice the calculations are usually stopped at the second order. Also, there are other ways of arriving at these equations, such as using geometric derivations; but the point is that they all lead to essentially the same result. The usual representation obtained is in terms of arc length, and the coordinates represent displacement and angles rather than momenta. Since the results are measured from the reference orbit, we have

$$\delta z_i(s) = \sum_j F_{ij}(s) \, \delta z_j^0 + \sum_{j,k} S_{ijk}(s) \, \delta z_j^0 \, \delta z_k^0, \tag{19.6}$$

where

$$\delta z = (x, \theta, y, \phi, \delta l, \delta p/p), \tag{19.7}$$

with x, y, θ, ϕ, giving the distance and angular deviations of the particle with respect to the reference particle, $\delta p/p$ the momentum perturbation (energy spread), and δl the difference in path length.

From the manner in which we have presented the matrix approach, it should be evident that we are essentially dealing with the first one or two terms of a Taylor's series expansion. The validity of the results, therefore, depends upon how well such a truncated series represents the problem. Experience has shown that for practical applications this approach is usually quite adequate and the truncated series can be used to predict the behavior of the beam. Programs using a matrix formalism have been very successful in systems in which the orbit displacements from the paraxial trajectory are not a significant fraction of the magnet apertures and the magnets are long compared to their apertures.

However, the accuracy of the results will depend upon the nonlinearities of the field and the extent to which the nearby particle maintains its proximity to the reference orbit.

The matrix method does cause the representation of the elements to become idealized. The idealized model of a particular element, for example a bending or quadrupole magnet, is formalized, and its transformation is then derived.

It should be evident that in this approach the electromagnetic element can be represented by the transformation that takes the beam through the element. Thus, we can build up the system using matrices to represent the various elements. Such matrices are derived once for each type of magnet in the system. Subsequently, whenever such an element appears in the transport system the idealized transformation matrix is applied to represent the element. One of the more critical and sometimes difficult tasks is to determine whether the idealized transformation used in a particular program accurately represents the real life element to which it is applied. Usually it does, but not always.

20. Program TRANSPORT

a. Introduction

This is perhaps the most versatile and general purpose program for the design of beam transport systems. Originally written in Balgol by Butler *et al.* [47]. It has been translated into Fortran by many experimenters in the field [48, 48a, 49] and many features have been added to increase its versatility and usefulness. Prominent features of this program include:

1. An array of flexible beam elements.
2. The availability of first-and second-order optics.
3. The capability to track simultaneously as many as 42 vectors.
4. Transformation of apertures forward and backwards in phase-space in order to create acceptance polygons; the calculations may be cycled over energy so as to obtain energy spectra.
5. Flexible output plotting which includes the beam line showing the ellipse projections, vector projections and apertures.
6. Second-order matrix effects [50].
7. The option to use space-charge and betatron function calculations.
8. An on-line interactive mode of operation, which is accomplished by a special version of this code.

20. Program TRANSPORT

Program TRANSPORT is available in Fortran for the CDC 6600/7600 computer systems. The program occupies approximately $75K_8$ memory locations and a typical run takes only a few seconds to solve.

b. Program Description

Program TRANSPORT facilitates the task of the beam designer by providing him with a computational model for most of the magnetic elements from which he may wish to develop his beam line. Many of the problem parameters may be varied according to specified constraints on the beam ellipsoid (or transformation matrix). Problems involving unknown quadrupole excitations, spacings, etc., may be solved. The user specifies the parameters that may be varied and the maximum and minimum excursions, as well as a list of conditions to be satisfied, such as focal points and beam size. The program then solves for the system parameters giving an optimum fit.

Basically, TRANSPORT is a matrix-representation program operating in a six-dimensional vector space with coordinates x, x', y, y', z and $\Delta p/p$. The second-order solutions can be linearized and written in matrix form by considering a higher-dimensional vector space where the six first-order terms plus the 36 second-order terms comprise a 42-dimensional vector space, for which the transformation matrix becomes a 42×42 matrix.

If m is the transformation matrix operating on a particle vector $v(0)$ located at $Z = (0)$ where Z is the beam direction, then $v(Z) = mv(0)$. By assuming that the particle space is bounded by an ellipsoidal countour, we can write the quadratic form of the equation of an ellipse

$$a_{11}x^2 + a_{12}xx' + a_{21}x'x + a_{22}x'^2 = 1.0, \qquad (20.1)$$

in matrix form as

$$(x, x') \begin{pmatrix} a_{11} & a_{12} \\ a_{21} & a_{22} \end{pmatrix} \begin{pmatrix} x \\ x' \end{pmatrix} = 1.0. \qquad (20.2)$$

If we set

$$\sigma^{-1} = \begin{pmatrix} a_{11} & a_{12} \\ a_{21} & a_{22} \end{pmatrix} \quad \text{and} \quad v = \begin{pmatrix} x \\ x' \end{pmatrix}, \qquad (20.3)$$

the equation of an ellipse in matrix form becomes

$$v^t \sigma^{-1} v = 1 \qquad (v^t = v \text{ transpose}). \qquad (20.4)$$

Note that if a $a_{12} = a_{21} = 0$, then $a_{11} = 1/x^2$ and if $x = 0$, then

$$a_{11} = 1/x^2 \quad \text{and} \quad a_{22} = 1/x'^2 \quad \text{and} \quad \sigma = \begin{pmatrix} x^2 & 0 \\ 0 & x'^2 \end{pmatrix}. \qquad (20.5)$$

The transformation properties of this matrix representation can be established by introducing the identity matrix $m^{-1}m = 1$ between the v's and the σ's and recalling that

1. The product of inverse matrices is the inverse of the product in reverse order, and
2. The product of transposed matrices is the transpose of the product in reverse order.

When one writes the equation of the ellipse at the beginning of the system as

$$v(0)^t \sigma^{-1}(0) v(0) = 1, \tag{20.6}$$

it follows that

$$v^t (m^{-1}m)^t \sigma^{-1} (m^{-1}m) v = 1, \tag{20.7}$$

$$v^t m^t m^{t-1} \sigma^{-1} m^{-1} m v = 1, \tag{20.8}$$

$$(v^t m^t)(m^{t-1} \sigma^{-1} m^{-1})(mv) = 1, \tag{20.9}$$

$$(mv)^t (m\sigma m^t)^{-1} (mv) = 1, \tag{20.10}$$

and

$$v^t(L) \sigma^{-1}(L) v(L) = 1, \tag{20.11}$$

where $mv(0) = v(L)$ is the transformation law of a particle vector and $\sigma(L) = m\sigma(0)m^t$ is the transformation law of the ellipse coefficient matrix representing the elliptical boundary of the particle beam. For example, consider an initially upright ellipse that has $a_{12} = a_{21} = 0$ and the matrix of the drift length

$$m = \begin{pmatrix} 1 & L \\ 0 & 1 \end{pmatrix}. \tag{20.12}$$

Then

$$\sigma(L) = \begin{pmatrix} 1 & L \\ 0 & 1 \end{pmatrix} \begin{pmatrix} x^2 & 0 \\ 0 & x'^2 \end{pmatrix} \begin{pmatrix} 1 & 0 \\ L & 1 \end{pmatrix}, \tag{20.13}$$

$$\sigma(L) = \begin{pmatrix} 1 & L \\ 0 & 1 \end{pmatrix} \begin{pmatrix} x^2 & 0 \\ Lx'^2 & x'^2 \end{pmatrix}, \tag{20.14}$$

$$\sigma(L) = \begin{pmatrix} x^2 + L^2 x'^2 & Lx'^2 \\ Lx'^2 & x'^2 \end{pmatrix}. \tag{20.15}$$

At $L = 0$, the beam has an extent given by $x = [\sigma_{11}(0)]^{1/2}$ at $x' = 0$. The extent of the beam is now the projection of the tilted ellipse on the coordinate axes (Fig. 66).

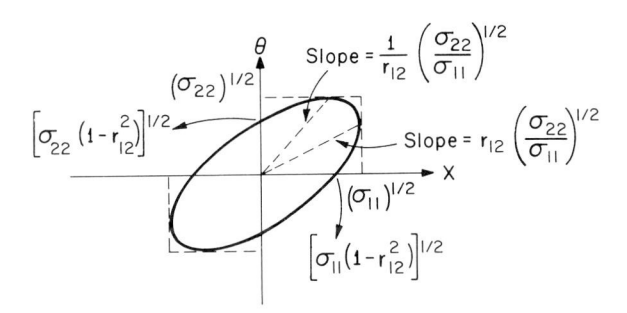

FIG. 66. Phase space ellipse showing the values of its intersection points with the coordinate axes and with the projection rectangle.

The maximum x excursion occurs at the point where $dx/dx' = 0$ and

$$x^2\sigma_{22} + 2xx'\sigma_{12} + x'^2\sigma_{11} = \sigma_{11}\sigma_{22} - \sigma_{12}^2. \tag{20.16}$$

Finding the value of x' where $dx/dx' = 0$ and substituting in Eq. (20.16), we find that $x = [\sigma_{11}(L)]^{1/2}$.

Program TRANSPORT obtains the projections of the beam ellipsoid onto the coordinate axes by taking the square root of the diagonal elements of the ellipse-coefficient matrix. If the transformations had been through a drift space Eq. (20.15) shows the resulting beam size would be

$$x = (\sigma_{11} + L^2\sigma_{22})^{1/2}.$$

Note that if simple ray-tracing had been used, the result would have been $x = (\sigma_{11})^{1/2}$ for an initial ray $(x_0, 0)$, and $x = L(\sigma_{22})^{1/2}$ for an initial ray $(0, x_0')$. Since σ_{11} and σ_{22} are always positive definite, the simple ray results are bounded by the elliptical result.

The transfer matrices simulating the various components of the beam line —bending magnets, quadrupoles, drift lengths, etc.—have been sufficiently described in the literature. For an excellent review of the considerations leading to these transformations the reader is directed to Penner [50a].

c. Program Organization

Program control is exercised by a sequence of "element-type data cards," which describe the ion optic system under consideration. Each element-type card describes a specific function to be performed and includes either the necessary parameters, or the output functions desired, or the various options that are provided by the program. The organization of TRANSPORT follows the simplified flow charts in Fig. 67a, 67b, and 67c, while Table 29 briefly describes the major subroutines included in the code. The flexible array of

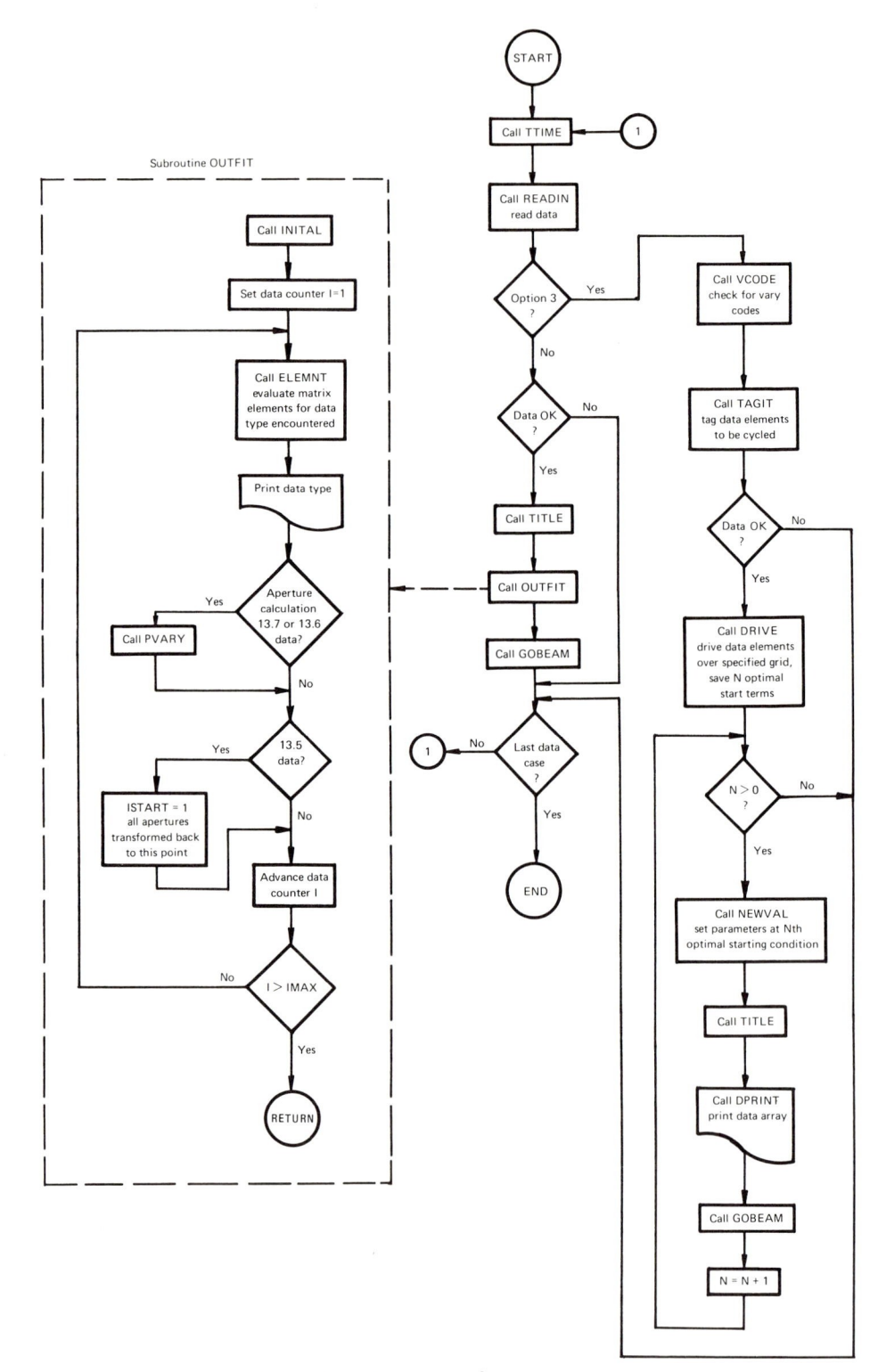

FIG. 67(a). Flow chart of program TRANSPORT.

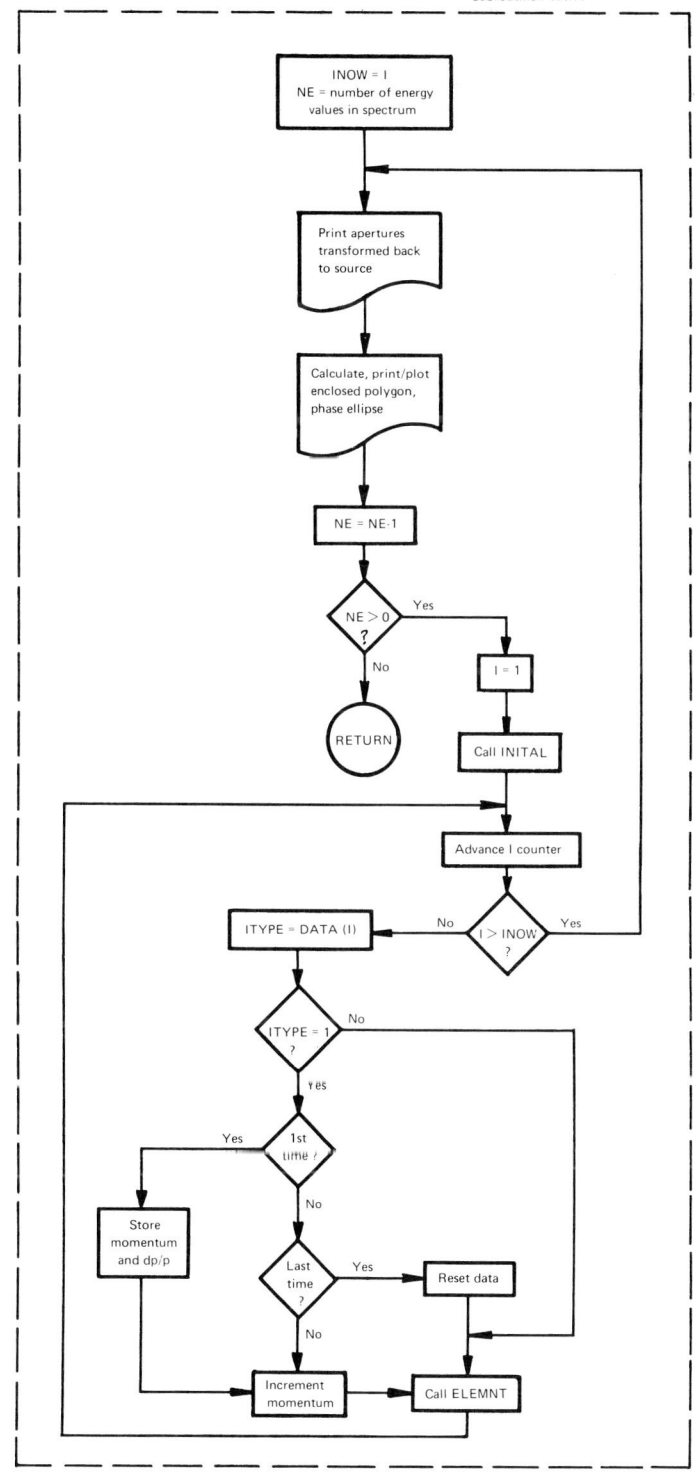

FIG. 67(b). Flow chart of subroutine PVARY.

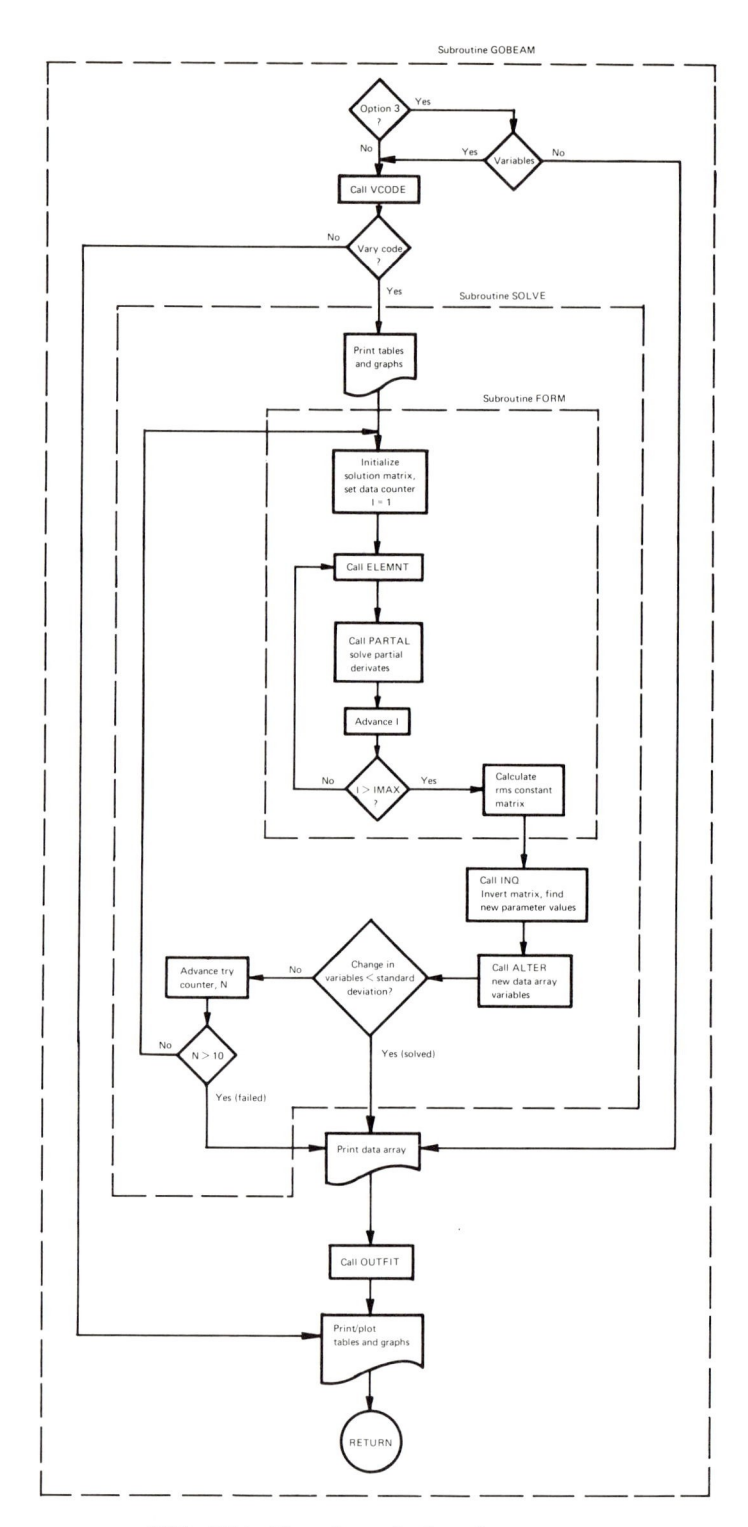

FIG. 67(c). Flow chart of subroutine GOBEAM.

Table 29

TRANSPORT Subroutine and Functions, Partial List

Name	Description
ALTER	Changes variables in the data array to values found during convergence
DRIVE	Cycles parameters and stores root-mean-square deviations for option 3
DPRINT	Prints data array
ELEMNT	Calculates the action of matrix for each data-type element
FORM	Sets up equations for solving variables
GOBEAM	Calls OUTFIT and SOLVE for calculation of beam transformation and prints beam tables
INITAL	Computes initialization of various parameters to standard values
INQ	Calculates an invert matrix of normal equations to find required values of variables
NEWVAL	Stores new values for parameters in data array during option 3
OUTFIT	Describes each type code encountered in input data and prints it at beginning of each job in the run
PARTAL	Calculates partial derivatives of constraints with respect to variables used
PVARY	Calculates aperture polygons to generate momentum acceptance spectra
READIN	Reads and stores data in the array DATA
SOLVE	Calls routines to solve for variables
TAGIT	Tags parameters to be cycled by option 3
VCODE	Checks for varycodes and stores in array TIE

beam elements, their interpretation, and their defining parameters are shown in Table 30. Even though this table includes only a few of the multitude of functions available to TRANSPORT, nevertheless, it indicates the extent of problem-solving capability.

The input/output functions are flexible, allowing the user to exercise considerable control in organizing his input for various runs, or to suppress selectively any part of undesirable printed output. TRANSPORT, if used correctly, is an indispensible program for the solution of problems relating to ion optics.

d. Preparation of Input Data

Every TRANSPORT data deck starts with a date card (format 8A10), followed by a case number card (format F10.0). Each data case within the deck begins with a title card (format 8A10), option card (format I5), and ends with a 73. terminator card (format 8F10.0). Between the option card and the terminator card are the data cards which describe the beam system and the calculations to be performed. Any number of data cases may be stacked with the last data case ending with a 73. 73. terminator card. A brief description of data follows.

1. The title card contains any alphanumeric identification by which the user desires to specify the problem he is attempting to solve.

2. The option card allows the user to exercise a certain degree of control in modifying the original input data. With a minimum of change, additional cases with modified parameters may be run sequentially. Specifically, TRANSPORT recognizes three distinct options:

   a. The standard option (option = 0) is used for single or multiple runs where no modification of data parameters is desired.
   b. The modification option (option = 2) is used to modify the preceding data deck by the proper specification of the desired change or changes.
   c. The initial conditions option (option = 3) is of particular value to problems sensitive to initial conditions, as well as to problems that have a large number of solutions, such as waist-to-waist transformation. It provides the mechanism by which the program may cycle over a range of values selected by the user. For example, one may attempt to discover the quadrupole fields required to produce waist-to-waist transformation, or he may desire to examine the effect on a beam line when two or more elements move in a correlated way.

3. The main data deck is composed of problem data cards that are arranged in a sequential order following precisely the order in which they appear in the beam line. Each card contains the quantities necessary either to properly define each beam line element, or to flag various input/output functions. The following three examples are typical TRANSPORT input:

   a. To define a quadrupole singlet lens the following required quantities may be punched: (See Table 30 for formats.)

$$5.0 \quad 2.0 \quad 2.2 \quad 5.0$$

The first entry, 5.0, refers to the element-type number peculiar to each beam element or input/output function. The second entry, 2.0, is the effective magnetic field length, along the paraxial trajectory, in the same units as the drift length. The third quantity, 2.2, refers to the magnetic field at the pole tip of the quadrupole, normally given in kilogauss. The last entry, 5.0, designates one-half the pole-to-pole gap.

Similarly, to describe a quadrupole doublet the following quantities would appear on the three cards:

5.02	54.0	−8.0	5.0	(first singlet)
3.0	4.5			(drift singlet)
5.03	54.0	10.0	5.0	(second singlet)

The quantities appearing on these cards have the same meaning as those

20. Program TRANSPORT   149

described earlier, with the addition of the new element-type number 3.0 which refers to a drift length of 4.5 units. However, notice that the code number is no longer 5.0, as in the example for the singlet quadrupole, but is 5.02 for the first and 5.03 for the second singlet. These extra digits, referred to as "vary codes," are of great significance to TRANSPORT and constitute, perhaps, the most outstanding feature of this program.

Vary codes make up the decimal part of the element-type number and may have the values 1 through 9. Each succeeding part of the decimal corresponds to the appropriate parameter in the parameter list for that particular code number. For example, the first decimal place (tenths), if not equal to zero would designate the first parameter as variable, and for the second decimal place, the second parameter as variable, and so on. Now consider the three cards for the quadrupole doublet previously mentioned. The element-type number 5.02 would imply that the second parameter of the list (magnetic field of −8.0 kG) could be employed as a variable since the second decimal place is nonzero. Similarly, 5.03 in the third card implies that the magnetic field of 10.0 kG is also a possible variable candidate.

It is often desirable to tie two or more variables together to effect the same change in each variable, as with a symmetric triplet where the first and third quadrupoles must have the same field. This is accomplished by using the same digit (other than 0 or 1) for the vary code of each quadrupole. A vary code of zero specifies that the corresponding parameter may not be varied, while a "1" specifies its possible use as a variable but does not couple it to any other vary code. A "2" in the first decimal part (tenths) will couple to any other vary code of 2 in the first decimal part of any element code number. This is true for all vary codes through 9. Codes 4 and 9 only will play a special role if each is used in the same decimal part of the element code number, in that the correction added to the variable with vary code 4 will be subtracted from the variable with vary code 9. As an example, consider finding the location of the horizontal waist following a bending magnet. This may be accomplished by sliding the waist constraint along the beam line while maintaining the total beam-line length. The following cards show that the drift length L1 will be varied so as to place the 10.0 data card at the horizontal waist while keeping L1 plus L2 constant

3.0	L.			
4.0	L.	B.	INDEX.	
3.4	L1.			
10.0	2.0	1.0	0.0	0.01
3.9	L2			
5.0	L.	B.	HALF-APERTURE	
3.0	L.			

Table 30

TRANSPORT Input for Elements and Parameters, Partial List

Element type number	Defining parameters	Description	
1	X,XP,Y,YP,DS,DP/P,P	Input beam	Parameters specify semiaxis of six-dimensional ellipsoid
2	T,B2	Pole-face rotation	T = angle (degree) between normal to paraxial trajectory and magnet face; B2 = field outside magnet (normally zero)
3	L	Drift space	L = length of drift space
4	L,B,INDEX	Bending magnet	L = effective length (meters); B = average magnetic field (kG); INDEX = magnetic field index, $R(dB/dx)B^{-1}$, where R = radius of curvature, and dB/dx = transverse gradient
5	L,B,APE	Quadrupole	L and B are defined above; APE = half-aperture or $\frac{1}{2}$ the pole-to-pole gap
6	J,X,Y	Slit	The three parameters have the following values: when J = 0 and X = 1, code updates beam J = 0 and X = 2, code initiates RC2 matrix J = 1 and X is the horizontal half-width of slit J = 3 and Y is the vertical half-width of slit J = 4, X and Y are the rectangular half-apertures J = 5, X and Y are the elliptical half-apertures J = 2, X and Y are rectangular half-apertures associated with the following element of length L
10	J,K,X,SD	Constraint	Code digits provide for varying physical parameters in the system to produce:
	J,K = Code digits X = Value SD = Standard deviation		J = 0, K = 0; constraints for system lengths -(J + 20),K = auxilary matrix RC2(J,K) constraint -(J + 10),K = Betatron phase-angle constraint -J, K = transformation matrix RC(J,K) constraint J, K>0 = beam matrix (σ) constraint SI(J,K) J + 10, K = correlation of the beam matrix RJK J>K<10 = covariance of beam constraint SI(J,K) When J = (K>0) and (K<10) these are the phase-projection constraint for matrix SQRT(SI(J,K)); J = ABCD, when K = Y, the storage-ring constraint solves: $Y \times R(A,B) + R(C,D) = X \pm SD$
14	X,X,X,X,X,X,J	Aribitrary matrix	The means by which user may enter his own transfer matrix; X = IJ matrix element, J = row number

Interpretation of these element-type numbers and their associated parameters is given in Table 30, while Table 31 shows the elements and the limits to which they may be varied.

b. In defining a constraint, TRANSPORT may restrict many quantities by either of two general means: constraint of nonmatrix system parameters and constraint on system matrices. In the latter type, TRANSPORT allows constraints on the beam matrix (σ-matrix) and the cumulative transformation matrices

20. Program TRANSPORT

Table 30 (*cont.*)

15	J,DIM,X	Unit change A set of these cards must be preceded by a 15.0, N, M card, where N = number of 15.0 cards following; N may be zero if M ≠ 0; units are restored to standard value before processing following 15.0 N cards; DIM = dimension; X = the value. Table that follows describes reference unit set

J	Description	Standard unit
1	Horizontal and vertical extent (x,y)	centimeters
2	Horizontal and vertical divergence (x')	milliradians
3	Vertical beam extent (y)	centimeters
4	Vertical beam divergence (y')	milliradians
5	Pulsed beam length (s)	centimeters
6	Momentum spread (dp/p)	percent
7	Not used	
8	Length (L)	meters
9	Magnetic field (B)	kilogauss
10	Mass	electron masses
11	Momentum or energy (p)	GeV/c

17	J,K,L	Second order Second, third, and fourth moment of the momentum distributions; all others are assummed Gaussian for which J = 1, K = 0, and L = 3 are required
18	L,B,APE	Sextupole magnet L = effective length; B = average field at pole tip; APE = half-aperture
19	L,B	Solenoid L = effective length; B = magnetic field on the axis of solenoid
22	X,XT,Y,YT,S,DP/P,COMP	Vector input Specifies particle to be traced with coordinates X,XT,Y,YT,S,DP/P; when COMP = 0, normal beam; COMP = β, beam to be deflected by separator; maximum number of input vectors = 42; COMP ≥ 2, second-order vectors
23	L,V,AH,AV,BETAO	Velocity separator A crossed magnetic- and electric-field element producing no deflection of $β_0$ particles when COMP = 0 but producing an angular deflection of particles with 0 < β ≤ 1. BETAO = velocity of vectors with COMP = 0; V = potential in energy units; and AH and AV are horizontal and vertical half apertures
25	MA,INPUTTYPE,PARM	Second-order matrix Allows input of first- and second-order matrices into four distinct arrays which can be used at other location in beam line when INPUTTYPE = 0
26	DL,AMPS,PRTFQ,RFFRE	Space charge Causes element types 3, 4, 5, and 19, to be divided into sublengths DL plus any remainder, and application of space charge impulse at end of each sublength; DL = distance interval at which space charge impulse will be applied; AMPS = current (amperes); PRTFQ = print frequency in units of DL; RFFRE = pulse frequency of bunched beam (megahertz)
73		End case A 73.0 card is required to separate multiple jobs in one computer run
73		End all cases A 73.0 73.0 card terminates computer run

(*RC* and *RC*2). The *RC* matrix is the cumulative transformation matrix from the previous beam update (See Table 30 for element-type 6.0), while the *RC*2 matrix is the cumulative transformation matrix from the point that has been specified by the user. A typical set of constraints that will produce both a horizontal and a vertical focus is shown in the following entries:

        10.0   −1.0   2.0   0.0   0.01
        10.0   −3.0   4.0   0.0   0.01

Table 31

TRANSPORT Internal Constraints on Variables[a]

Element type	Variables	Limits Lower	Limits Upper	Limits Lower	Limits Upper
Beam	1.111111	.01	1000.	-	-
Pole rotation	2.1	-60.	60.	-	-
Drift	3.1	0.1	1000.	-	-
Bend	4.011	-18.	18.	-500.	50.
Quad	5.110	.01	10.	-20.	20.
Alignment	8.111111	-1.	1.	-50.	50.
Auxiliary matrix	14.111111	-	-	-	-
Solenoid	19.01	-	-	-	-
Beam rotation	20.1	-360.	360.	-	-

[a] Denotes adjustment of various physical parameters of the elements to produce the desired fitting requirements imposed by a 10. element type card. To accomplish this, the element types listed above consist of a floating point number whose exponent specifies the code number of the element, and the mantissa, the number of variables associated with the element type. A "zero" in the mantissa portion specifies "no variation", and a "one" specifies "variation", e.g., a quadrupole magnet specified by 5.000 indicates that none of the defining parameters (length, field, and half-apeture) may be varied. However, a 5.010 specifies that the second parameter (field) may be varied to produce constraints imposed upon the system. Similarly a 3.1 card indicates a drift length that may be varied between the upper and lower limits given.

The quantity 10.0 is the element-type number for constraint, the −1.0 and 2.0 in the first card are the J, K elements of the RC matrix $[RC(J, K)]$ to be constrained, while 0.0 and 0.01 specify the value of the constrained matrix element and its limits, that is $RC(J, K) = 0.0 \pm 0.01$.

c. In defining an output function, TRANSPORT uses a 13. element-type card. The flexibility of this program with its capability to generate or suppress normal output or to initiate output not normally generated, is controlled by the appropriate value of a control digit associated with this 13. card. For example, a 13. 2. code will suppress the beam ellipsoid output, a 13. 4.

code will print output for transfer matrix *RC*, and a 13. 7. code will print output for the phase-space aperture polygons. For a complete description of all the output functions and the element types not listed in Table 30 the reader is referred to the work of Paul [48].

e. Illustrative Example

The example in Table 32 will be analyzed in detail with a description of the input, to further illustrate the manner in which data is prepared for a successful TRANSPORT run, to establish criteria of performance and to permit comparison with the other beam transport programs described in later sections.

The data in Table 32 correspond to a simple beam transport system consisting of a bending magnet and a quadrupole doublet. The first quadrupole converges vertically, the second converges horizontally. The system requires both a horizontal and a vertical focus at the beam end, to be obtained by proper adjustment of the quadrupole magnetic field which, therefore, constitute the variables of the system. The interpretation of the input data for this example is straight forward. After the date, case, title, and option cards, a specification of the changes in the internal unit reference system is made (15.0, 3.0), followed by three cards specifying the unit changes desired. Next, a specification is made of the vectors to be traced (six 22. cards), followed by a description of the beam system, interspersed with output option cards (13.). Note the manner in which the quadrupole doublets have been specified: The first doublet (5.02) has an effective length of 27.0 cm, an estimated field of -6.0 kG (the minus sign indicates the quadrupole is horizontally diverging), and a half-gap of 5.0 cm. The possibility of adjustments of the quadrupole field is indicated by the nonzero unit position of the decimal portion (hundredths) or the quadrupole element-type number 5.02. A similar argument describes the second quadrupole magnet. The requirement of both a horizontal and a vertical focus is specified by the two 10.0 cards with the appropriate values as specified in earlier paragraphs (also see Table 30). The last card of the deck (73. 73.) terminates the run.

f. Output Description

The output appearing in Fig. 68 is obtained from subroutine OUTFIT and reflects the element-by-element parameter list, followed by the accumulated length. Unless suppressed, the printout of the beam ellipsoid, centroid, and vectors will occur after each element that possesses a transformation matrix. The printout of several transformation matrices will occur only if it is explicitly specified. Some extra output generated by bending magnets, quadru-

Table 32

TRANSPORT Sample Input Data

			Card Column				
1	11	21	31	41	51	61	71
DATE CARD							
1000.0							
N PI PI BEAM.							
0							
15.0	3.0						
15.0	1.0	IN	2.54				
15.0	6.0	PC	1.0				
15.0	8.0	IN	0.0254				
22.0	0.0	0.0	0.0	0.0	0.0	4.0	
22.0	0.2156	−10.0	0.22726	−6.0	0.0	4.0	
22.0	0.25	0.0	0.25	0.0	0.0	−4.0	
22.0	−0.2165	10.0	−0.22726	6.0	0.0	−4.0	
22.0	0.25	0.0	0.25	0.0	0.0	0.0	
22.0	−0.25	−0.0	−0.25	−0.0	0.0	0.0	
1.0	0.25	20.0	0.25	12.0	0.0	0.0	2.5
13.0	4.0						
24.0	0.0	0.0	0.0	0.0	1.0	6.0	
24.0	4.0	2.4	2.6	0.1			
24.0	1.0	0.25	1000000.	0.25	1000000.		
13.0	5.0						
13.0	2.0						
3.0	10.0						
6.0	2.0	3.0	6.0				
2.0	−8.11						
4.0	44.0	−17.115	0.0				
13.0	4.0						
2.0	−5.07						
3.0	40.0						
5.02	27.0	−6.0	5.0				
5.02	27.0	−6.0	5.0				
13.0	4.0						
3.0	4.5						
5.03	27.0	6.0	5.0				
5.03	27.0	6.0	5.0				
13.0	4.0						
3.0	50.0						
13.0	1.0						
10.0	−1.0	2.0	0.0	0.01			
10.0	−3.0	4.0	0.0	0.01			
13.0	6.0						
13.0	4.0						
24.0	1234.0						
3.0	50.0						
73.0	73.0						

poles, unit changes, angles of bend, radius of curvature, and focal length of quadrupoles, etc., are easily identifiable.

It was mentioned earlier that TRANSPORT is extremely flexible in controlling or suppressing output based on the requirements of the user, which are exercised by the proper choice of the element-type number 13.0. In the example, the first reference of an output function is given by the 13.0   4.0 card which specifies the printout of the *RC* matrix. The next data entry, 13.0   5.0, indicates the beginning of the phase-space aperture calculation while the 13.0   6.0 card specifies the end of this calculation (see Fig. 69.) These apertures are

20. Program TRANSPORT

```
UNIT()=  7.615E-04  1.000E-03  7.615E-04  1.000E-03  2.998E-04  1.00GE-02  1.000E+00  7.615E-04  1.000E+00  5.108E-04  1.000E+00
                                                   .....IN .....MR .....IN .....MR .....CM .....PC .....COM
                                          VECTOR  1(A)   0.        0.        0.        0.        0.       4.000   -0.
                                          VECTOR  2(B)   .216   -10.000    .227    -6.000    0.       4.000   -0.
                                          VECTOR  3(C)   .250      0.        .250      0.        0.      -4.000   -0.
                                          VECTOR  4(D)  -.216    10.000   -.227     6.000    0.      -4.000   -0.
                                          VECTOR  5(E)   .250      0.        .250      0.        0.        0.      -0.
                                          VE----------------------------------------------------------------------- 19
BEAM  1.000000    2.50 GEV/C
                                                        0.        .250 IN                     LENGTH=    0.     IN
                                                        0.      20.000 MR       0.
                                                        0.        .250 IN       0.       0.
                                                        0.      12.000 MR       0.       0.
                                                        0.        0.   CM       0.       0.       0.
                                                        0.        0.   PC       0.       0.       0.       0.
                                                   .....IN .....MR .....IN .....MR .....CM .....PC .....COM
                                          VECTOR  1(A)   0.        0.        0.        0.        0.       4.000   -0.
                                          VECTOR  2(B)   .216   -10.000    .227    -6.000    0.       4.000   -0.
                                          VECTOR  3(C)   .250      0.        .250      0.        0.      -4.000   -0.
                                          VECTOR  4(D)  -.216    10.000   -.227     6.000    0.      -4.000   -0.
                                          VECTOR  5(E)   .250      0.        .250      0.        0.        0.      -0.
                                          VECTOR  6(F)  -.250      0.       -.250      0.        0.        0.      -0.
TRANSFORM  1
     0.        0.        0.        0.        0.        0.
     0.        0.        0.        0.        0.        0.
     0.        0.        0.        0.        0.        0.
     0.        0.        0.        0.        0.        0.
     0.        0.        0.        0.        0.        0.
     0.        0.        0.        0.        0.        0.
PLOTS 24.0      0       0.     IN      0.     IN      0.     IN      1.0000    6.0000
PLOTS 24.0      4     2.4000 GEV    2.6000 GEV    .1000 GEV/   -0.       -0.
PLOTS 24.0      1      .2500 IN *0000.0000 MR      .2500 IN *0000.0000    -0.
APERTURE   TRANSFORMATION POINT                                                          LENGTH=     10.000 IN       54
DRIFT 3.0    12.000 IN                                                                   LENGTH=     10.000 IN       56
SLIT  6.0     2    3.0000 IN    6.0000 IN                                                LENGTH=     10.000 IN       60
ROTAT 2.0    -8.110 D    -0.    KG                                                       LENGTH=     54.000 IN       63
BEND  4.000  44.000 IN  -17.115 KG    .00    ( -13.142 D)
                                             ( -191.826 IN )
TRANSFORM  1
     1.0062E+00  5.3677E-02  0.           0.           0.          -5.0241E-02
    -4.6188E-01  9.6919E-01  0.           0.           0.          -2.2737E+00
     0.           0.          9.6731E-01  5.3673E-02  0.           0.
     0.           0.         -7.4286E-01  9.9257E-01  0.           0.
     5.8700E-01  1.8631E-02  0.           0.           1.0000E+00 -9.7742E-03
     0.           0.          0.           0.           0.           1.0000E+00             LENGTH=     54.000 IN       69
ROTAT 2.0    -5.070 D    -0.   KG                                                        LENGTH=     94.000 IN       72
DRIFT 3.0    40.000 IN                                                                   LENGTH=    121.000 IN       74
QUAD  5.02   27.000 IN    -8.2721 KG    5.00 IN   ( -69.18 IN )
QUAD  5.02   27.000 IN    -8.2721 KG    5.00 IN   ( -69.18 IN )                          LENGTH=    148.000 IN       78
TRANSFORM  1
     1.8411E+00  2.3874E-01  0.           0.           0.          -4.1669E-01
     3.4608E+01  5.0310E+00  0.           0.           0.          -9.0880E+00
     0.           0.          2.7313E-01  7.2791E-02  0.           0.
     0.           0.         -1.9750E+01 -1.6023E+00  0.           0.
     5.8700E-01  1.8631E-02  0.           0.           1.0000E+00 -9.7742E-03
     0.           0.          0.           0.           0.           1.0000E+00             LENGTH=    152.500 IN       84
DRIFT 3.0     4.500 IN                                                                   LENGTH=    179.500 IN       86
QUAD  5.03   27.000 IN   10.0932 KG    5.00 IN   (  64.98 IN )
QUAD  5.03   27.000 IN   10.0932 KG    5.00 IN   (  64.98 IN )                           LENGTH=    206.500 IN       90
TRANSFORM  1
     1.8171E+00  2.5751E-01  0.           0.           0.          -4.6182E-01
    -4.0239E+01 -5.1522E+00  0.           0.           0.           8.9551E+00
     0.           0.         -1.0394E+00  1.8900E-01  0.           0.
     0.           0.         -3.2147E+01 -3.7755E-01  0.           0.
     5.8700E-01  1.8631E-02  0.           0.           1.0000E+00 -9.7742E-03
     0.           0.          0.           0.           0.           1.0000E+00             LENGTH=    256.500 IN       96
DRIFT 3.0    50.000 IN
                                                        0.        .049 IN                     LENGTH=    256.500 IN
                                                        0.     103.533 MR      .137
                                                        0.        .662 IN       0.       0.
                                                        0.       9.226 MR       0.       0.      .871
                                                        0.        .400 CM     -.403   -.962    0.       0.
                                                        0.        0.   PC       0.       0.       0.       0.
                                                   .....IN .....MR .....IN .....MR .....CM .....PC .....COM
                                          VECTOR  1(A)  -.056    35.820     0.        0.        .033     4.000   =0.
                                          VECTOR  2(B)  -.097    78.630    -.602    -5.040    -.098     4.000   -0.
                                          VECTOR  3(C)   .008   -45.880    -.662    -8.037     .106     4.000   -0.
                                          VECTOR  4(D)   .097   -78.630     .602     5.040     .098    -4.000   -0.
                                          VECTOR  5(E)  -.049   -10.060    -.662    -8.037    -.147     0.      -0.
                                          VECTOR  6(F)   .049    10.060     .662     8.037    -.147     0.      -0.
FIT  10.0    -1  2    0.           /   .010000
                     -.000098
FIT  10.0    -3  4    0.           /   .010000
                      .000022
```

FIG. 68. Partial output for the sample case appearing in Table 32. The several transformation matrices printed occur at points specified by a 13.   4.   input card.

transformed backward from their location to the 13.0   15.0 card. The aperture calculations apply only to systems in which the horizontal and vertical phases are decoupled, which restricts this type of calculation to beam lines that do not contain elements that couple the horizontal and vertical planes such as solenoids, misaligned quadrupoles, etc.

156　IV. Programs Employing Matrix Formalism

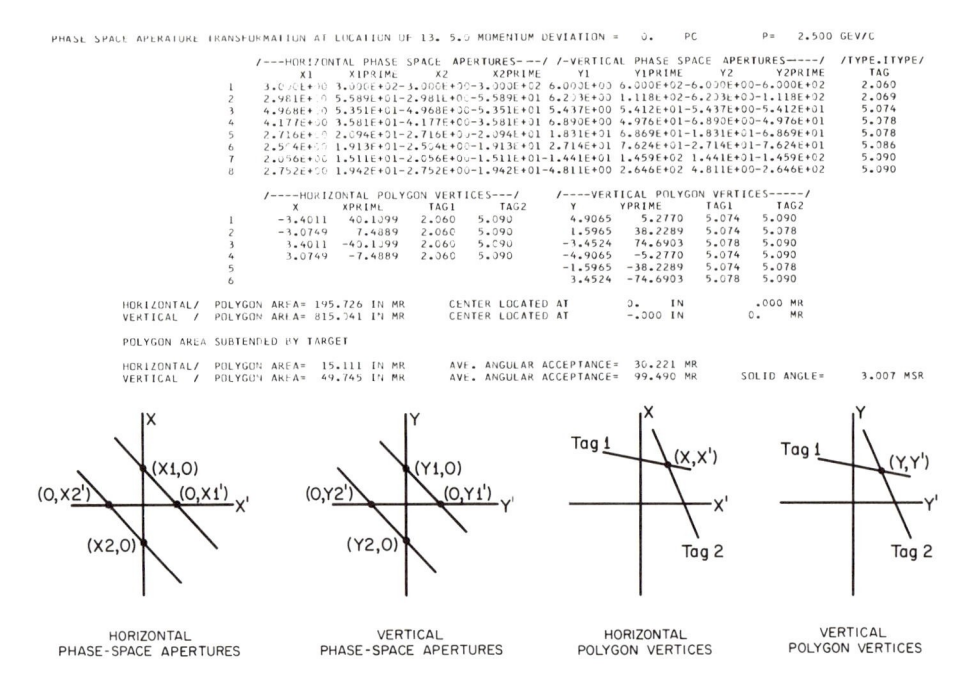

```
PHASE SPACE APERTURE TRANSFORMATION AT LOCATION OF 13. 5.0 MOMENTUM DEVIATION =   0.     PC       P=   2.500 GEV/C
         /---HORIZONTAL PHASE SPACE APERTURES---/ /--VERTICAL PHASE SPACE APERTURES----/ /ITYPE.ITYPE/
              X1      X1PRIME      X2      X2PRIME      Y1      Y1PRIME      Y2      Y2PRIME        TAG
    1    3.000E+00  3.000E+02-3.000E+00-3.000E+02  6.000E+03  6.000E+02-6.000E+00-6.000E+02       2.060
    2    2.981E+00  5.589E+01-2.981E+00-5.589E+01  6.203E+03  1.118E+02-6.203E+00-1.118E+02       2.069
    3    4.968E+00  5.351E+01-4.968E+00-5.351E+01  5.437E+00  5.412E+01-5.437E+00-5.412E+01       5.074
    4    4.177E+00  3.581E+01-4.177E+00-3.581E+01  6.890E+00  4.976E+01-6.890E+00-4.976E+01       5.078
    5    2.716E+00  2.094E+01-2.716E+00-2.094E+01  1.831E+01  6.869E+01-1.831E+01-6.869E+01       5.078
    6    2.544E+00  1.913E+01-2.544E+00-1.913E+01  2.714E+01  7.624E+01-2.714E+01-7.624E+01       5.086
    7    2.056E+00  1.511E+01-2.056E+00-1.511E+01-1.441E+01  1.459E+02  1.441E+01-1.459E+02       5.090
    8    2.752E+00  1.942E+01-2.752E+00-1.942E+01-4.811E+00  2.646E+02  4.811E+00-2.646E+02       5.090

             /----HORIZONTAL POLYGON VERTICES---/        /----VERTICAL POLYGON VERTICES----/
                 X      XPRIME     TAG1    TAG2              Y      YPRIME     TAG1    TAG2
    1        -3.4011    40.1099   2.060   5.090          4.9065    5.2770   5.074   5.090
    2        -3.0749     7.4889   2.060   5.090          1.5965   38.2289   5.074   5.078
    3         3.4011   -40.1099   2.060   5.090         -3.4524   74.6903   5.078   5.090
    4         3.0749    -7.4889   2.060   5.090         -4.9065   -5.2770   5.074   5.090
    5                                                   -1.5965  -38.2289   5.074   5.078
    6                                                    3.4524  -74.6903   5.078   5.090

HORIZONTAL/  POLYGON AREA= 195.726 IN MR     CENTER LOCATED AT        0. IN       .000 MR
VERTICAL  /  POLYGON AREA= 815.041 IN MR     CENTER LOCATED AT       -.000 IN      0.    MR

POLYGON AREA SUBTENDED BY TARGET

HORIZONTAL/  POLYGON AREA=  15.111 IN MR    AVE. ANGULAR ACCEPTANCE=  30.221 MR
VERTICAL  /  POLYGON AREA=  49.745 IN MR    AVE. ANGULAR ACCEPTANCE=  99.490 MR       SOLID ANGLE=   3.007 MSR
```

```
        X                        Y                       X                        Y
        |                        |                       |  Tag 1                 |  Tag 1
        |(X1,0)                  |(Y1,0)                 |\                       |\
 (0,X2')\|    (0,X1')     (0,Y2')\|    (0,Y1')           | \   (X,X')             | \   (Y,Y')
 --------+--------- X'    --------+--------- Y'   -------+--\-------- X'    ------+--\-------- Y'
         |\                       |\                     |   \                    |   \
         | (X2,0)                 | (Y2,0)               |    \                   |    \
         |                        |                      |   Tag 2                |   Tag 2

     HORIZONTAL              VERTICAL               HORIZONTAL              VERTICAL
  PHASE-SPACE APERTURES  PHASE-SPACE APERTURES  POLYGON VERTICES       POLYGON VERTICES
```

FIG. 69. Phase-space aperture transformation table.

The quantities printed in Fig. 69 correspond to the sketches drawn in the lower portion of the same figure, while the two polygons in Fig. 70 show a plot of the horizontal and vertical transformed apertures. In relation to these two polygons, the solid angle, Ω, and the average angular acceptance is calculated by finding the area overlap of the target and the beam-line acceptance polygon, A_0, divided by the target size, W, with $\Omega = A_{0X} A_{0Y} / W_X W_Y$. If one wishes to perform the acceptance calculations at several different momenta, then the 13.0　6.0 card is replaced by a 13.0　7.0 card. Now the polygon transformation will be performed at each momentum specified by a 24.0　4.0 card and will include the effect of dispersion and chromatic aberration. The result of such a calculation is the determination of the first-order momentum acceptance of the beam line.

The 13.0　2.0 card suppresses the beam ellipse output on the first run, while later in the example the 13.0　1.0 card temporarily overrides this ellipse suppression.

Summary tables giving accumulated length, apertures, beam projections, etc., may also be obtained by the proper choice of the 13.0 card, and the user may usually alter the output requirements at his discretion.

The 24.0　1234.0 card allows the plotting of the phase-space ellipses at the projected planes 12 and 34, respectively (see Fig. 71.). Plane 12 specifies the

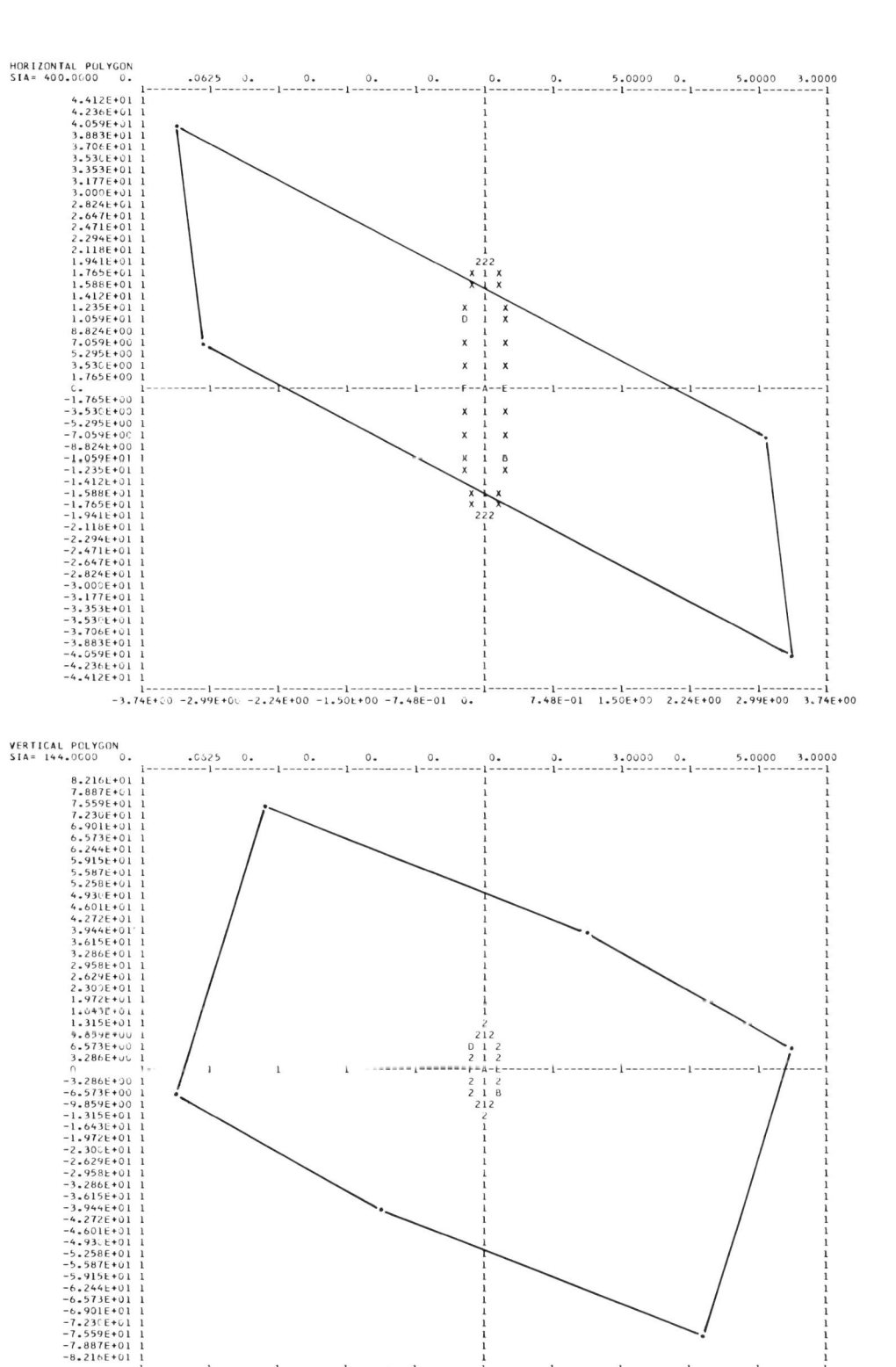

FIG. 70. Printer plot of the transformed apertures in the horizontal (upper) and vertical (lower) planes.

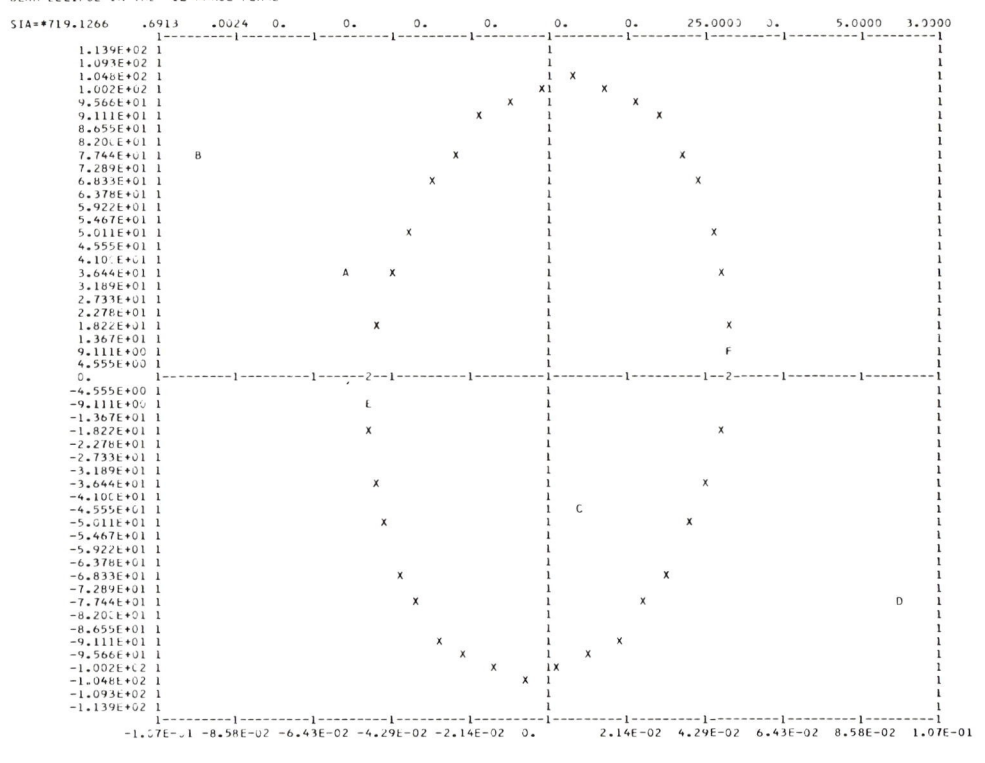

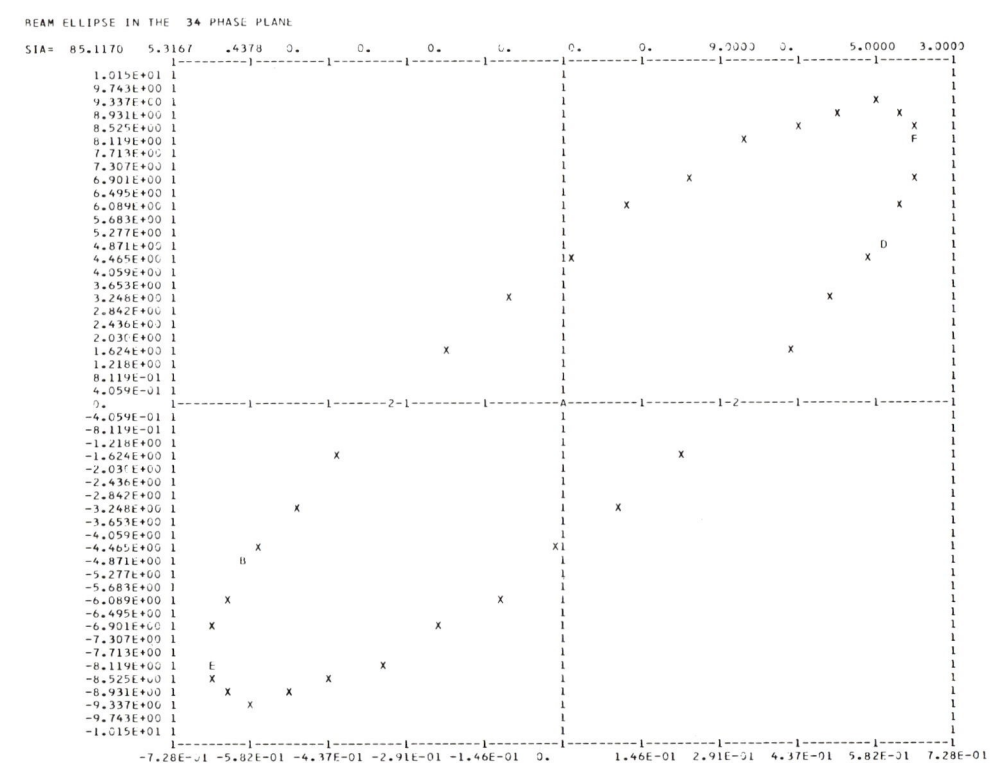

FIG. 71. Phase-space ellipse projected at the requested planes. The x, x' plane (upper) and the y, y' plane (lower).

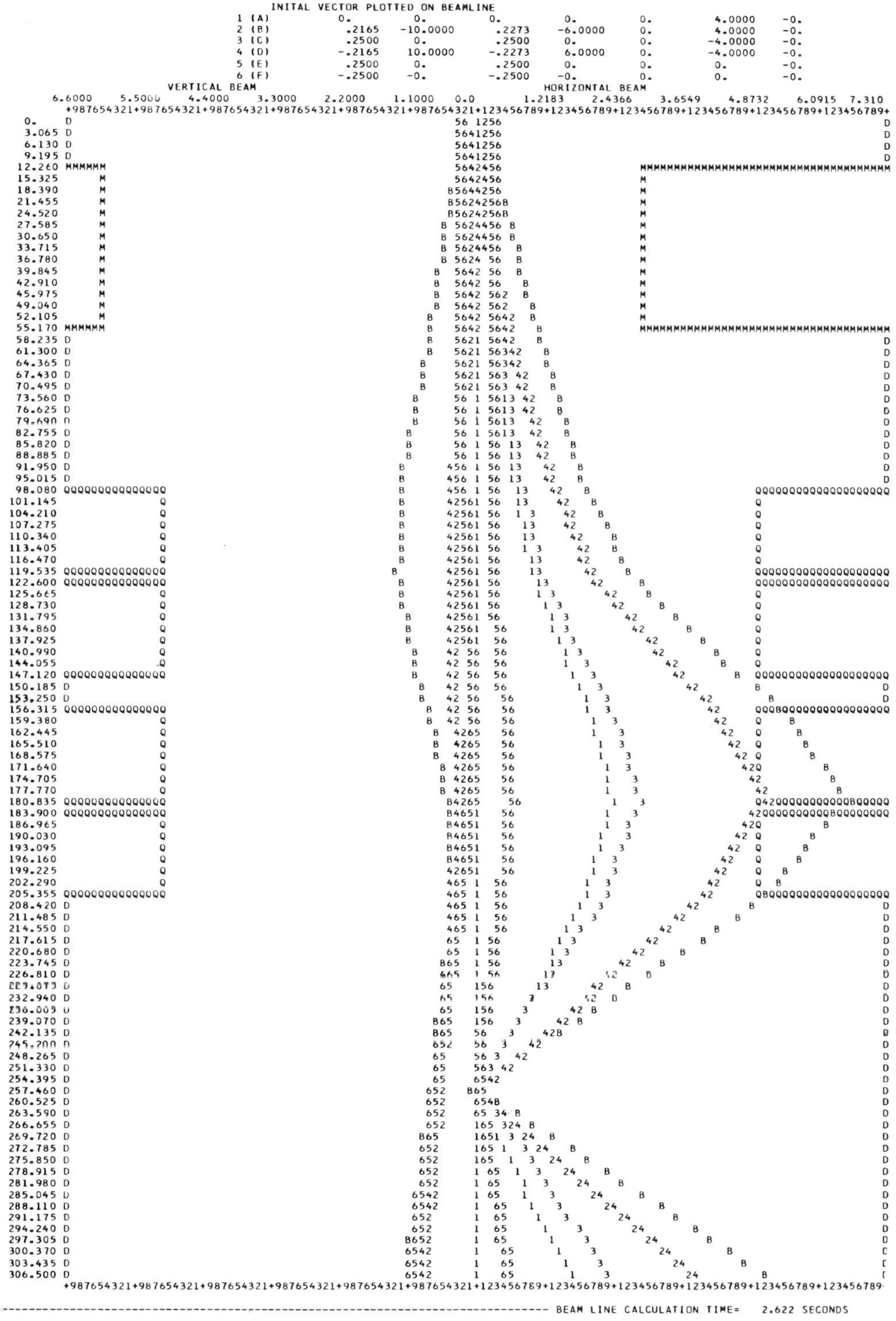

FIG. 72. Printer plot of the beam envelope, and ray trace of six vectors.

x, x' plane while plane 34 specifies y, y' plane. It should be seen that any combination of the six-dimensional space coordinates by which the beam is specified may be projected; thus, a 24.0   1216.0 card would specify the x, x' and $x, \Delta p/p$ planes, etc. Figure 72 shows the printed plot of the beam envelope and a ray-trace of the specified vectors as produced by the 24.0   0.0 card.

g. Further Development of TRANSPORT

The application of graphic display devices [51, 51a, 52] to a computer system has opened new horizons to man-machine communications. It is no longer necessary for the user to be an observer, tacitly accepting the idiosyncrasies of his computational device. Graphic displays provide the means of interacting with the computational device, permitting the user to modify or augment the information held in the computer memory, select calculation procedures, or direct the program to alternate modes of operation.

These characteristics have been implemented in program TRANSPORT and add flexibility to the program in the sense that input and output, as well as information flow, is at the discretion of the experimenter. This special version of TRANSPORT has been implemented on the CDC 6600 computer system using a CDC 252 display console,[†] which provides for instantaneous visual readout of selected portions of the computer memory in the form of alphanumeric symbols vectors, or points.

Incoming data are displayed directly on the console's 19-inch CRT within a 1024×1024 horizontal- and vertical-position raster. An attached teletype also provides for data input or external program interrupts.

The existing software package provides for the utilization of programmable interrupts, use of teletype to enter or extract data and for performing various plotting functions.

A session begins with the user viewing on the cathode ray tube screen the data for the problem under consideration (Fig. 73). The data correspond to the previously-run simple beam (Table 32) and the user begins the interactive process by exercising any of the options listed in the lower part of the screen, which are explained in Table 33. These options by no means exhaust those that the user may want to incorporate, and which are accommodated by the modularity of program TRANSPORT. To exercise an option the user simply selects the desired option on the screen by an entry on the teletype.

Upon initiation of the option, SOLVE, the executive program of TRANSPORT, decodes the action and signals the beginning of execution resulting in Fig. 74,

[†] A special version using only teletype has been implemented and used successfully through remote communication links.

20. Program TRANSPORT

Table 33

Available Options for On-Line TRANSPORT

Option name	Description
ALTER	Performs data alteration by entering appropriate address in data array
DLINE	Deletes one line of input or one element of beam system
ALINE	Adds one line of input; allows user to enter all input via teletype, or to add new element at any location along beam system
SNAPB	Records beam line display on microfilm
SNAPD	Records data display on microfilm
MDATA	Allows user to view more data; often one frame is not enough to display all input data
SAVE	Saves current data in special array for future use
RECAL	Restores data previously saved
CANCL	Cancels any selected option user has initiated
VECT	Performs changes on displayed vectors; vector parameters may be added or removed if already specified
BEAM	Reviews beam that is already calculated
RAYS	Provides for display of vector(s)
SCALE	Provides for longitudinal and transverse scaling of beam line; via teletype user enters region desired and scaling factor
MATRX	Allows for display of phase-space ellipses and permits selected matrices
GO	Permits program to execute without optimization
ITER	Permits program to execute with optimization
NCASE	Reads data for next case
FIN	Terminates run
SEGMT	Separates beam data into segments
FIX	Removes vary codes and negates constraints

which displays the beam envelope. Program TRANSPORT again waits (releases CPU) for the user to resume the interacting cycle based on the information displayed.

At this point, the user may decide to observe the particle phase-space distribution at some location along the beam line. To do so, he enters on the teletype the option MATRX, the location on the beam line where he desires the matrix and the planes on which the six-dimensional phase-space ellipse is to be projected. Upon completion of this input specification (see the lower half of Fig. 74) the ellipse is displayed. Here, as before, the beam envelope is shown which is defined as the plot of beam size versus position along the system. Interpretation of the trace is as follows. A rectangle is drawn to represent a magnet, the size of which reflects the magnet length and aperture. The beam size is interpolated linearly in drift spaces. Both the horizontal and vertical planes are displayed (half of each because of symmetry) along with the cumulative beam length (see the upper half of Fig. 74.) Below the beam

```
CURRENT OPT. SI. EXMPLE                                CASE 10.455  5
 1 UNIT CON.    15    5   ;;;;;;    1.0000
 3 UNIT CON.    15    1   IN        2.5400
 9 UNIT CON.    15    6   PC        1.0000
15 UNIT CON.    15    8   IN         .0254                              LENGTH
17 VECTORS     22.00000  6.00000
               VECTOR NO. 1   0.       0.       0.       0.      0.    4.000  -0.
               VECTOR NO. 2    .250    0.        .250    0.      0.    0.     -0.
               VECTOR NO. 3   0.      20.000    0.      12.000   0.    0.     -0.
               VECTOR NO. 4   -.216   10.000   -.227     6.000   0.    0.     -0.
               VECTOR NO. 5   -.250    0.       -.250    0.      0.    0.     -0.
               VECTOR NO. 6   0.     -20.000    0.     -12.000   0.    0.     -0.
19 BEAM         1.00000   .25000  20.00000   .25000  12.00000  0.      0.    2.50000
27 I/O CON.    15.00000  4.00000
29 PLOT        24.00000  0.       0.        0.       0.       1.00000  6.00000
35 PLOT        24.00000  4.00000  2.40000   2.60000   .10000  -0.     -0.
43 PLOT        24.00000  1.00000   .25000*1000.00000   .25000*1000.00000  -0.
50 I/O CON.    15.00000  5.00000
52 I/O CON.    15.00000  2.00000
54 DRIFT        5.00000  10.00000                                             1.000E+01
56 RESOLUTION   6.00000   2.00000  5.00000   6.00000
60 POLE F. ROT  2.00000  -8.11000 -0.
63 BEND         4.00000  44.00000 -17.11500 0.                                5.400E+01
67 I/O CON.    15.00000  4.00000
69 POLE F. ROT  2.00000  -5.07000 -0.
72 DRIFT        5.00000  40.00000                                             9.400E+01
74 QUAD         5.02000  27.00000 -6.00000  5.00000                           1.210E+02
78 QUAD         5.02000  27.00000 -6.00000  5.00000                           1.480E+02
82 I/O CON.    15.00000  4.00000
```

ALTER	OLINE	ALINE	SWPB	MDATA	GO	ITER	SAVE	RECL	SEAT
JECT	RAYS	BEAM	SCALE	MATRX	CANCL	SWPD	FIX	NOISE	FIT

FIG. 73. Input data displayed on a CDC 252 console.

envelope the cumulative transfer matrix (RC) is displayed along with the beam matrix σ.

Adjacent to RC and σ matrices, the projection of the phase-space ellipsoid on the chosen coordinate axes is displayed in both the horizontal and the vertical planes; also displayed is the location of the tracked vectors if any.

Now, upon examination of the various plots and parameters displayed, the experimenter proceeds to exercise any of the options shown in Table 33 either to obtain a ray-tracing (option RAY) as shown in Fig. 75, or to alter his initial design and thus repeat the computational loop. This loop may be repeated

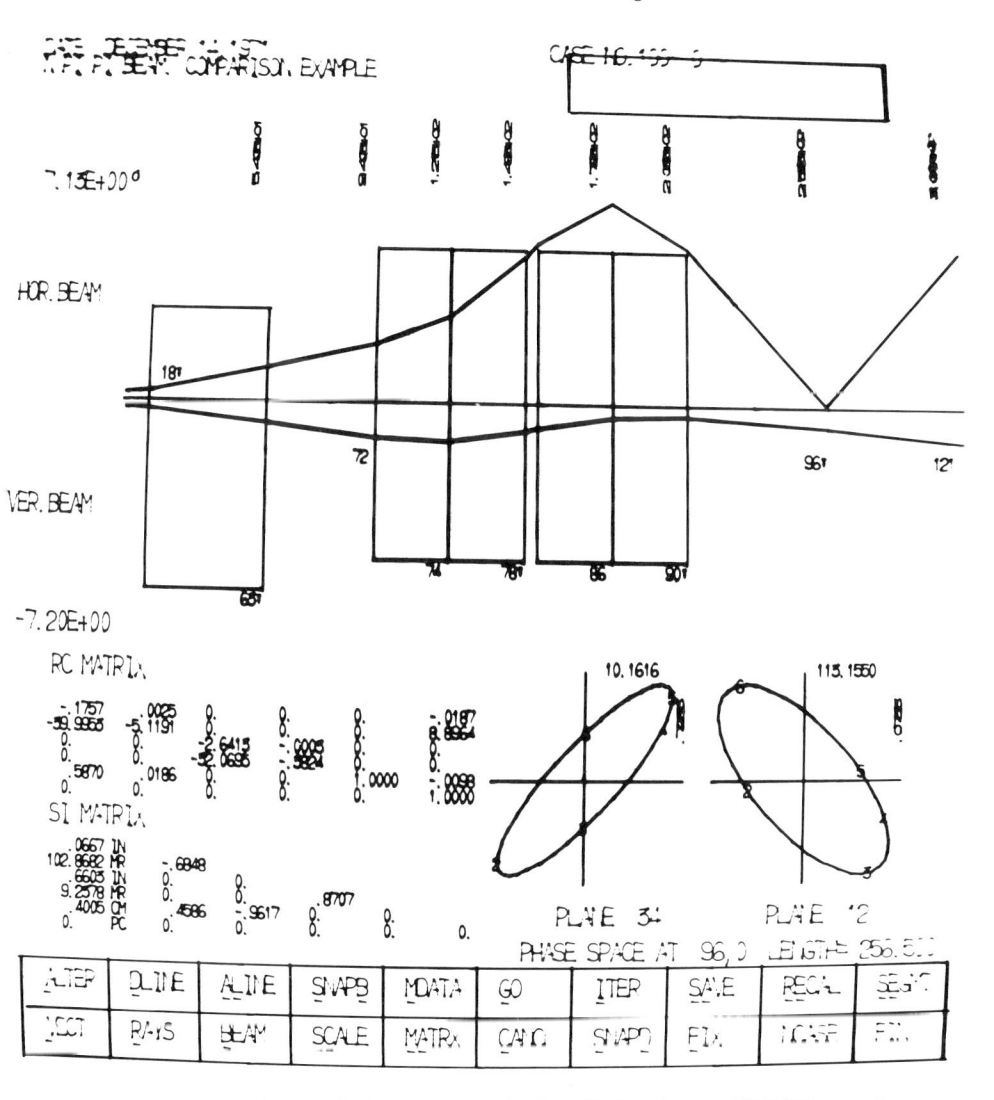

FIG. 74. Beam envelope and phase-space projection displayed on a CDC 252 console.

as many times as the experimenter feels is necessary to optimize his beam design. At all times he may exercise the options SNAPD or SNAPB to record the CRT display on microfilm, which later may be developed to a hard copy for a permanent record.

This interacting mode of operation is extremely desirable since the operator can quickly examine the overall characteristics of his system and take immediate action. Errors are detected easily, and time loss because of erroneous data is minimized. During the course of the session, warning messages are available to guide the operator in avoiding invalid conditions.

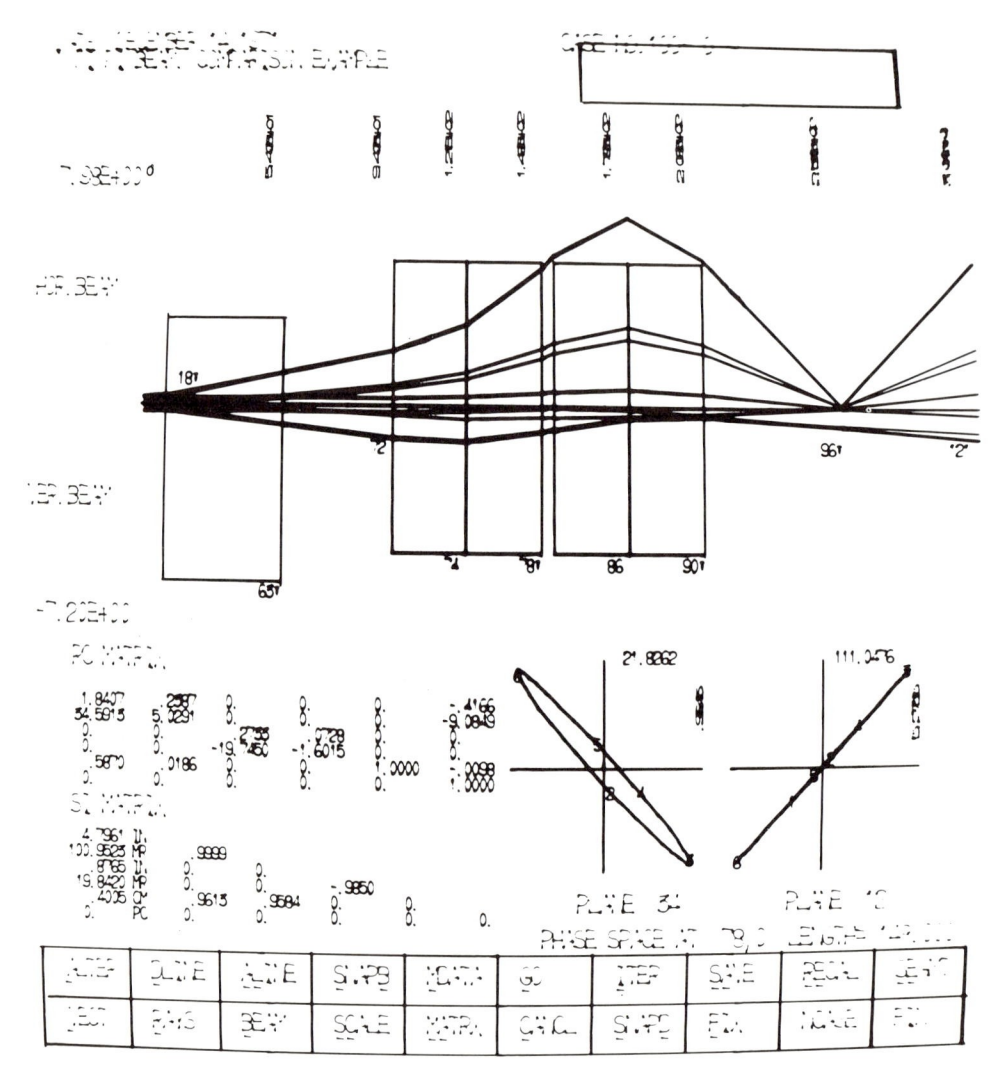

FIG. 75. Beam envelope with ray trace displayed at a different location along the beam line.

21. Program OPTIK

a. Introduction

Program OPTIK as well as program TRANSPORT, provide the experimenter with the means to handle problems encountered in the optical design of high-energy particle beams. It was originally written by Devlin [53] for the IBM 709 computer system and it has been modified considerably by Chaffee [54] and Kane [55] to increase its usefulness and flexibility.

The present version of OPTIK allows the experimenter to follow particles through a beam transport system consisting of: field-free regions, bending magnets (either wedge-shaped or rectangular region of homogeneous magnetic field), quadrupole lenses with one or two degrees of freedom, and velocity separators.

b. Mathematical Development

The mathematical model of OPTIK provides for a beam defined by a five-dimensional vector space consisting of the following coordinates:

x = horizontal displacement from the beam line (inches)

x' = horizontal divergence of the beam (radians)

y = vertical displacement from the beam line (inches)

y' = vertical divergence of the beam (radians)

$\Delta p/p = (p' - p)/p$ = fractional difference from desired momentum (p')

and a projection operator, m_1 used with a velocity separator. The magnet system may consist of a sequence of bending magnets, quadrupole lenses, velocity separators, and field-free drift spaces. The magnet system is represented by 5×5 transfer matrices which multiply the initial beam vectors to determine new vectors.

The phase space passed by any aperture is bounded in the xx' plane by straight lines fixed in x (see Fig. 76). The effect of these boundaries on the

FIG. 76. Phase-space area accepted by target; this area is divided into 40 equal strips and the effect of each aperture is transformed back to the target. These effects are calculated in terms of the intercepts with the x and x' or y and y' axes. Thus, all the effects of the apertures can be shown in one graph.

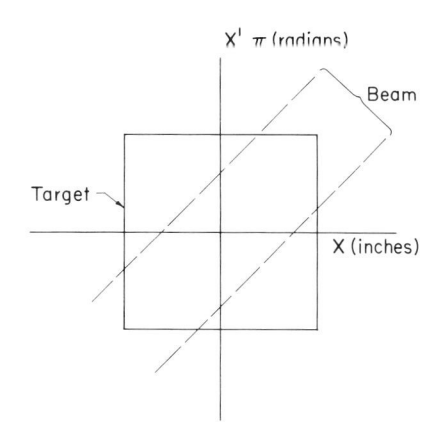

total phase space is found by transforming them back to the source with the inverse of the transformation matrix. These lines remain straight lines in xx' space but with slope and intercept that depend on the transformation matrix. The effect of each aperture is obtained by estimating the phase-space area enclosed by the transformed boundaries.

c. Program Organization

OPTIK consists of a main program and an executive routine (EXEC) which controls the flow of information based upon the input. EXEC routes control based on an identification number in the data to one of the 12 main routines whose function is shown in Table 34.

Table 34

OPTIK Subroutines[a]

Name	Description
START	Reads particle momentum and "N" initial vectors for starting a new problem. Forms 6 × N vector matrix that transfers control to the main program
FREE	Forms matrix for a field-free region
QUADU	Solves quadrupole problem as follows: When an unknown quadrupole is encountered, a temporary matrix is calculated that represents the effect of the transport system between the lens and the condition to be satisfied. A small change in the lens strength is made, and the effect is calculated on the matrix element specified by the condition, with the resulting derivative being used to compute a new value of the lens strength. This process is repeated until the change is smaller than the value specified on the condition card or until a maximum of 15 iterations has been reached; in either case the lens is treated as a fixed lens. Similar procedure is used for two variables where field strength is represented by a two-dimensional vector
BEND	Forms matrix of bending magnet. (See Table 36.) Calculates and prints magnetic field values and $\int H \cdot dl$ through the magnet
SEP	Uses maximum voltage and physical dimensions to calculate the deflection induced on particles of different masses. OPTIK then constructs the corresponding matrix
CONDTN	Reads data; makes it available to QUADU and SING as a condition on the beam
QUADK	Forms matrix for doublet or triplet with a known solution. Output is the same as QUADU without going through the iteration process to arrive at a solution
ODD	Reads in an arbitrary matrix
SING	Finds gradients for one or two singlets to satisfy one or two conditions (similar to QUADU)
FLUX	Calculates phase-space projections
REPEAT	Permits cycling of program to produce successive runs
QUADM	Acts as QUADU the first run and as QUADK on subsequent runs; only difference lies in the phase-space calculation since it allows splitting the doublet into singlets and drifts; thus the effects of stops inside the lens may be calculated

d. Input Preparation

Preparation of data for a successful run follows the same pattern as that of TRANSPORT. Data are punched format-free and the resulting deck is ordered as follows:

1. Title card
2. Plot cards (if any)
3. Beam-line elements
4. End of job (two blank cards).

A list of all possible beam-line elements, plotting option, input/output flags, etc. will be found in Table 35.

Numbers on data cards are punched in sequence anywhere in columns 1 to 80 with one or more blanks inserted between each entry. All parameters that follow the last nonzero entry on the card are considered to be zero. For example, a START card punched as: 1 0.25 0.25 710. implies $\delta p/p = N$ = XGRF = DXGRF = 0 (see Table 35). Comments on a data card may be enclosed within $ signs; program OPTIK ignores all intervening characters. Note that a number followed by Rm is equivalent to repeating the number m times (e.g., to repeat 2.54 seven times enter 2.54R7.)

All lengths and widths are in inches. The beam divergence is in radians, bend angles are in degrees and $\Delta p/p$ is dimensionless. The magnetic field gradient is in gauss per inch. Table 35 gives the array of elements used in program OPTIK along with the parameters required. The following two examples of input preparation will direct the experimenter in the proper usage of Table 35 and the sequence of cards required for a successful run:

Example 1. The data shown in Table 36 describe a beam which would give the necessary field strengths in the symmetric triplet to focus particles from the target to the final focus. No output will be shown for this example.

Example 2. The data shown in Table 37 allow the experimenter to avail himself of most of the potential power of this program. This problem is the same one run with program TRANSPORT. The code will produce a ray-trace of six vectors in a beam system consisting of a bending magnet and a symmetric quadrupole doublet, in which both a horizontal and a vertical foci are desired.

e. Output Description

Program OPTIK provides printed output controlled by a 14 element-type card, and CalComp plots controlled by a 13 element-type card. The output

168 IV. Programs Employing Matrix Formalism

Table 35

OPTIK Input Specifications

Element type	Parameters	Description
1	HH,HV,P,DP/P,N,XGRF,DXGRF	START. HH and HV are the symmetric apertures at the end of the element; P = momentum; N = number of vectors (if any) to be traced; if N = -1, refer to element type 17; if XGRF and DXGRF $\neq$ 0, and ray trace mechanism not already started (See element 13.), CalComp routine calls for phase-space plot of region X = ±XGRF, $\dot{X}$ = ±DXGRF and Y = ±XGRF, $\dot{Y}$ = ±DXGRF; setting XGRF $\neq$ 0 and DXGRF $\neq$ 0 is the only change required for plotting
	X,DX,Y,DY,DP/P,M	VECTORS. Initial vectors for ray tracing; if used, must follow START card Maximum number of vectors = 10
2	HH,HV,LENGTH	DRIFT SPACE. HH and HV are the entrance apertures for the next element, and L is the LENGTH in inches
3	HH,HV,IDQ,IDP,L1,L2,D,THETA1, THETA2,DB/DR1,DB/DR2	QUADU. HH and HV (See element 1.); if IDQ = 1, doublet quad; if = 2, symmetric triplet quad; if IDP = 1, first element of lens converges in horizontal plane; if = 2, lens diverges; L1 and L2 are the effective lengths; D = center-to-center distance; THETA1 and THETA2 = quadrupole strengths, OPTIK will search input until two condition cards (See element 6.) are found, then the code will vary THETA1 and THETA2 to satisfy the condition $\theta = (1/B_\rho \; dB/dA \;)^{\frac{1}{2}} L$; DB/DR1 and DB/DR2 = field gradients; user need not specify both θ and DB/DR, if one is set equal to zero the program uses the other
4	HH,HV,E1,E2,THETA,RHO,L,N	BENDING MAGNET. HH and HV (See element 1.); E1 and E2 = entrance and exit angles; THETA = E1 + E2 (for a rectangular bending); also, THETA = total bend of central ray for wedge magnet; RHO = radius of curvature; L = effective length; generally with a constant field, radial dependence can be supplied by specifying nonzero field index N
5	HH,HV,MV,MD,V,L,D	SEPARATOR. Velocity separator. It is not necessary to include separate field-free regions before and after the separator card; HH and HV (See element 1.); MV and MD = masses (MeV) of undeflected and deflected particles; V = potential difference between electrodes (millivolts); L = length (inches) of separator; D = gap
6	HH,HV,I,J,COND,DEL	CONDITION. Sets the appropriate matrix element to the desired beam condition; OPTIK will vary strength of the unknown lens until COND ± DEL has been achieved; HH and HV are irrelevant in this type, may be ignored
7	HH,HV,IDQ,IDP,L1,L2,D,THETA1, THETA2,DB/DR1,DB/DR2	QUADK. Form the matrix of a known quadrupole (either doublet or symmetric triplet); parameters have same meaning as element type 3
8	HH,HV,DZ $\begin{bmatrix} M_{11} & M_{12} & M_{13} & M_{14} & M_{15} & M_{16} \\ \cdot & \cdot & \cdot & \cdot & \cdot & \cdot \\ \cdot & \cdot & \cdot & \cdot & \cdot & \cdot \\ M_{61} & M_{62} & M_{63} & M_{64} & M_{65} & M_{66} \end{bmatrix}$	ODD. Arbitrary external matrix; HH and HV (See element 1.); DZ = scaling factor for ray trace; if DZ = 0, no ray trace is made; if a ray trace is made, DZ = physical length before scaling; if = blank, a value of 3.0 × scale is used
9	HH,HV,ISOLVE,IDP,L,	

Table 35 (cont.)

Element type	Parameters	Description
		SING. Solves quadrupole singlet problem for either one or two singlets (similar to QUADU); if ISOLVE = 1, OPTIK solves for DB/DR; if ISOLVE = 2, OPTIK uses the given value of DB/DR; if ISOLVE = 3, OPTIK solves for DB/DR for first problem but for all repeated runs it will use the value of DB/DR found in first problem; remaining parameters similar to element 3
10	I,VALUE(1),VALUE(2),......	REPEAT. Cycles program for successive problems; for first solution with REPEAT card, the I th position on the card following the REPEAT card is replaced by VALUE(1); for second solution, replaced by VALUE(2); this process continued until final number on first REPEAT card is processed, REPEAT ≤ 101; OPTIK can count two or more cards as one if $ symbol is punched in column 80 of first card and column 1 of second card; additional cards used in similar $ punch sequence
11	(Quantities same as QUADK)	QUADM. Acts as QUADU the first run and as QUADK in all subsequent runs, uses field values found in the first run and splits lens into its component singlets and drifts; in this way, effect of stops inside lens is found in phase-space calculations
12	(No parameters)	NIL or SKIP. Do nothing card; TRACE card could be turned into a NIL card for REPEATed problems in which a ray trace not wanted
13	ID,HMAX,VMAX,SCALE	TRACE. Provides for ray trace plots; if ID = 13.0, ray trace; if ID = 13.1, suppress ray trace; TRACE initiates CalComp for ray trace of region ±HMAX, ±VMAX in horizontal and vertical plane with z dimension scaled by SCALE, i.e., plot length = physical length/SCALE; draws rays specified by START card and following VECTOR cards
14	CODE	OUTPUT. Printout options: if CODE = 0, prints all information; if = 1, prints information in abbreviated format; if = 2, prints all information for one element then makes code equal to original value; if = 3, intermediate printout (default option)
15	XMAX,XMIN,YMAX,YMIN,CODE	APERTURE. Provides for asymmetric nontilted rectangular apertures; if CODE = 0, aperture acts as collimator (flux stopped by aperture, not passed on to other elements); if CODE = 1, apertures act as counter (flux stopped by aperture and not included in phase-space calculation is restored at end of problem) remaining parameters are coordinates of the aperture
17	DXMAX,DYMAX,DXMIN,DYMIN	VECSET. Permits entering initial vectors into the system; precedes START card; if N on the START card = -1, OPTIK will generate internally 10 vectors; the first 5 vectors in x direction will be spaced between DXMIN and DXMAX and the other 5 in the y direction will be between DYMIN and DYMAX

Table 36

Optik Input Data for Example 1

		Card Column	
1	10	20	40
SIMPLE BEAM.	EXAMPLE NO.1		$ TITLE CARD $
1 0.25 0.25 716 0 0 0 0			$ START CARD $
2 4 4 100			$ DRIFT SPACE UP TO LENS $
3 4 4 2 1 20 36 33.3 0 0 2000 2000			$ CDC SYMMETRIC TRIPLET LENS $
2 4 4 100			$ ANOTHER DRIFT $
6 4 4 1 2 0 0.01			$ TWO CONDITIONS FOR THE $
6 4 4 3 4 0 0.01			$ LENS TO SATISFY $
			$ BLANK CARD $
			$ BLANK CARD $

in Fig. 77 shows Example 2 in an abbreviated form. The first part lists the input data, while the second part lists the transformation matrix at the end of each element, the condition imposed on the beam, and the computed fitted value. The program then prints the transformation matrix at the end of each beam element encountered. Upon reaching the unknown quadrupole doublet, the analysis is made and pertinent information relating to this analysis is printed in detail. Since CalComp plots were requested, the appropriate flags (element-type 10) are activated. Some acceptance polygon plots are shown in Fig. 78. Figure 79a shows a beam envelope while Fig. 79b shows a ray tracing plot. An examination of the output shown in Fig. 79 and the output produced by program TRANSPORT on the same problem (Fig. 68) will show that both programs produce similar answers.

f. Further Developments of OPTIK

Perhaps the most significant additions to program OPTIK are the graphic displays for on-line interaction. Following the same logic of program TRANSPORT it was possible to alter the off-line version of OPTIK to include the desirable characteristics of an on-line computer program.

This version allows the user to view on a cathode ray tube the input data (Fig. 80), the resulting product and vector matrices (Fig. 81), and the beam

Table 37

OPTIK Input Data for Example 2

Card Column		
1 10 20 30	48	67 78
N PI PI BEAM.EXAMPLE NO.2 SHOWING CALCOMP USE	$ TITLE $	
14 0	$ PRINT ALL OUTPUT $	
10 1 13 12R5	$ PRODUCE RAY-TRACE THE FIRST $	
13	$ TIME,THEN DO NOTHING FOR THE	
	REST OF THE PROBLEMS$	
10 8 0 0.05 R5	$ MAKE PHASE-PLOTS $	
10 5 0 -4 -2 0 2 4	$ SET THE MOMENTUM - DEVIATION $	
1 0.25 0.25 2500 0 6 2 0.05	$ START CARD $	
0.25 0 0.25	$ 1ST VECTOR $	
0.25 0.02 0.25 0.012	$ 2ND VECTOR $	
0.25 -0.02 0.25 -.012	$ 3RD VECTOR $	
-.25 0 -.25	$ 4TH VECTOR $	
-.25 0.02 -0.25 0.012	$ 5TH VECTOR $	
-0.25 -0.02 -0.25 -.012	$ 6TH VECTOR $	
2 6 4 10	$ DRIFT FROM TARGET TO BEND $	
4 6 4 -8.11 -5.07 -13.18 0 44	$ BENDING MAGNET $	
2 5 5 40	$ DRIFT SPACE $	
11 5 5 1 2 54 54 58.5 0 0 1000 1000	$ DOUBLET LENS TO BE SOLVED $	
2 6 6 50	$ DRIFT SPACE $	
6 4 4 1 2 0 0.01	$ CONDITIONS (A HORIZONTAL AND $	
6 4 4 3 4 0.01	$ VERTICAL FOCUS) TO BE SATI-	
	SFIED BY THE DOUBLET $	
2 6 6 50	$ DRIFT BEYOND FINAL FOCUS $	
	$ BLANK CARD.END OF JOB $	
	$ BLANK CARD.END OF JOB $	

```
                                                                   P=   2500.00 MEV/C DP/P= 0.        JOB  2  REPEAT  1 PAGE 1
N PI PI BEAM.EXAMPLE NO.2 SHOWING CALCOMP USE   $ TITLE $                                               REPEAT NUMBER   1

P =   2500.00 MEV/C           DELP/P =  0.                SOURCE IS  .2500 BY  .2500 INCHES   (HALF-WIDTHS)

   INITIAL VECTORS
    .250000     .250000     .250000     -.250000    -.250000    -.250000
   0.           .020000    -.020000     0.           .020000    -.020000
    .250000     .250000     .250000     -.250000    -.250000    -.250000
   0.           .012000    -.012000     0.           .012000    -.012000
   0.          0.          0.          0.          0.          0.
   0.          0.          0.          0.          0.          0.

FIELD FREE REGION OF LENGTH =     10.000 INCHES
    HORIZ )   1.000   10.00          VERTICAL)  1.000   10.00          MOMENTUM  )     0.            MASS        )    0.
    MATRIX) 0.        1.000          MATRIX  ) 0.       1.000          SEPARATION)  0.               SEPARATION)  0.
    SOLID ANGLE =960.000000       HORIZ. ANGLE = 1200.00000      VERT. ANGLE =  800.00000

BEND = -13.18 DEGREES   ENTRANCE=  -8.11 DEGREES   EXIT=  -5.07 DEGREES   RADIUS=  -191.698 IN.   EFF.LENGTH=  44.000 IN.
    AVERAGE FIELD = -17.115 KILOGAUSS     (EFFECTIVE LENGTH)(AVERAGE FIELD) =   -753.08 KILOGAUSS-INCHES
    HORIZ )   1.006   53.77          VERTICAL)   .967   53.77          MOMENTUM  )   -5.05           MASS        )    0.
    MATRIX) -.00000    .994          MATRIX  ) -.00119   .968          SEPARATION)  -.2303           SEPARATION)  0.
    SOLID ANGLE = 33.203966       HORIZ. ANGLE =  223.16952      VERT. ANGLE =  148.78361
    CENTER-TO-CENTER DISTANCE FROM THE LAST BEND =   32.15 INCHES.

FIELD FREE REGION OF LENGTH =     40.000 INCHES
    HORIZ )   1.006   93.53          VERTICAL)   .920   92.48          MOMENTUM  )  -14.26           MASS        )    0.
    MATRIX) -.00000    .994          MATRIX  ) -.00119   .968          SEPARATION)  -.2303           SEPARATION)  0.
    SOLID ANGLE = 11.562036       HORIZ. ANGLE =  106.92181      VERT. ANGLE =  108.13544

DOUBLET        ENTRANCE DIVERGING  HORIZONTALLY                        L1= 54.00 INCHES THETA1=1.21194 DB1/DR= 1653.70 GAUSS/INCH
   (UNKNOWN)                               D=     58.50INCHES,     L2= 54.00 INCHES THETA2=1.33881 DB2/DR= 2018.07 GAUSS/INCH

**********OPTIK WILL NOW ANALYZE THIS DOUBLET IN DETAIL
SINGLET DIVERGING  HORIZONTALLY
   (KNOWN)     LENGTH = 27.00 INCHES       THETA = .60597       FIELD AT POLE TIP = 6.615 KILOGAUSS      DB/DR = 1653.70 GAUSS/INCH
    HORIZ )   1.197  139.74          VERTICAL)   .726  100.57          MOMENTUM  )  -23.57           MASS        )    0.
    MATRIX)  .01454   2.533          MATRIX  ) -.01273   -.387         SEPARATION)  -.4800           SEPARATION)  0.
    SOLID ANGLE =  7.115819       HORIZ. ANGLE =   71.56289      VERT. ANGLE =   99.43448
SINGLET DIVERGING  HORIZONTALLY
   (KNOWN)     LENGTH = 27.00 INCHES       THETA = .60597       FIELD AT POLE TIP = 6.615 KILOGAUSS      DB/DR = 1653.70 GAUSS/INCH
    HORIZ )   1.840  238.85          VERTICAL)   .273   72.85          MOMENTUM  )  -41.80           MASS        )    0.
    MATRIX)  .03458   5.032          MATRIX  ) -.01974  -1.603         SEPARATION)  -.9114           SEPARATION)  0.
    SOLID ANGLE =  4.163071       HORIZ. ANGLE =   41.86747      VERT. ANGLE =   99.43448
FIELD FREE REGION OF LENGTH =     4.500 INCHES
    HORIZ )   1.996  261.49          VERTICAL)   .184   65.63          MOMENTUM  )  -45.90           MASS        )    0.
    MATRIX)  .03458   5.032          MATRIX  ) -.01974  -1.603         SEPARATION)  -.9114           SEPARATION)  0.
    SOLID ANGLE =  3.802593       HORIZ. ANGLE =   38.24219      VERT. ANGLE =   99.43448
SINGLET CONVERGING HORIZONTALLY
   (KNOWN)     LENGTH = 27.00 INCHES       THETA = .66941       FIELD AT POLE TIP = 8.072 KILOGAUSS      DB/DR = 2018.07 GAUSS/INCH
    HORIZ )   2.430  330.99          VERTICAL)  -.346   34.30          MOMENTUM  )  -58.81           MASS        )    0.
    MATRIX) -.00359   -.077          MATRIX  ) -.02104   -.804         SEPARATION)  -.0086           SEPARATION)  0.
    SOLID ANGLE =  3.004135       HORIZ. ANGLE =   30.21220      VERT. ANGLE =   99.43448
SINGLET CONVERGING HORIZONTALLY
   (KNOWN)     LENGTH = 27.00 INCHES       THETA = .66941       FIELD AT POLE TIP = 8.072 KILOGAUSS      DB/DR = 2018.07 GAUSS/INCH
    HORIZ )   1.816  257.63          VERTICAL) -1.038   18.91          MOMENTUM  )  -46.33           MASS        )    0.
    MATRIX) -.04020  -5.153          MATRIX  ) -.03212   -.378         SEPARATION)   .8980           SEPARATION)  0.
    SOLID ANGLE =  3.004135       HORIZ. ANGLE =   30.21220      VERT. ANGLE =   99.43448
**********END OF DOUBLET
    CENTER-TO-CENTER DISTANCE FROM THE LAST BEND =  118.40 INCHES.

FIELD FREE REGION OF LENGTH =     50.000 INCHES
    HORIZ )   -.194    .00           VERTICAL) -2.644    .01           MOMENTUM  )   -1.43           MASS        )    0.
    MATRIX) -.04020  -5.153          MATRIX  ) -.03212   -.378         SEPARATION)   .8980           SEPARATION)  0.
    SOLID ANGLE =  3.004135       HORIZ. ANGLE =   30.21220      VERT. ANGLE =   99.43448

BEAM CONDITION          PRODUCT MATRIX (1,2) =        0.    +/-   .010          FITTED VALUE =        .0000
    CENTER-TO-CENTER DISTANCE FROM THE LAST BEND =  224.65 INCHES.

BEAM CONDITION          PRODUCT MATRIX (3,4) =          .010+/-   .000          FITTED VALUE =        .0100
    CENTER-TO-CENTER DISTANCE FROM THE LAST BEND =  224.65 INCHES.

FIELD FREE REGION OF LENGTH =   100.000 INCHES
    HORIZ )  -4.214 -515.26          VERTICAL) -5.856  -37.79          MOMENTUM  )   88.37           MASS        )    0.
    MATRIX) -.04020  -5.153          MATRIX  ) -.03212   -.378         SEPARATION)   .8980           SEPARATION)  0.
    SOLID ANGLE =  2.315747       HORIZ. ANGLE =   23.28917      VERT. ANGLE =   99.43448

- - - - - - - - - - - - - - - - - - - - - - - - - - - - - - - - - - - - - - - - - - - - - - - - - - - - - - - - CP=  4961  PP=      72
A CAL-COMP RAY-TRACE WAS MADE DURING THIS PROBLEM.
```

FIG. 77. Partial printout obtained by program OPTIK.

line in both the horizontal and vertical planes (Fig. 82). In Fig. 82 the bending magnets are shown with their equivalent optical lenses, the apertures are shown by square blocks and the cumulative length is indicated along the beam line.

The program accommodates interaction by allowing the user to control the progress of the solution at his discretion. Some of the options that may be exercised include the ability to alter input data, to scale the resulting beam line, to display up to ten vectors, and to take pictures of the information displayed on the screen.

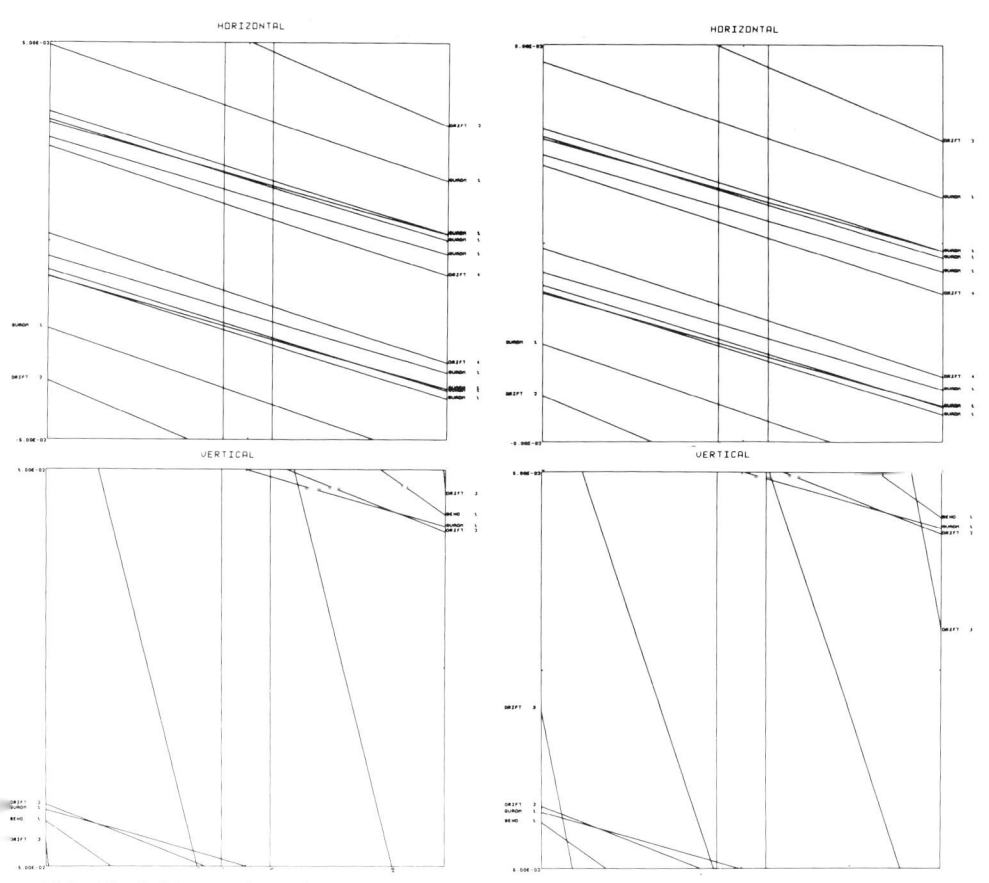

FIG. 78. CalComp plots of acceptance polygons at various locations along the beam line.

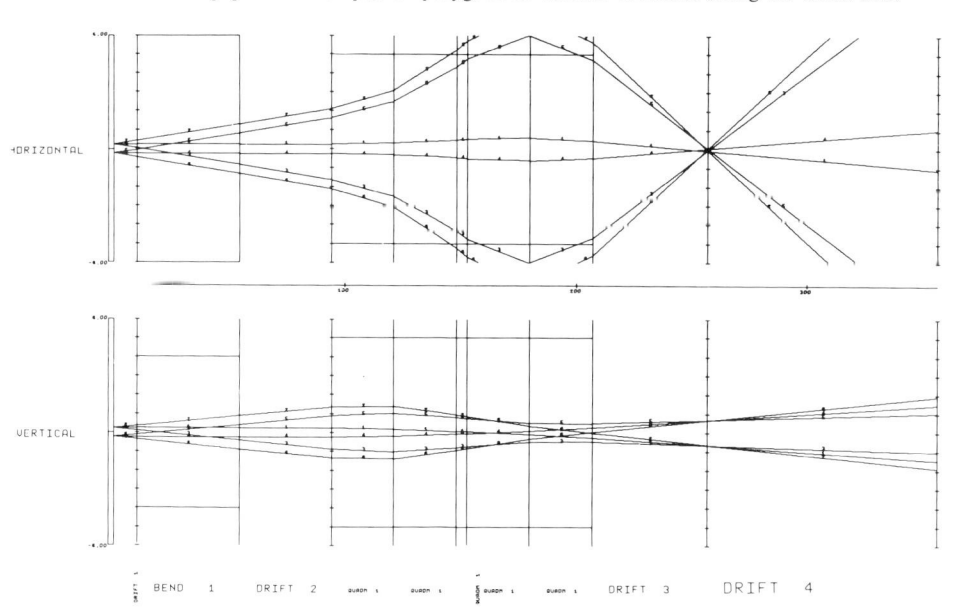

FIG. 79. Beam envelope for the test problem.

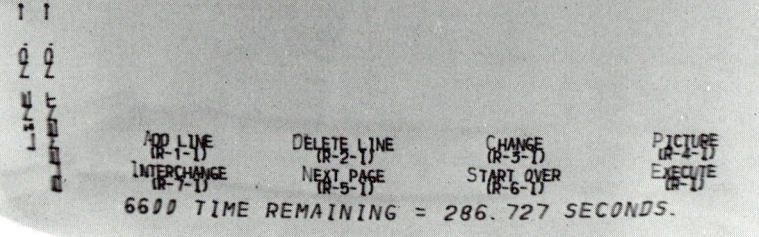

FIG. 80. OPTIK input data displayed on a CDC 252 console.

```
08/19/71, 14.19.09, [N PI PI BEAM.EXAMPLE], RUN  1
        RESULTS AT ELEMENT   3
              BENDING MAGNET


BEND = -15.18 DEGREES,          ENTRANCE = -8.11 DEGREES
EXIT = -5.07 DEGREES            RADIUS = 191.698 INCHES
EFFECTIVE LENGTH = 44.00000 INCHES    AVERAGE FIELD = 17.115 KILOGAUSS
              PATH LENGTH = 44.097 INCHES
(AVERAGE FIELD)X(EFFECTIVE LENGTH) = 753.04 KILOGAUSS - INCHES
```

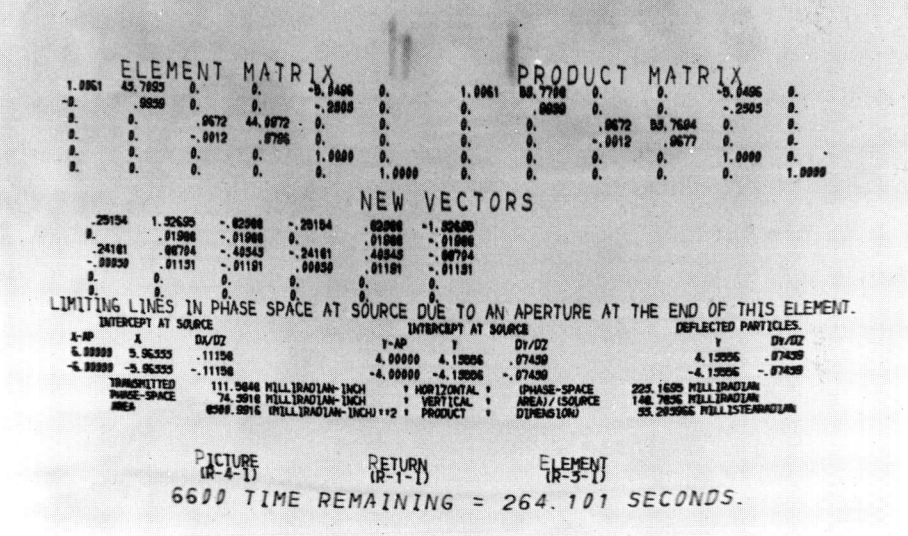

FIG. 81. Display of partial output on a CDC 252 console.

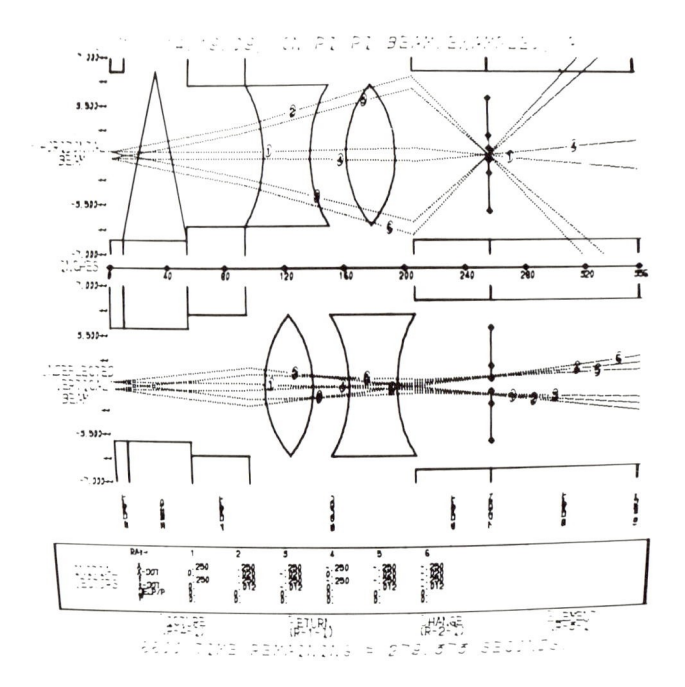

FIG. 82. Beam envelope display on a CDC 252 console.

22. Program TRAMP

a. Introduction

Program TRAMP was developed at Rutherford High Energy Laboratory by Gardner and Whiteside [56, 57] to provide solutions to problems encountered in beam transport design. It has been extensively modified by various experimenters (see Kane [55], Butler [58]), to fit the needs and the computer facilities of their respective laboratories.

The version described herein is capable of tracking and matching trajectories, beam profiles, or phase-space ellipses through a given beam transport system. The program is written in Fortran, and it is operational on the CDC 6600 computer system at Lawrence Berkeley Laboratory. The program occupies $66K_8$ memory locations, and a typical run takes less than one minute.

b. Beam Elements

The elements that may be used in TRAMP are represented by a 2×2 matrix for each plane, transforming the displacement-divergence vectors from beginning to end. The following elements are recognized:

1. Drift length is identified by a element-type number 1.0 with the parameters L, L_h, L_v, F, b, A_h, A_v in a card format of 8F8.3 (All of the following element cards are 8F8.3.) The length L (meters) is measured between centers, *not the edge*, of any adjacent nondrift element. The horizontal and vertical lengths L_h and L_v (meters), are entered as the third and fourth parameters. These lengths may be different, allowing cases of virtual focus in one plane, but then consecutive drift lengths are necessary so that the distance between real elements is the same in both planes. This equality is maintained as a result of any change of drift lengths during matching. The fifth parameter F is used to describe variation of the parameters that proceed it, e.g., if $F = 0$, all parameters are fixed; if $F = 1$, the second parameter L, will be varied by TRAMP, if $F = 2$ the third parameter L_h, will be varied, etc. The sixth parameter b, indicates blank field, i.e., not used in this card and any following element-type card. The seventh and eighth parameters A_h and A_v (centimeters) are the half-horizontal and half-vertical apertures; if either aperture is left blank or zero, the value will be set to 15 cm to insure similar treatment of all elements in certain tracking routines.

2. Quadrupole lenses are coded by a element-type number 2.0 for horizontal focusing, and 3.0 for vertical focusing; both with the parameters L, dB/dx, F, b, b, A_h, A_v. The effective length L (meters) is given by the second entry. The third parameter specifies the field gradient divided by the design momentum in units (gauss/centimeters per GeV/c).

3. Bending magnets are designated by an element-type number 4.0 with the parameters L, B, θ_0, ϕ_0, b, A_h, A_v. Where L is the effective length (meters), B specifies the magnetic field divided by momentum (kilogauss per GeV/c). The fourth parameter θ_0 (degrees) gives the entrance angle, and the fifth entry ϕ_0 (degrees) specifies the sector angle.

4. Velocity separators are identified by either of two element-type numbers, 5.0 or 5.5, both with parameters L, E, F, m_u, m_d, A_h, A_v. The type-5.0 separator is normally used and indicates that the separation of the beam is influenced only by the separator through which it is passing and the separation effect is not cumulative. In contrast, a type-5.5 separator accumulates the effects of previous separations. Following the effective length L (meters) is the electric field strength, E (kV/cm). If the fourth entry F is 1.0 or 2.0 respectively for anti-proton-pion or kaon-pion separations, then one can omit the fifth and sixth parameters; these give the rest masses m_u and m_d (GeV/c²) of the undeflected and deflected particles.

5. Sextupole magnets are designated by a element-type number 6.0 with the parameters L, d^2B/dx^2, dx, K, b, A_h, A_v. The third entry, d^2B/dx^2 (gauss/centimeter²), is the second derivative of the magnetic field. The fourth parameter, dx, is reserved for the integration step; this is required since the sextupole field is integrated to determine the dispersion and separation. The fourth entry is also required for tracking vectors, tracing rays, or finding phase-space

area. If the fifth entry, K, is nonzero the progress of the ray-vector through the sextupole is printed out. For example, if 200.0 and 10.0 appear at the fourth and fifth entry respectively, the integration is carried out in 200 steps; the initial step and every tenth step thereafter are printed in detail.

c. Matching Beam Conditions

The program recognizes the following conditions:

1. Focal and phase-space conditions. To achieve the requirements imposed by the beam designer, program TRAMP breaks the beam into sections between positions where the beam requirements are known. Matching involves making systematic changes in specified variable elements, setting up a matrix for each section, and by some iterative procedure determining the value of the variables so that the product matrix conforms to the specified focal conditions. Mathematically this is equivalent to setting a specific matrix element to zero as follows:

$a_{11} = 0$ Parallel-to-focus matching

$a_{12} = 0$ Focus-to-focus matching

$a_{21} = 0$ Parallel-to-parallel matching

$a_{22} = 0$ Focus-to-parallel matching.

Table 38 describes the data needed for these matching beam conditions.

2. The focal conditions specified above match a single particle trajectory and are valid whenever the phase-space extent of the beam is small. Since the focus is not the place at which one beam has the smallest physical size, an investigation of the waist (which does have this property) is desirable. TRAMP varies parameters until a waist is produced at the desired position, however, since the code does not distinguish between a broad and a narrow waist, it is advisable to use focal matching first and then use the resulting parameters in the waist match. Table 38 shows the quantities necessary for waist-matching input.

3. Waist and magnification matching. In the above two types of matching it is impossible to specify the physical size of the beam, the magnification at a focus or the extent at a waist, since all use only two variables. When a third variable in the section is allowed, then it is possible to specify the physical

Table 38

TRAMP Beam-Matching Options

IT	Beam condition	Parameters[a]
0	No matching desired	
1	Parallel-to-focus matching	
2	Focus-to-focus matching	$M(H)$ = initial value of element in horizontal plane
3	Parallel-to-parallel matching	$N(H)$ = final value of element in horizontal plane
4	Focus-to-parallel matching	$KT(H)$ = the KT value in column 1
5	Waist matching	a = semiaxis phase-space ellipse displacement[b]
1	Parallel-to-focus matching with end extent given	b = semiaxis phase-space ellipse divergence[b]
2	Focus-to-focus matching with end extent given	a' = terminal extent in terms of a final displacement of semiaxis[b]
3	Parallel-to-parallel matching with end extent given	$M(V)$, $N(V)$, and $KT(V)$ refer to a similar set of
4	Focus-to-parallel matching with end extent given	parameters for the vertical plane
5	Waist matching with end extent given	

extent of the beam at the end of the section. The five types of matching which match the final beam extent as well are:

a. Parallel-to-focus matching with end extent given
b. Focus-to-focus matching with end extent given
c. Parallel-to-parallel matching with end extent given
d. Focus-to-parallel matching with end extent given
e. Waist matching.

The variables referred to may be drift lengths or quadrupoles. They are designated by a variable number, F, on the element-type card (see Section b. Beam Elements). If two element-types have the same number, F, they are varied together, such as the two outside quadrupoles of a triplet. If one of the split drift lengths is varied (e.g., horizontal and vertical lengths differ implying an adjacent complementary length), the complementary one is varied to maintain separation of the physical elements. The data for a section are contained in two sets on one card: one set for the horizontal plane and one for the vertical. Input data necessary for these matching conditions are also given in Table 38.

d. Dispersion and Separation Conditions

It is possible, with program TRAMP, to match conditions on dispersion and separation. The data necessary are listed on one card following an element-type 989 card and a printing requirement card (PRC). If the PRC card is not blank, the product matrices, the dispersion and separation, and the phase-space projections are printed at the end of each iteration. If the PRC card is blank, these quantities are printed initially and again at the end. The parameters necessary for dispersion and separation conditions have a format of 3I3,1X, 3(I3,F8.3), and are explained as follows:

NI, NP, IRS, (LMV, FINC, LMC, DESR, ISP, HATS).

The quantities listed in parenthesis are repeated for the vertical plane.

NI is the number of iterations, and is limited to five if NP $>$ 10; or is limited to 10 if NP $<$ 10.

NP is the plane number: if NP $=$ 1, dispersion matching is attempted; if NP $=$ 2, separation matching is attempted; if NP $=$ 21, or 12, dispersion-and-separation matching is attempted.

IRS is a flag: if IRS $=$ 0, values of variables after successful match are retained; if IRS $\neq$ 0, initial values of variables are retained.

LMV is the number of variable elements.

FINC is a fractional increase used with an associated variable that essentially forms the derivatives for a matrix inversion. Values of 0.005 to 0.05 are generally appropriate.

LMC is the number of elements at the end of which beam conditions are to be matched.

DESR is the desired result. When IRS $=$ 0, these values are taken to be the dispersion or separation at the end of the appropriate LMC (in units of centimeters); for IRS $\neq$ 0, the value of DESR has the following interpretation:

1. For the horizontal plane, DESR is the irreducible momentum bite, β, at the element in question and is essentially the fraction of off-momenta whose central orbits are overlapped by the on-momentum-source image. It is defined by

$$\beta = \frac{2 \times R_{11} \times \text{HATS} + R_{12} + \theta_H}{2 \times \text{Dispersion for } +1\% \text{ at end of LMC}}, \quad (22.1)$$

where R_{11} and R_{12} are elements of the product matrix in the appropriate plane, θ_H is the full-acceptance angle, and HATS is the half-target size (centimeters) and may be equal to zero.

2. For the vertical plane, DESR is interpreted as the separation ratio, η, defined by

$$\eta = \frac{\text{Separation at end of LMC}}{2 \times R_{11} \times \text{HATS} + R_{12} \times \theta_V}, \quad (22.2)$$

If HATS is nonzero, the tracking which determines the phase-space limitations due to all beam apertures is followed by a calculation of the average solid angle over the face of this source size. If neither the vertical or horizontal HATS is zero the solid angle printed in the output is simple $\theta_H \times \theta_V$.

e. Tracking Facilities

The capability of tracking particle trajectories enables the user to track certain rays through the entire beam line. Program TRAMP tracks single-particle trajectories or phase ellipses with profiles (for any specified momentum) relative to a central orbit for that momentum. The input necessary for tracking purposes consists of the following parameters (format I3, 4F8.3): KT, V(1), V(2), V(3), V(4). Interpretation of these parameters depends on the value of the flag KT whose following types are recognized:

KT = 1: the V array is a four-vector array; x, x', y, y'. The units are cm and crad.

KT = 3: the V array is the displacement and divergence semi axes for the initial upright horizontal and vertical phase-space ellipses.

KT = 6: a printout of product matrices for the elements given from V(1) to the end of the element given by V(2).

KT = 7: a printout of product matrices is made as required.

KT = 8: phase-space and acceptance calculations are performed depending upon the values of V(1) to V(4). Four distinct cases are recognized:

1. KT = 8 and V(1) = V(3) = V(4) = 0.0
The apertures for all elements are projected back to the source and the intercepts for these lines are printed.

2. KT = 8 and V(1) = V(3) = V(4) ≠ 0.0
The program determines, but does not print, information of case 1 above. But it uses the acceptance angles in the following way: $x = v(1)$, $x' = 0.50\theta_H/v(3)$, $y = v(2)$, $y' = 0.50\theta_V/v(4)$; and then this ray is tracked as in KT = 1.

3. KT = 8 and V(1) = 0.0, V(3) = V(4) ≠ 0
Determines and prints information of case 1, then interprets V(3) and V(4) as half the horizontal and vertical source sizes (cm). The program calculates the phase-space by integrating the acceptance angle versus distance plots, and finally obtains the average solid angle.

4. KT = 8 and V(1) ≠ 0.0, V(3) and V(4) = 0.0
The program determines, but does not print, information of case 1 above. However, it uses the acceptance angles in the following way: $a_H = v(1)$ and $a_V = v(2)$ with both in units of centimeters; $b_H = \theta_H/2$ and $b_V = \theta_V/2$ in units of crads. Then these two phase-space ellipses, which represent the true accepted phase-space of the beam, are tracked in KT = 3 above.

182   IV. Programs Employing Matrix Formalism

KT = 9: The bending-magnet tolerances and quadrupole aberrations are listed.

KT = 10: the change of tracking momentum: v(1) is the fraction of momentum used; if v(1) = 0.01, then tracking momentum is 1.01 times the design momentum used; if v(1) = −0.01 then it is 0.99 times the design momentum, etc. v(2) and v(3) give the horizontal and vertical divergence at the source, in centiradians.

KT = 11: allows N = v(1) ≤ 24 tracking cards (with KT ≤ 10) to be stored for reusing purposes, provided it is followed by a PRC card, and v(1) ≠ 0. In the following example, the program encounters KT = 11 and v(1) = 6, which permits the storing of the six tracking cards following, to be used later in the problem.

Examples

11	6.0					
		"Blank card"		(PRC card) Print everywhere		
9.08		3222.21		0.3	0.1	
8.001		0.001		0.08		
8.3		0.1				tracking cards to be stored
10	−0.01					
7						
8						
20	1.0		1.	2.0	0.6	move source
20	1.0		1.	3.0	0.6	
11	0.0					Track previously stored tracking cards (no PRC card follows and v(1) = 0.0)

KT = 20: any of the first four variables on an element-type card can be changed; v(1) = design momentum, v(2) = number of elements to be changed, and v(3) = number of variables to be changed, and v(4) = new value.

KT = 21: the N = v(1) ≤ 3 elements are replaced, N cards must follow in the element data format. v(2), v(3), and v(4) are the numbers of the elements to be replaced.

KT = 22: elements numbered v(2) and v(1) are interchanged.

KT = 23: the N = v(1) ≤ 4 elements on the following N cards are added as a group after the element numbered v(2).

KT = 24: the N = v(1) elements following the element numbered v(2) are deleted as a group.

In both cases where KT = 23 or 24, all matrices, dispersion, separation, and matching sections are updated to comply with the new numbering.

f. Preparation of Input Data

The organization of input follows the same pattern as that for programs previously described, in the sense that each beam element is described in terms of its defining parameters. Specifically, a data deck has the following structure:

Group 1. The first card format of F8.3,I3, constitutes the first group of the data input deck. It contains the design momentum (GeV/c), and the number of elements (subsequent cards in Group 2) in the beam system.

Group 2. The second group consists of I-element cards ($I \leq 80$), which give the basic structure of the beam system in terms of its defining parameters.

Group 3. Here, we list the sections of the beam in which matching is to be specified. No more than nine sections are permitted. Data for a section are contained in two sets on one card: one set for the horizontal plane, and one for the vertical plane. For all types of matching, the initial element and the final element of the section must be given. Matching is made for the product matrix from the beginning of the initial element to the end of the final element in the section. It is not necessary for the initial and final sections of the horizontal and vertical planes to be the same; however, care must be exercised because interference with the matching of another section might occur.

For all matching types it is necessary to specify the upright phase-space ellipse at the beginning of the section by the two semi-axes. displacement in centimeters and divergence in centiradians.

For beam types 11 through 15 it is necessary to give the terminal extent in terms of a final displacement semi-axis (a') in centimeters.

For matching on separation and dispersion, the data are listed on one card placed behind the PRC following the 989 element-type card (see Group 4.) If the PRC is not blank, the product matrices (dispersion and separation) and the phase-space passed are printed after each iteration; if the PRC is blank these quantities are printed only initially and terminally.

Group 4. This group consists of one card with one of two possible option flags in format I3. The first option flag is 999 and indicates termination of the reading of matching section, followed by a PRC. The second option flag is 989 which terminates reading of matching sections and then enters separation and or dispersion matching. This flag is followed by a PRC and then additional separation and or dispersion cards if desired.

Group 5. This group consists of one card with two parameters in a format of I3,F8.3. The first parameter (KT) indicates the utilization of the tracking facilities of the program or if not exercised, KT indicates future action. The second parameter (L) is used by the program when KT = 30. A summary of the values of KT follow:

1. If KT = 30, the program will retain and store the first L sections and expect new matching sections. When a terminator card is reached (999 or 989), the reading of matching sections stops and all currently stored sections may be used in any additional problems. If the stored sections are not being retained, new sections must be read in again, since matching always follows the encounter of a terminator card.
2. If KT = 40, the program expects new matching sections.
3. If KT = 50, the program expects a complete set of data.
4. If KT = 300, the program will terminate.

g. Illustrative Example

Application of the foregoing instructions is demonstrated in the example whose data appear in Table 39. This example was also run by programs TRANSPORT and OPTIK so that the potential user may compare input data and observe the necessary conversions from one program to the other.

The first card of the data deck shows a 2.5-GeV/c-design momentum entry followed by the numeral 8, which indicates the number of elements that follow.

The next eight cards comprise the beam elements of the specific example, following the exact formats and units required by program TRAMP.

The next two cards indicate the matching requirements of the problem, namely, a request for a focus-to-focus matching (KT = 2) from the first through the seventh element in both planes and a request for no matching for the seventh and eight elements. Note that since the matching type is less than four, no specification of the phase ellipse is necessary.

The next card (999) indicates the end of reading matching sections.

The blank card following the (999) card corresponds to the PRC, requesting a printout of all program information.

Next we track four vectors through the system. The next two cards (KT = 5) request a printout of the individual matrices for the sixth and seventh elements.

The next card (KT = 7) requests the printout of the product matrix.

The next card (KT = 8) produces printout relating to projection of the apertures back to the source, since v(1) = v(2) = v(3) = 0.0.

Table 39

TRAMP Input Data for Example 1

			Card Column			
1	11	20		34	42	50
2.5	8					
1.0	0.8128	0.3182	0.0			
4.0	1.1176	6.846	8.11	0.0		
1.0	2.2606	2.2606	0.0			
2.0	1.3715	314.9607	1.0		5.0	5.0
1.0	1.4359	1.4359	0.0			
3.0	1.3715	+314.961	2.0		5.0	5.0
1.0	1.9558	1.9558	0.0			
1.0	2.540	2.540	0.0			
1 7 2					1 7 2	
7 8 0					7 8 0	
999						
	← Blank card (PRC, Print Requirement Card)					
1	0.250	0.0	0.25	0.0		
1	0.250	0.002	0.25	0.0012		
1	0.250	−0.002	0.25	−0.0012		
1	−0.250	0.0	−0.25	0.0		
5	1.0	2.0				
5	6.0	7.0				
7						
8						
9						
10	0.01	0.0	0.0			
	← Blank card (PRC, Print Requirement Card)					
300						

186   IV. Programs Employing Matrix Formalism

The card with KT = 9 is intended to print the bending magnet tolerances and quadrupole aberrations if any.

The card with KT = 10 allows the change of momentum by $0.01 \times$ design momentum.

The next to the last card, corresponding to the PRC, while the last card (300) signals the end of input data for this test case.

h. Output Description

Output from program TRAMP is determined by the proper use of the PRC and the appropriate value of KT (described in the tracking facilities paragraphs). Each column in the 80-column PRC stands for an element (this alone imposes a limit on the number of elements in the beam). Thus a nonzero digit in the seventh column causes a printout for the seventh beam element, etc. However, if no digits are punched on this card, printout will occur for all elements.

Figure 83 shows the partial printout obtained from the illustrative example listed in Table 39. The program lists the initial input parameters, and tracking information, followed by a printout of the individual or product matrices at various locations of the beam line. If rays are to be traced in the system, the program prints the projections in the horizontal and vertical plane for each ray. Depending on the input data, the program gives various printouts pertaining to the KT value entered. An examination of the printout of Fig. 83, with that obtained from TRANSPORT (see Fig. 68) and OPTIK (see Fig. 79) will show that TRAMP produces similar results.

No CalComp plots are available with this version of the program.

i. Further Improvement of Program TRAMP

In order to solve beam-transport problems most efficiently, Kane [55] has combined program TRAMP with program OPTIK (described previously) and produces program TRAMTIK. The combination of these two programs is accomplished through control cards to run under the CDC 6600/7600 computer systems.

The first part of TRAMTIK is basically program TRAMP as modified by Butler [58]. It serves primarily to solve the beam-transport problem, and is oriented towards using sections rather than elements as the basic unit of beam design. Sections are delimited by beam conditions that are solved by iteration on a section-by-section basis. Focusing and phase-space conditions are handled exactly as in program TRAMP. The second part of TRAMTIK is basically program OPTIK with its solving facilities removed and its output facilities supplemented. Its ordinary use is to output ray-trace or phase-space beam elements.

```
DESIGN MOMENTUM =    2.5000 GEV/C,    8 ELEMENTS
                           INITIAL SYSTEM PARAMETERS
         AST                                   XTRA                    WID
   1.000    .813     .818    0.         -0.       -0.        15.000   15.000
   4.000   1.118    6.846    8.110       0.       -0.        15.000   15.000
   1.000   2.261    2.261    0.         -0.       -0.        15.000   15.000
   2.000   1.372  314.961    1.000      -0.        5.000      5.000   15.000
   1.000   1.486    1.486    0.         -0.       -0.        15.000   15.000
   3.000   1.372  314.961    2.000      -0.        5.000      5.000   15.000
   1.000   1.956    1.956    0.         -0.       -0.        15.000   15.000
   1.000   2.540    2.540    0.         -0.       -0.        15.000   15.000

MATCHING ALL AVAILABLE SECTIONS

 SECTION 1, SUM OF SQ ERRORS= 4E-18
HORIZONTAL MATCHING  1 TO  7  F-F    VERTICAL  MATCHING  1 TO  7  F-F

     SYSTEM PARAMETERS IN TRACKING    FULL BEND ENTR-EXIT    DELX
LMN        METERS                        DEG      DEG          CM
 1 L         .813         .818 METERS
 2 BM       1.118      17.115 KGAUSS       13.18    3.045     2.9699
 3 L        2.261       2.261 METERS
 4 FH       1.372     654.846 G/CM
 5 L        1.486       1.486 METERS
 6 DH       1.372     795.073 G/CM
 7 L        1.956       1.956 METERS


 SECTION 2, SUM OF SQ ERRORS= 0
HORIZONTAL MATCHING  7 TO  8           VERTICAL  MATCHING  7 TO  8

     SYSTEM PARAMETERS IN TRACKING    FULL BEND ENTR-EXIT    DELX
LMN        METERS                        DEG      DEG          CM
 7 L        1.956       1.956 METERS
 8 L        2.540       2.540 METERS

                               PRODUCT MATRICES                              DISPERSION        SEPARATION
              DIMENSIONS FOR 12 AND 21 ARE METERS/RADIANS,RADIANS/METERS    (PER +1PERCENT)  (   I VS   I )

LMN    (H11)    (H12)    (H21)    (H22)    (V11)    (V12)    (V21)    (V22)      CM    CRAD     CM   CRAD
 1 L   1.000     .254    0.       1.000    1.000     .259    0.       1.000    0.      0.      0.    0.
 2 BM  1.006    1.366    -.000     .994     .967    1.371    -.047     .967     .127    .228   0.    0.
 3 L   1.006    2.376    -.000     .994     .920    2.354    -.047     .967     .359    .228   0.    0.
 4 FH   .350    1.878    -.836   -1.628    1.605    5.997    1.167    4.982     .366   -.219   0.    0.
 5 L    .255    1.692    -.836   -1.628    1.739    6.566    1.167    4.982     .341   -.219   0.    0.
 6 DH -1.002     .487   -1.263    -.384    1.562    6.473   -1.385   -5.097     .297    .146   0.    0.
 7 L  -2.606    -.000   -1.263    -.384    -.196    -.000   -1.385   -5.097     .483    .146   0.    0.
 8 L  -5.814    -.975   -1.263    -.384   -3.713  -12.946   -1.385   -5.097     .853    .146   0.    0.

A RAY IS TRACED THROUGH THE SYSTEM
             PROJ  IN HORIZONTAL PLANE      PROJ  IN  VERTICAL  PLANE
LMN           X(CM)     DX(CRAD)              Y(CM)    DY(CRAD)
INITIAL        .250      0.                    .250     0.
 1 L           .250      0.                    .250     0.
 2 BM          .252     -.000                  .242    -.012
 3 L           .252     -.000                  .230    -.012
 4 FH          .088     -.209                  .401     .292
 5 L           .064     -.209                  .435     .292
 6 DH         -.251     -.316                  .391    -.346
 7 L          -.652     -.316                 -.049    -.346
 8 L         -1.453     -.316                 -.928    -.346

                               INDIVIDUAL MATRICES
              DIMENSIONS FOR 12 AND 21 ARE METERS/RADIANS,RADIANS/METERS

LMN    (H11)    (H12)    (H21)    (H22)    (V11)    (V12)    (V21)    (V22)
 1 L   1.000     .254    0.       1.000    1.000     .259    0.       1.000
 2 BM  1.006    1.111    -.000     .994     .967    1.120    -.047     .980
 3 L   1.000    1.016    0.       1.000    1.000    1.016    0.       1.000
 4 FH   .348    1.058    -.831     .348    1.834    1.735    1.362    1.834
 5 L   1.000     .114    0.       1.000    1.000     .114    0.       1.000
 6 DH  2.039    1.820    1.735    2.039     .229     .997    -.950     .229
 7 L   1.000    1.270    0.       1.000    1.000    1.270    0.       1.000
 8 L   1.000    2.540    0.       1.000    1.000    2.540    0.       1.000

            SYMMETRIC LIMITING APERTURES PROJECTED BACK TO SOURCE GIVING X,DX OR Y,DY INTERCEPTS FOR +/- HALF APERTURE
            HORIZONTAL CLIPPING FOR CENTRAL RAY OF CURRENT MOMENTUM, VERTICAL CLIPPING FOR CENTRAL RAY OF OFF MASS PARTICLES

LMN   HGAP/2  +TO  XI  +TO DXI    -TO  XI  -TO DXI   CRAY   VGAP/2 +TO  YI   +TO DYI    -TO  YI  -TO DYI   CRAY
       (CM)    (CM)   (CRAD)       (CM)    (CRAD)            (CM)   (CM)    (CRAD)      (CM)    (CRAD)
 1 L  15.00   15.000   59.06     -15.000   -59.06           15.00  15.000   57.83     -15.000  -57.83
 3 L   5.00    4.969    2.10      -4.969    -2.10           15.00  16.311    6.37     -16.311   -6.37
 4 FH  5.00   14.282    2.66     -14.282    -2.66           15.00   9.343    2.50      -9.343   -2.50
 5 L   5.00   19.643    2.95     -19.643    -2.95           15.88   8.626    2.28      -8.626   -2.28
 6 DH  5.00   -4.989   10.26       4.989   -10.26           15.00   9.600    2.32      -9.600   -2.32

IN THE FOLLOWING, A HORIZONTAL ACCEPTANCE 2* 2.104 CRAD AND A VERTICAL ACCEPTANCE 2* 2.284 CRAD ARE USED


            SYMMETRIC LIMITING APERTURES PROJECTED BACK TO SOURCE GIVING X,DX OR Y,DY INTERCEPTS FOR +/- HALF APERTURE
            HORIZONTAL CLIPPING FOR CENTRAL RAY OF CURRENT MOMENTUM, VERTICAL CLIPPING FOR CENTRAL RAY OF OFF MASS PARTICLES

LMN   HGAP/2  +TO  XI  +TO DXI    -TO  XI  -TO DXI   CRAY   VGAP/2 +TO  YI   +TO DYI    -TO  YI  -TO DYI   CRAY
       (CM)    (CM)   (CRAD)       (CM)    (CRAD)            (CM)   (CM)    (CRAD)      (CM)    (CRAD)
 1 L  15.00   15.000   59.06     -15.000   -59.06           15.00  15.000   57.83     -15.000  -57.83
 3 L   5.00    4.969    2.10      -4.969    -2.10           15.00  16.311    6.37     -16.311   -6.37
 4 FH  5.00   14.282    2.66     -14.282    -2.66           15.00   9.343    2.50      -9.343   -2.50
 5 L   5.00   19.643    2.95     -19.643    -2.95           15.00   8.626    2.28      -8.626   -2.28
 6 DH  5.00   -4.989   10.26       4.989   -10.26           15.00   9.600    2.32      -9.600   -2.32

IN THE FOLLOWING, A HORIZONTAL ACCEPTANCE 2* 2.104 CRAD AND A VERTICAL ACCEPTANCE 2* 2.284 CRAD ARE USED
QUADRUPOLE ABERATIONS AND BENDING MAGNET TOLERANCES
        LATTER ASSUMES ABERATION AT SOURCE OF=0.    CM
 1 BM ENTR   0.             PER CM
 2 BM EXIT   0.             PER CM
 3 FH ENTR  -1.040E-02 CM
 4 FH EXIT  -8.736E-02 CM
 5 DH ENTR  -1.146E-01 CM
 6 DH EXIT   7.558E-03 CM
```

FIG. 83. Partial printout obtained from successful execution of program TRAMP.

23. Program 4P

This program will evaluate the linear properties of beam transport systems and calculate the third-order aberration properties of a system consisting of quadrupole and octupole magnets [59]. The code uses one of two algorithms to achieve maximum weighted error in meeting the set of conditions or the minimization of the weighted sum of errors in meeting the conditions. The linear programming method used in 4P automatically incorporates constraints upon the minimum and maximum values of adjusted parameters; thus, solutions are restricted to physically reasonable parameters.

This program is difficult to implement and to use even though a maximum of 12 linear beam conditions may be met by this linear programming method. Preparation of input data is cumbersome and requires specific instructions applicable to the IBM 7094 computer. The code is written in Fap and Fortran languages and is adapted for use only with the IBM 7090 and IBM 7094 computers.

Execution of the program is controlled by input cards specified in a mnemonic format similar to mnemonic computer instructions. Input-output control is exercised through internal switches that are set in one card at the discretion of the user. The output of the program is particularly flexible, providing cathode-ray tube plots of the optimal effects of the aberrations. Various programmed options allow the user to direct the code to plot beam profiles for the entire system or for phase ellipses, trajectories, and other options.

24. Naval Research Laboratory Beam Transport Programs

The Naval Research Laboratory has developed a set of Fortran subroutines and functions to facilitate calculations of the first-order properties of a beam transport system [60]. This set of routines may be used with a "driver program" that is specifically written by the user to meet the requirements of the particular problem making use only of the routines needed. The technique of writing such driving programs is simple, since they require only a sequence of call statements that assemble the required subroutines in a form suitable to calculate the optical properties of the system under consideration.

This method of approach is desirable for small computer installations where the available computer memory is limited. It also provides the user with the added flexibility to incorporate his own programs in special cases where the routines available do not include the problem under consideration. An excellent description regarding the use of this set of subroutines is given by Shapiro [60].

25. Program BOPTIC

This program was developed by Dangerfield and Walsh [61] of the Australian Atomic Energy Commission for the purpose of analyzing the behavior of monoenergetic charged particles in post-acceleration beam transport systems. Basically, the program includes the various beam elements found in all programs of this type and offers various parameter optimization schemes to satisfy the specified constraints on the system. In particular, BOPTIC can handle beam systems comprised of any number of bending magnets, quadrupoles, drift lengths or beam collimators, in a manner similar to that of TRANSPORT, and can produce solutions for various types of calculations. BOPTIC traces trajectories for single particles. A beam envelope is obtained by tracing a number of trajectories, and the profile is drawn by inspection. The program is written in Fortran and a very good description of the code is given by Dangerfield and Walsh [61].

26. Program BEATCH

Developed at CERN by Burton *et al.* [62], this program evaluates particle motion in beam transfer channels. Some of the outstanding characteristics are calculations of

1. Three-dimensional transfer channels
2. Orbit trajectories with respect to the central orbit
3. Beam envelope along the channel
4. Betatron parameters
5. Matching of betatron phase-space ellipses at one or two points
6. Matching of momentum compaction at a predetermined point.

Program control is exercised by appropriate control cards, each of which is followed by the appropriate data which perform the designated functions. Data cards may be stacked to produce runs for successive transfer-channel modifications or to calculate new beam transfer channels. The flexibility of the input allows the user to prepare only the data necessary for his particular run and to ignore those functions not needed.

Various options in the program have been incorporated in subroutines which the user may call with appropriate control cards. The beam transfer channel itself may be made of any sequence of elements that the program can handle, such as drift lengths, quadrupole lenses (horizontally focusing or defocusing), and bending magnets. Each beam element is represented by a 3×3 horizontal and vertical transfer matrix with no coupling between the horizontal and vertical planes.

190   IV. Programs Employing Matrix Formalism

The program output is determined also by the requirements of the user. For example, in the tracking option, in addition to the printout of the initial particle coordinates, the program may print the particle coordinates at the exit of each beam element and in special instances various normalized quantities relative to amplitudes, displacements or slopes, may be printed. Printed output for the matching calculation consists of two parts: that relating to the minimization process and that relating to various matching quantities, such as mismatch factors, the maximum allowed beam size and discharge.

Program BEATCH is excellently documented and is operational on the CDC 6600/7600 computer systems. It is included in the CERN library of programs.

27. Program SYTRAN

This program was developed at CNEN by Bassetti *et al.* [63] to allow for the determination of various parameters necessary to fulfill requirements that arise during the design of beam transport systems. The system may consist of bending magnets, quadrupoles, and drift spaces, with each beam element represented by a 3×3 transfer matrix. The beam is represented by an ellipsoid with coordinates x, x', y, y', $\Delta p/p$ in a similar fashion outlined in program TRANSPORT.

SYTRAN is capable of varying the values of certain parameters to produce desired beam conditions at the end of the beam channel or at some intermediate section. These conditions may be a focus, an image point, waist-to-waist transfer, etc. The parameters being varied are the quadrupole strengths and the bending magnet field index, since these parameters exert a considerable influence on the system and have a wide range of variations. The system of equations that result are solved by a modified Newton-Raphson method.

28. CERN Beam Transport Programs

a. Introduction

The programs listed under the CERN Program Library series W-101 and W-102 are written for the study of particle transmission through a linear magnetic system [64, 65, 66]. These codes have the facility of calculating the transmission of either primary particles or neutrinos and charged secondary particles produced by the decay of the primary particles.

Various extensions of the basic program permit the automatic calculation of the momentum spectrum of primary and secondary particles transmitted by the system. The transmission of particles through scattering media may also be calculated.

The assorted writeups for these programs refer to pions and muons, although these codes will calculate the transmission for any primary beam particle. However, a restriction for secondary particles, is that they must be produced via the two-body decay mode, $S \to T + V$, unless the program is otherwise instructed by entering explicitly desired momenta and corresponding angles.

There are extensive writeups of the programs in the CERN "reports" cited below which heavily cross-reference one another and presume some familiarity with the CERN programming system. For this reason, in this section the function of each program will be briefly described along with its relationship to other existing codes. The beam lines handled by these CERN programs are made up of elements of the following type: Drift spaces (E), thin lenses (L), quadrupoles (Q), Bending magnets (A), and arbitrary matrices (M).

These are represented by horizontal and vertical 3×3 matrices. Additionally one may specify limiting apertures (Diaphrams, D), polygons through which the beam must pass, etc.

The programs calculate and print the transformation matrices, forward and backward polygons, and beam transmission through the system. The programs handle seven basic types of problems.

b. Program B-50

This program calculates problem types 1, 2, and 3 (see Fronteau and Hornsby [64]):

1. Dispersive systems whose transmission and pass band are determined by the linearized matrix calculated for the central momentum;
2. Nondispersive system, calculating the transmission of charged secondaries emitted in the phase space including the effect of π decay on the transmission;
3. Nondispersive neutral particle beam (neutrinos) emitted in phase space by the primary target.

c. Program B-51

This program calculates the Coulomb scattering deviation $\sigma = (\langle y^2 \rangle)^{1/2}$ and $\sigma' = (\langle \theta^2 \rangle)^{1/2}$ in a focusing medium as a function of momentum (see Banaigs et al. [66]). A biconvergent lens is described and allowed as a beam element in this program.

There are various combinations of these programs to provide alternate modes of operation and calculations of other quantities. For example, a combination of B-51 and B-52 exists (through a linking program B-58) that calculates transmission spectra through absorbers.

d. Program B-52

This program is an extension of the B-50 code and handles two additional types of problems, problems 4 and 5 (see Banaigs *et al.* [66]):

4. Transmission and pass band for a broad momentum band by calculating all matrices at the appropriate momentum and referencing the results to the central momentum;
5. Dispersive system (variant of problem type 2), where the transmission is calculated for the π's in systems containing bending magnets.

e. Program B-56/57

This program is an extension of the B-52 code and introduces two additional types of problems, problems 6 and 7 (see Citron *et al.* [65], Banaigs *et al.* [66]). The code treats the entire beam complex from target to experiment. Any of the four models of particles production—Cocconi, Van Dowdel, Uniform, or Trilling—may be specified for the particle production distribution $d^2N/dp\,d\Omega$.

The effect of the accelerator guide field on the particles in the beam line is introduced by the use of the appropriate Michaelis plots for that beam on the accelerator. With this additional data, program B-56/57 is able to calculate the momentum spectra of the pions and muons at specified points along the beam line and particularly at the downstream end of the system.

V | Programs Employing Integration of Equations of Motion

29. Introduction

As we have noted previously, the other conventional approach to solving orbit problems is numerical integration of the equations of motion. Orbit codes such as GOC and CYDE base their results on orbits calculated by numerically integrating through a given magnetic field, which can be measured empirically or derived analytically. A program such as SOTRM uses numerical integration to generate a transfer matrix which represents the effect of the field on a beam of charged particles passing through the field. Again, the field can be measured or calculated.

Which differential equations are to be integrated depends upon how the problem is formulated and upon the choice of different coordinate systems. These details can be found by consulting a particular program. The method of integration is also different for the variety of codes available. Quite often, Runge–Kutta methods are employed because of the ease with which they can handle arbitrary fields, various geometrical regions (through which the integration can be carried out), and different step sizes. This is not a necessary choice and any suitable scheme can be used. For example, SOTRM uses a predictor-corrector scheme with automatic step-size adjustment. But, for some applications this scheme might not be suitable, so, in that case the program would be modified without affecting the basic code that generates the matrix elements. This is more or less true of most programs based on integration; the integrator is essentially a "black box" that numerically solves the equations of motion.

Obtaining results by numerical integration has many advantages: formost would be the ability to utilize essentially arbitrary electromagnetic field elements at whatever precision is justified by the data. No simplifications are necessary; the equations of motion can be solved as they stand, and the field can be supplied in any manner that is suitable and consistent with its specification as an electromagnetic field. For example, we can supply measured values, numerically calculated fields, approximation formulae, exact analytical values; the equations of motion allow a completely free specification.

There are disadvantages, however, and probably the chief one is the time and work required to solve the problem. While it is possible to supply a detailed field description, this usually will require considerable data input with perhaps, preprocessing, and possible large amounts of internal storage. To integrate along a path, we must specify the field along this path, and thus the boundaries and the field values in specific regions, the description of the geometric shape of the system can become a complicated feat in itself. Finally, integrating through the field can consume copious amounts of computer time. So, it is possible to grind to a virtual halt. For example, an integration of 20 to 30 orbits through a reasonable length of a few meters can take a few seconds of central processor time on a test field which is supplied analytically. Yet, when a measured three-dimensional field is used in the same program, and these field values are interpolated in three dimensions at every integration step along the orbit, then the same set of orbits can take a few *hundred* seconds of central processor time. Obviously then, time—meaning dollars and cents—is a critical element.

Another disadvantage with numerically integrating to obtain the orbits is the inclusion of "round-off" errors that can accumulate when the integration is carried over a large number of steps. This situation can be easily illustrated: A typical cyclotron, such as the 88-inch cyclotron, can accelerate a beam of particles to full energy in 500 turns. Therefore, the crude step-size of 16 steps per revolution would lead to 8000 steps. Thus, care would have to be exercised to insure that round-off errors were not unduly influencing the answers.

One might also have to consider the numerical stability of the integration scheme, and whether the scheme itself introduces systematic errors.

It has been my experience that integration methods—while having the potential ability to yield detailed information about the particle behavior in arbitrary electromagnetic fields—are more complicated than matrix transformations and quite often require for their effective use, a good knowledge or feeling of the problem in order to obtain meaningful results.

30. Program SOTRM

a. Introduction

In the design of beam transport systems, it is often desirable to generate transformation elements from a magnetic field by numerically integrating the orbits through the field. Such a transformation matrix is needed when only a measured field is available or when the effect of various trial magnetic fields is being investigated. Essentially, program SOTRM, by Close [67] produces first- and second-order elements when an arbitrary magnetic field is given. The resulting transformation matrix is readily applicable to beam transport programs such as TRANSPORT.

SOTRM formulates a system of equations which, when integrated, produces the coordinates of the reference particle and of any nearby particle(s) specified. Once this is completed, the program calculates (if requested) the first- and second-order transformation matrix elements using the reference orbit as the origin in a suitably choosen coordinate system. The program is written in Fortran and is operational in the CDC 6600/7600 computer systems. SOTRM occupies 60K_8 memory locations and a typical run takes less than one minute on the CDC 6600 system.

b. Program Organization

The program is organized in a fashion that allows the user to arrange his data in a very flexible way for a variety of cases. Depending on the data setup, SOTRM will ray-trace and/or produce first- and second-order transformation matrices. The organization of the program is shown in the simplified flow chart appearing in Fig. 84, while the main subroutines are briefly outlined in Table 40. The integration "package" and output routine (OUTRT) may be substituted to fit the requirements of the user. In general, SOTRM may be used with very little effort as a "black box"; it may be adapted to run as a subroutine provided sufficient knowledge has been accumulated by the user.

The mathematical model derived to simulate the equation for the reference orbit and the nearby orbits, as well as the formalism to find the transformation elements for a particle moving in a specified magnetic field, have been adequately described in the cited references [67, 68]. The program has been tested in a variety of cases involving analytically defined magnetic fields, and has produced results well within the error limits of the integration routines. However, for practical problems where the field is known from measured field data, the user must find some independent criteria by which to judge the significance of the generated transformation elements.

c. Preparation of Input Data

The ability of SOTRM to perform a variety of tasks is demonstrated by the flexible manner in which the input data may be prepared. Specifically, SOTRM

V. Integration of Equations of Motion

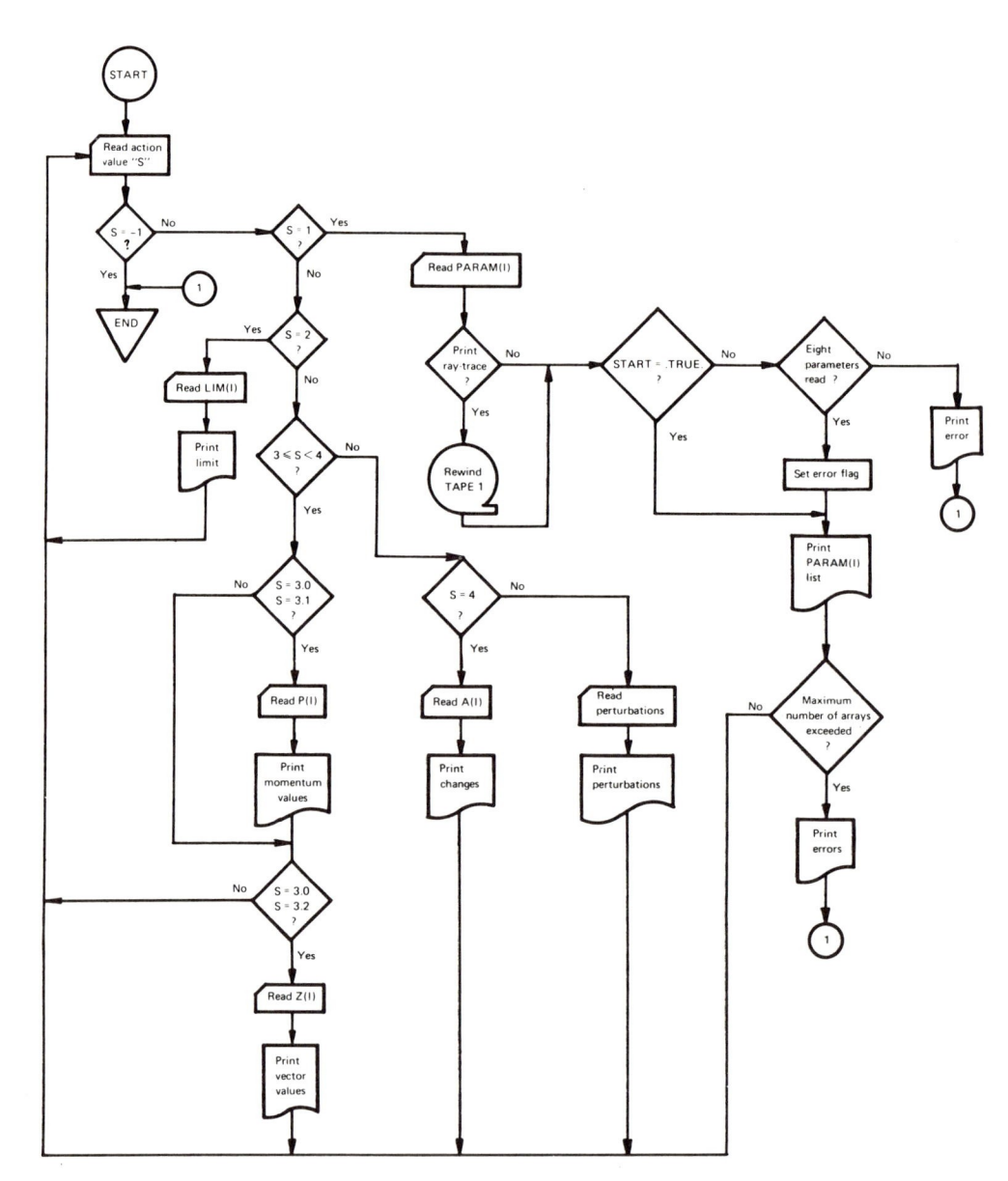

FIG. 84. Flow chart for program SOTRM.

is controlled by an action card that tells the code the specific set of data that will follow, and the portion of the program to be executed. Upon the completion of this task, SOTRM returns and reads another action card and performs the indicated functions. In this manner, the user may process as many cases as desired by the appropriate choice of an action card. The following remarks

30. Program SOTRM

Table 40

SOTRM Subroutines and Functions

Name	Description
SOTRM	The main control program: it reads input data, calls appropriate subroutines, and provides control for various functions for which the program is capable
SOTRM1	Uses values of coordinates of reference ray and nearby rays, initial value of independent variable, and momentum perturbation $\delta P/P_0$ of nearby particles to compute first- and second-order transfer matrix elements. Secondary function is to supply ray trace information for one reference ray and up to thirty nearby rays
INTO	The main integration routine
ZINIT	Initializes values of nearby rays
ZFINAL	Calculates final values of any nearby ray; is the inverse of ZINIT except that it handles one ray at a time
SF	Obtains off-diagonal elements of second-order transfer matrix
SETZ	Sets initial vectors for nearby rays in order to generate first-order matrix F and second-order matrix S
FUNC	Computes right side of differential equations set up for reference orbit, as well as those for nearby rays
SAVEZP	Saves initial values of momentum and orbit information used to start integration loop
OUTRT	Provides printed output of various quantities
APTRAN	Applies appropriate transfer matrix to nearby rays
FIELD	Establishes field (supplied by the user) at current position of reference particle and nearby particles; origin, that of reference particle. Actual field in gauss is supplied by subroutine GAUSS
GAUSS	Furnishes actual magnetic field to be used by FIELD

are intended to clarify utilization of the action cards and their associated parameters shown in Table 41.

When the value of the action card = 1.0, the program expects a redefinition of some of the parameters [PARAM(I)]. An action card with a value of 3.1 is for ray-tracing. The cards that follow carry the momentum of the reference particle and the momenta of the remainder of the nearby particles (one card per momentum). An action card with a 3.2 value is used to furnish a set of parameters, $z(1)$ to $z(4)$ for the reference particle and NUM-1 sets of particles showing the parameters $z(5)$ to $z(8)$. When the value of the action card = 4.0, the program expects to read an externally defined magnetic field. With the action card = 6.0, the standard value of the perturbations may be changed. An action card with a value of 99.0 directs the program to perform the cal-

V. Integration of Equations of Motion

Table 41

SOTRM Input Specifications

Action card	Parameters	Format	Description
3.2		8E10.0	Specify conditions for central (reference) and nearby particle
	$\{K, [I, Z(I), I = 1, K]\}$	[I5/(I5,5X,E15.0)]	K = number of Z(I) values to follow
	Z(1)		Parameter for reference ray momenta (MeV)
	Z(2)		Particle angle (degree)
	Z(3)		Starting x_1 parameter (meter)
	Z(4)		Starting x parameter (meter)
			Parameters Z(1) to Z(4) must always be present
	Z(5)		Fifth parameter δx_1 (centimeter)
	Z(6)		Sixth parameter θ_1 (milliradian)
	Z(7)		Seventh parameter δx_2 (centimeter)
	Z(8)		Eight parameter θ_2 (milliradian)
			There are (NUM-1) entries of last four parameters when ray-tracing; vector Z(I) contains values of coordinates of reference ray and nearby rays corresponding to value of initial arc length (S01) upon entry and exit
4.0		8E10.0	Allows input of externally defined magnetic field
	RMIN, RMAX, DR	3E10.0	RMIN and RMAX = beginning and ending radius of magnetic field; DR = radius interval
	B(I), I = 1, N	8F10.0	Magnetic field values, entries N = 1 + (RMAX - RMIN) / DR
6.0		8E10.0	Allows change of perturbations used to generate transformation matrix elements
	$\{K, [I, A(I), L = 1, K]\}$	[I5/(I5,5X,E15.0)]	K = number of A(I) values to follow
	A(1)		First parameter δx_1, standard value = 0.01 centimeter
	A(2)		Second parameter $\delta x_1'$, standard value = 0.1 milliradian
	A(3)		Third parameter δx_2, standard value = 0.01 centimeter
	A(4)		Fourth parameter $\delta x_2'$, standard value = 0.1 milliradian
	A(5)		Fifth parameter $\delta P/P_0$; standard value = 0.0001
99.0		8E10.0	End of data input; initiates a call to subroutine SOTRM1 to begin execution
-1.0		8E10.0	Terminates program

culations indicated (ray trace or matrix generation action cards), using the current parameter values.

d. An Illustrative Example

The performance of this program was tested by evaluating the first-order transfer matrix for the "fringe field" of the 184-inch cyclotron at the Lawrence Berkeley Laboratory. With this test case we will show data organization,

Table 41 (cont.)

Action card	Parameters	Format	Description
1.0		8E10.0	Initializes, changes PARAM array values
	$\{K, [I, PARAM(I), L = 1, K]\}$	[I5/(I5,5X,E15.0)]	K = number of PARAM(I) values to follow
	PARAM(1) = Q		Signed number of charges of reference particle and of nearby particles (coulomb)
	PARAM(2) = DELSIN		Maximum step size (centimeter)
	PARAM(3) = SOMAX		Maximum arc length (meter)
	PARAM(4) = S01		Initial Arc length (meter)
	PARAM(5) = OPTION		Options available: 0 = ray trace; 1 = first-order matrix; -1 = first-order matrix with a second-order trace; 2 = first- and second-order matrices that are asymmetric; 3 = first- and second-order matrices that are symmetric
	PARAM(6) = NUM		Number of rays, including center reference ray; may be set to 1 if generating matrix elements; maximum NUM $\leq$ 31.0
	PARAM(7) = MAX		Maximum number of steps that can be taken; includes all step adjustments of the variable step-size integrator
	PARAM(8) = NPRINT		Printout has three choices: if 0, no printout of ray trace or matrix is done; if > 0, printout of every NPRINT step is generated, also initial and final values are printed; if < 0, output is same as > 0 except that matrix elements will be stored on Tape 1 if generating elements, or will be read from Tape 1 and applied to initial conditions if ray-tracing
2.0		8E10.0	Optional, specify geometric field limits
	$[LOWLIM(I), UPLIM(I), I = 1, 3]$	8E10.0	
	LOWLIM(I)		Lower limit of reference particle (meter)
	UPLIM(I)		Upper limit of reference particle (meter)
			In figure to the left, reference particle s_0 moves in 1,3 plane
3.1		8E10.0	Specify momentum of nearby particle $\delta P/P_0$
	$\{K, [I, P(I), L = 1, K]\}$	[I5/(I5,5X,E15.0)]	K = number of P(I) values to follow
	P(I)		This card required for ray-tracing; vector P(I) contains momentum perturbation $\delta P/P_0$ of nearby particles. (NUM-1) values must be read in at least once

sequential use of action cards, output options, and additional comparison of results. In the sample test data (see Table 42) eight parameters are initialized by the 1.0 action card; the 3.2 action card then specifies the coordinates of the reference ray; the 4.0 action card inputs the magnetic field values from 1.0 to 208.0 inches in 1.0 inch increments; and finally, the 99.0 action card initiates execution for this case. The parameters are now reset by another 1.0 action card to generate the first-order matrix case PARAM(5) = 1.0. SOTRM continues reading cards until again a 99.0 action card starts execution. When the com-

200 V. Integration of Equations of Motion

Table 42

SOTRM Sample Input Data

1	5	11	26	36	46	56	66	76
			Card Column					
1.0								
	8							
	1	-1.0						
	2	7.01095						
	3	140.219						
	4	0.0						
	5	0.0						
	6	1.0						
	7	2000.0						
	8	2.0						
3.2								
	4							
	1	182.275122						
	2	70.0						
	3	28.0456518						
	4	-77.054795						
4.0								
1.		208.	1.					
		23347	23327	23306	23284	23261	23238	23217	23198
		23181	23164	23147	23130	23112	23094	23075	23056
		23037	23020	23004	22990	22977	22964	22952	22940
		22928	22915	22902	22888	22874	22860	22845	22830
		22816	22802	22789	22776	22764	22752	22741	22730
		22720	22709	22698	22687	22676	22665	22654	22642
		22630	22618	22606	22594	22583	22572	22561	22551
		22541	22531	22520	22510	22503	22498	22493	22487
		22479	22470	22459	22447	22435	22423	22412	22403
		22395	22387	22379	22371	22362	22354	22346	22338
		22326	22298	22236	22116	21916	21618	21209	20685
		20043	19265	18347	17327	16256	15180	14136	13133
		12169	11250	10380	9560	8792	8070	7389	6745
		6135.00	5560.00	5021.55	4517.00	4061.00	3613.00	3191.00	2795.00
		2429.00	2093.00	1787.000	1510.00	1247.00	1001.00	775.00	576.00
		414.00	230.000	121.00	14.0	5.	4.	3.	2.
		414.00	230.000	121.00	14.0	-5.	-97.0	-146.	-166.
		-169.	-168.	-160.	-150.	-140.	-130.	-120.	-110.
		-100.	-90.	-80.	-70.	-60.	-50.	-40.	-30.
		-20.	-15.	-12.	-10.	-8.	-6.	-5.	-4.
		-3.	-2.	-1.	0.	0.	0.	0.	0.
		0.	0.	0.	0.	0.	0.	0.	0.
		0.	0.	0.	0.	0.	0.	0.	0.
		0.	0.	0.	0.	0.	0.	0.	0.
		0.	0.	0.	0.	0.	0.	0.	0.
99.0								
1.0								
	3							
	4	0.0						
	5	1.0						
	8	20.0						
3.2								
	4							
	1	182.275122						
	2	70.0						
	3	28.0456518						
	4	-77.054795						
99.0								
-1.0								

```
START OF SOTRM1

 THE FOLLOWING HEADINGS ARE TO BE USED FOR COL. 1-7

                  K= 0              PX(MEV/C)     PY(MEV/C)        X(IN)        Y(IN)     THETA(DEG)    FUNC CAL_S
                  K=(1,NUM-1)        X(IN)          X'(MR)         Y(IN)        Y'(MR)     DP/PO            DPX/PO        DPY/PO
                   U(I,J)    / J=     1               2              3            4           5
                   S(I,J,K)  / K=     1               2              3            4           5

    I    SO(METER)   K           1             2              3              4              5             6            7

    0      0.00000   0  1.7128259E+02  6.2341763E+01  2.8045652E+01 -7.7054795E+01  7.0000000E+01          1
    2     14.02190   0  1.1129993E+02 -1.4434872E+02  4.0565285E+01 -8.0882642E+01  1.4236602E+02        142
    4     28.04380   0 -1.0188757E+01 -1.8199009E+02  4.3827520E+01 -9.4218587E+01 -1.7679563E+02        246
    6     42.06570   0 -5.0571975E+01 -1.7511907E+02  4.1137553E+01 -1.0795011E+02 -1.6389201E+02        334
    8     56.08760   0 -5.5919595E+01 -1.7348549E+02  3.6930051E+01 -1.2132530E+02 -1.6213431E+02        473
   10     70.10950   0 -5.5954179E+01 -1.7347434E+02  3.2598307E+01 -1.3466131E+02 -1.6212288E+02        729
   12     84.13140   0 -5.4896210E+01 -1.7381203E+02  2.8342526E+01 -1.4802175E+02 -1.6247197E+02        768
   14     98.15330   0 -5.4775800E+01 -1.7385001E+02  2.4126332E+01 -1.6139476E+02 -1.6251166E+02        831
   16    112.17520   0 -5.477559-E+01 -1.7385008E+02  1.9912602E+01 -1.7476855E+02 -1.6251173E+02        871
   18    126.19710   0 -5.477559-E+01 -1.7385008E+02  1.5698874E+01 -1.8814233E+02 -1.6251173E+02        881
   20    140.21900   0 -5.477559-E+01 -1.7385008E+02  1.1485145E+01 -2.0151612E+02 -1.6251173E+02        888
CONTROL=  3.200
NUMBER OF PARAMETERS READ =    4

Z(  1)=   182.27512200(MEV) Z(  2)=    -0.00000000(DEG) Z(  3)=    28.04565180(IN)  Z(  4)=   -77.05479500(IN)
Z(

CONTROL= 99.000


 THE FOLLOWING HEADINGS ARE TO BE USED FOR COL. 1-7

                  K= 0              PX(MEV/C)     PY(MEV/C)        X(IN)        Y(IN)     THETA(DEG)    FUNC CALLS
                  K=(1,NUM-1)        X(IN)          X'(MR)         Y(IN)        Y'(MR)     DP/PO            DPX/PO        DPY/PO
                   U(I,J)    / J=     1               2              3            4           5
                   S(I,J,K)  / K=     1               2              3            4           5

    I    SO(METER)   K           1             2              3              4              5             6            7

    0      0.00000    0  1.7128259E+02  6.2341763E+01  2.8045652E+01 -7.7054795E+01  7.0000000E+01          1
                  U(1,J)     1.0000000E+00  0.             0.             0.             0.
                  U(2,J)     0.             1.0000000E+00  0.             0.             0.
                  U(3,J)     0.             0.             1.0000000E+00  0.             0.
                  U(4,J)     0.             0.             0.             1.0000000E+00  0.
                  U(5,J)     0.             0.             0.             0.             1.0000000E+00

   20    140.21900    0 -5.4775465E+01 -1.7385012E+02  1.1485201E+01 -2.0151614E+02 -1.6251177E+02        984
                  U(1,J)    -1.4271207E+01  2.1840791E-02  0.             0.            -2.6117285E+02
                  U(2,J)    -1.2135812E+02  1.1566146E-01  0.             0.            -2.1966608E+03
                  U(3,J)     0.             0.            -1.1891118E+00  5.4119084E-02  0.
                  U(4,J)     0.             0.            -2.2203332E+01  1.6955778E-01  0.
                  U(5,J)     0.             0.             0.             0.             1.0000000E+00
```

FIG. 85. Partial output from program SOTRM; printout is obtained by appropriate use of print flags as outlined in the text.

202   V. Integration of Equations of Motion

putational cycle is completed, it may be terminated by a −1.0 action card, or a new cycle may be initiated with new parameters.

The output from SOTRM is controlled by the appropriate choice of print flags [PARAM(8)]. In the solution for the reference ray, NPRINT = 2.0, the matrix printout was in 2.0-inch increments. The numbers under the so heading

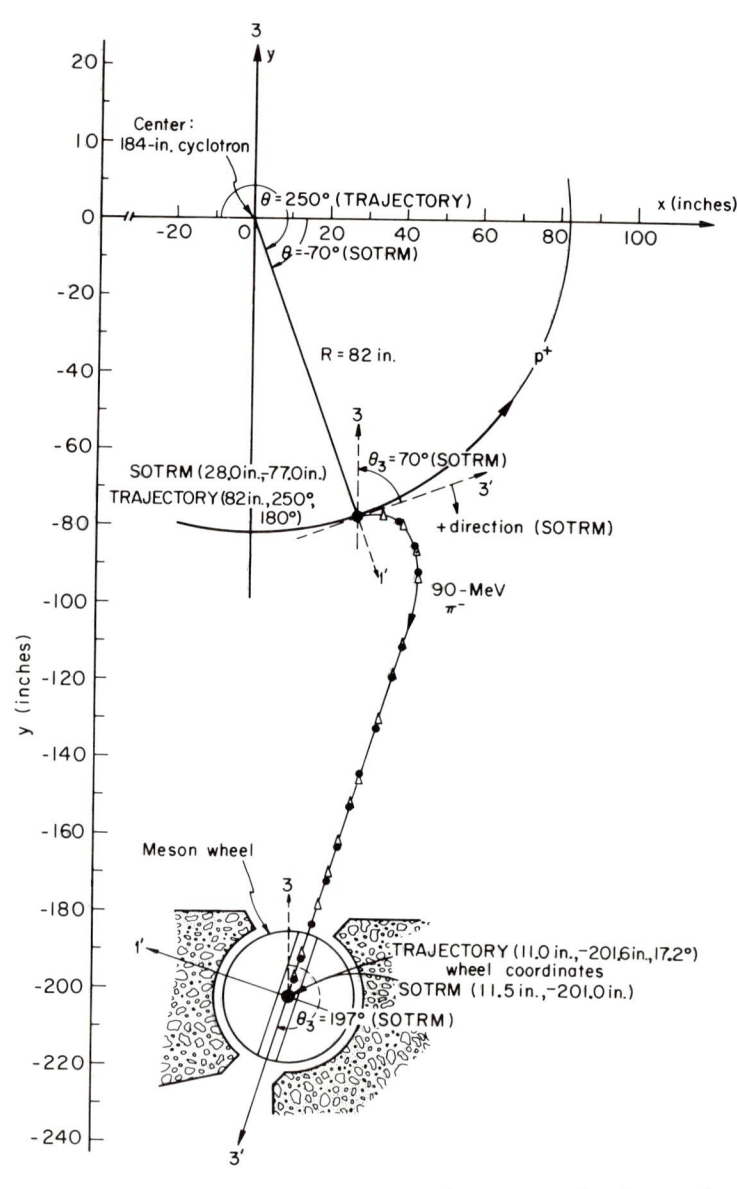

FIG. 86. Orbits computed by program SOTRM and TRAJECTORY for the sample case described in the text. △ = computed SOTRM orbit; ● = computed TRAJECTORY orbit.

indicate the paraxial trajectory orbit (see Fig. 85). This trajectory has been plotted (Fig. 86) along with the trajectory obtained from program TRAJECTORY; both programs arrived at similar solutions.

The second half of the partial printout lists the transformation matrix, U(I, J), at the beginning of the system and at the location desired (in this case the center of the meson-wheel). However, it is possible to print the transfer matrix at any location along the particle trajectory by use of PARAM(8). A comparison of the transformation matrices of SOTRM and TRAJECTORY indicates both programs produced virtually identical matrix solutions. These calculations were verified also by results obtained with program TRANSPORT.

31. Program TRAJECTORY (TRAJ)

a. Introduction

This program by Paul [69], calculates particle trajectories in a given magnetic field, and simultaneously integrates the differential equations of motion to construct the first-order ion-optic matrix representing particle transformations close to this trajectory. One of the desirable characteristics of this program is that the ion-optic matrix produced may be directly used with programs TRANSPORT or OPTIK previously described. This direct utilization is made possible because all the transformations (coordinate, unit, etc.) are performed internally in the code, resulting in a matrix conforming to the peculiarities of the designated ion-optic program.

The program will track protons and positive or negative pions, in either the forward or reverse directions. The horizontal matrix and orbit calculations are performed in the median plane, with the first-order Taylor series terms retained for calculation of the horizontal and vertical transformation matrix. The code is written in Fortran and is operational on the CDC 6600/7600 computer systems. It occupies approximately $50K_8$ memory locations and a typical case runs in less than one minute in the CDC 6600 computer.

b. Program Organization

TRAJECTORY is organized as outlined in the block diagram appearing in Fig. 87 while the functions of the major subroutines are outlined in Table 43. The magnetic field may be read in either rectangular or polar coordinates by the appropriate subroutines provided by the user. The program performs calculations for each control card read and produces the corresponding output, if any. This cycle is repeated until a 10.0 card is read, which signals the program to end the calculation loop.

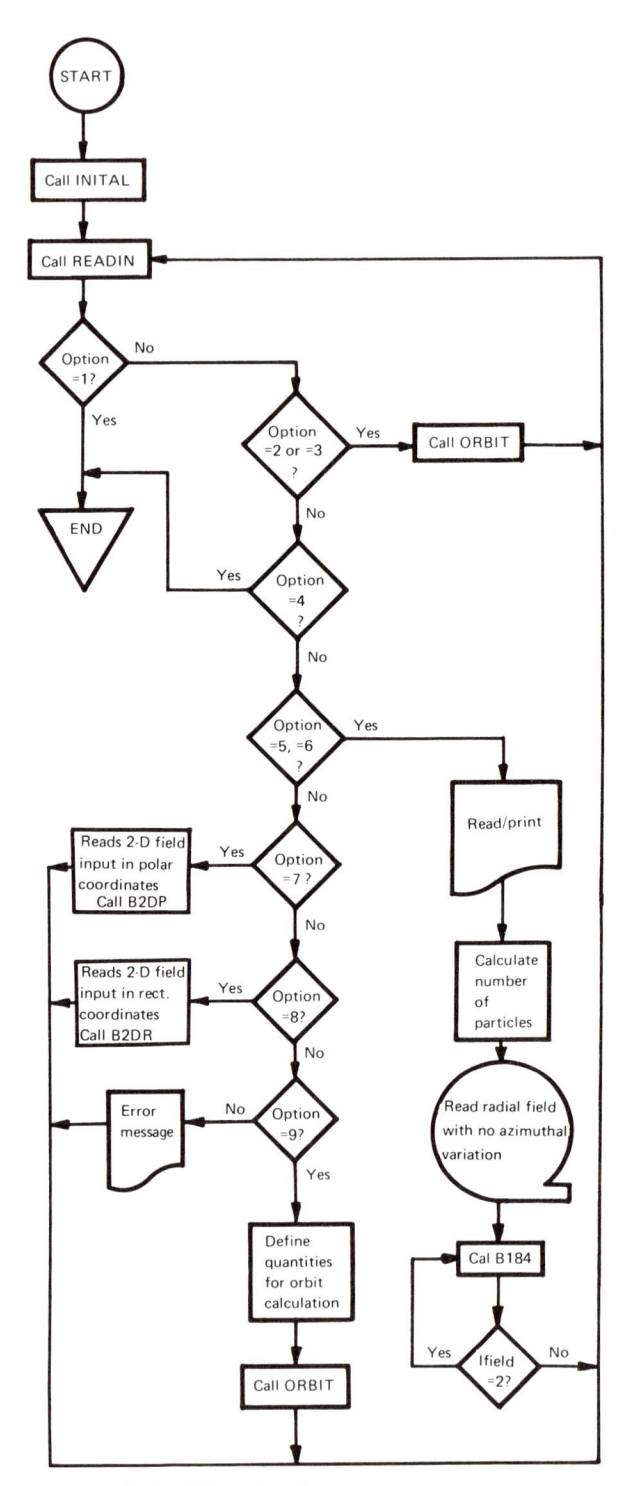

FIG. 87. Flow chart for program TRAJECTORY.

Table 43

TRAJECTORY Subroutines and Functions

Name	Description
TRAJ	Main control program
INITAL	Initialize program parameters
B2DP, B2DR	Dummy routines (supplied by user) for reading two-dimensional field in cylindrical (B2DP) or rectangular (B2DR) coordinates
DISCOT, DISSER, LAGRAN, UNS	Three-dimensional Lagrangian interpolation routines
FIELD	Calculates the magnetic field B_z and gradient DB/DR for a given input rectangular coordinate x, y, z
OUTFIT	Prints the calculated forward and reverse matrices
B184	Perturbation field input
CROSS	Finds point which has been passed
BAKSTEP	Backward integration to point which has been passed

c. Preparation of Input Data

Input is prepared in a straightforward manner, and is controlled by a function code that specifies either the parameters associated with the type of calculation, or the input/output function to be performed; these are listed in Table 44. The program reads the function code and the associated parameters, performs the indicated operations, and returns to read the next function code. The process is continued until the function code equal to 10.0 is encountered which terminates the program. Although Table 44 should be self-explanatory, the following remarks will elaborate in some areas in which questions might arise.

Since the appropriate program function is directed by corresponding choice of the function code, the user must stack the necessary cards in appropriate sequence to produce the results desired. For example, to use code = 1 or 2 with a 730-MeV proton beam starting tangentially at a 70-inch radius and 172 degrees azimuthally, the user would input a code = 1 card (8F10.0) as follows:

$$1.0 \quad 70.0 \quad 172.0 \quad 0.0 \quad 0.73$$

The remaining quantities necessary for this function code need not be included if the standard values are to prevail.

Table 44
TRAJECTORY Input Specifications

Function code	Parameter list			Format	Description
1.0	R,THETA,PHI,E,DS,SMAX,SOUT			8F10.0	Particle input in polar coordinates; R = particle radius (inches); THETA = particle angle (degrees); PHI = outward angular direction (degrees); E = particle energy (GeV); DS = orbit integration step size (inches), positive for forward and negative for reverse calculation; SMAX = maximum trajectory length (inches); SOUT = interval of ion matrix printed along trajectory (inches)
2.0	X,Y,PHI,E,DS,SMAX,SOUT			8F10.0	Particle input in rectangular coordinates (X,Y); remainder same as above
3.0	PARM,VALUE			8F10.0	Parameter specification
3.0	1.0	NC2			Order of Lagrangian interpolation for field values
3.0	2.0	E0			Rest mass of particle (GeV)
3.0	3.0	SMAX			Maximum trajectory length (inches)
3.0	4.0	SOUT			Distance between matrix printouts (inches)
3.0	5.0	JMAX			Maximum number of integration steps
3.0	6.0	SPRINT			Prints step size
3.0	7.0	MATRIX			MATRIX = 0, no matrix calculation; if = 1, compute matrix for TRANSPORT; if = 2, compute matrix for program OPTIK
3.0	8.0	PUNCH			PUNCH = 0, no punch; if = 1, punch matrix elements on cards
3.0	9.0	PHIIN			PHIIN = 0, input in degrees; if = 1, input in p_r/p
3.0	10.0	DS			Integration step size (inches)
3.0	11.0				Ejects page
4.0	XFIELD,BSCALE,PSCALE			4F10.0	Magnetic field input specifications
4.0	1.0	1.0	1.0		User must insert following sequence for radial magnetic field Title card (8A10) RMIN,RMAX,DR,BSCALE (4F10.0) . . . . (Radial field in ascending order of radius extending from RMIN to RMAX in steps of DR, format 8F10.0) . . . . Blank card
4.0	2.0	1.0	1.0		Sequence for radial field plus perturbation over a sector Title card (8A10) RMIN,RMAX,DR,BSCALE (4F10.0) Radial field as above Blank card RMIN,RMAX,DR,BSCALE (4F10.0) . . . . [Perturbations in ascending order of azimuth for each radius; R,B(1),B(2) . . . B(15), format I3,2X,15F5.0] . . . . Insert 3 blank cards
4.0	3.0	1.0	1.0		Two-dimensional polar magnetic field over a sector
4.0	4.0	1.0	1.0		Two-dimensional rectangular magnetic field; subroutines called by 4.0 3.0 and 4.0 4.0 (FIELD, GAUSS) must be supplied by user
5.0	Select,value			3F10.4	Allows interruption of calculations, permits subsequent printout of ion optic matrix at specified value for either particle radius, azimuth, or trajectory length

Table 44 (cont.)

Function code	Parameter list	Format	Description
5.0	1.0 R		Prints matrix at radius, R
5.0	2.0 THETA		Prints matrix at azimuth, THETA
5.0	3.0 S		Prints matrix at trajectory length, S
5.0	4.0 1.0		Stops calculations at desired R
5.0	4.0 2.0		Stops calculations at desired azimuth, THETA
5.0	4.0 3.0		Stops calculations at desired trajectory length, S
5.0	5.0 0.0		Turns off matrix output
6.0	N	8A10	Reads N comment cards
7.0	CODE,V1,V2,DV,W1,W2,DW	8F10.4	Phase-plot options; CODE selects phase plot on V,W axis: if CODE = 1, radius; if = 2, angle THETA; if = 3, angle PHI; if = 4, momentum or energy, e.g., CODE = 13, radius vs. angle PHI plotted V1,V2 and W1,W2 are coordinates of particle
8.0	INIT	2F10.4	If INIT = 0.0, orbit and matrix calculations resumed after previously terminated by 5.0-type card; if = 1.0, matrix initialized to unit matrix
9.0	XW,YW,RLINE,TLINE,PHILINE	6F10.4	XW, YW = center coordinates of meson wheel (inches); RLINE, TLINE, and PHILINE = new coordinate system for origin location, and direction of external beam or center for Michaelis plot
10.0		F10.4	Terminates job

If now we wish to read some magnetic field related to the particle input specified, we would input a code = 4 card (4F10.0) as follows:

$$4.0 \quad 3.0 \quad 1.0 \quad 1.0.$$

This card specifies that the field that follows is a two-dimensional field in polar coordinates (XFIELD = 3.0). BSCALE = 1.0 and PSCALE = 1.0 specifies that no scaling is to be imposed upon the field.

To continue this imaginary example we would probably desire to print the ion-optic matrix at a particular radius (90 inches), azimuth (150 degrees), and trajectory length (11.5 inches). These options would be accomplished by the following 3 cards (format 3F10.4):

$$5.0 \quad 1.0 \quad 90.0$$
$$5.0 \quad 2.0 \quad 150.0$$
$$5.0 \quad 3.0 \quad 11.5.$$

In a similar fashion the user may stack as many cards as necessary for the problem solution and then complete the deck with a function code 10.0 card that terminates the problem. A complete set of data required to produce specific results from an actual experiment will be described in detail in the following section.

208   V. Integration of Equations of Motion

d. An Illustrative Example

The performance of TRAJECTORY was tested with the calculation of the first-order fringe-field transfer matrix for an internally produced π beam in the 184-inch cyclotron at the Lawrence Berkeley Laboratory (see also program SOTRM). The field is assumed to fall radially and without azimuthal variation. Figure 86 shows a schematic arrangement of the contemplated problem. The beam is created by bombarding a beryllium target from 2- to 3-inches thick at a radius of 82 inches from the cyclotron, and at an angle of 250 degrees. The π^- beam generated will describe a radius of curvature of opposite sign to that of the proton beam, therefore it will cross the fringe field of the cyclotron.

Table 45

TRAJECTORY Sample Input Data

				Card Column				
1	10	20	30	40	50	60	70	80
4.		1.		-1.		1.		
MEASURED RADIAL FIELD, 184 INCH CYCLOTRON, LAWRENCE BERKELEY LABORATORY								
73.		272.		1.		1.		
	22395	22387	22379	22371	22362	22354	22346	22338
	22326	22298	22236	22116	21916	21618	21209	20685
	20043	19265	18347	17327	16256	15180	14136	13133
	12169	11250	10380	9560	8792	8070	7389	6745
	6135.00	5560.00	5021.55	4517.00	4061.00	3613.00	3191.00	2795.00
	2429.00	2093.00	1787.000	1510.00	1247.00	1001.00	775.00	576.00
	414.00	230.000	121.00	14.0	-5.	-97.0	-146.	-166.
	-169.	-168.	-160.	-150.	-140.	-130.	-120.	-110.
	-100.	-90.	-80.	-70.	-60.	-50.	-40.	-30.
	-20.	-15.	-12.	-10.	-8.	-6.	-5.	-4.
	-3.	-2.	-1.	0.	0.	0.	0.	0.
	0.	0.	0.	0.	0.	0.	0.	0.
	0.	0.	0.	0.	0.	0.	0.	0.
	0.	0.	0.	0.	0.	0.	0.	0.
	0.	0.	0.	0.	0.	0.	0.	0.
	0.	0.	0.	0.	0.	0.	0.	0.
	0.	0.	0.	0.	0.	0.	0.	0.
	0.	0.	0.	0.	0.	0.	0.	0.
	0.	0.	0.	0.	0.	0.	0.	0.
	0.	0.	0.	0.	0.	0.	0.	0.
	0.	0.	0.	0.	0.	0.	0.	0.
	0.	0.	0.	←Blank card				
3.	2.		.139579					
3.	3.		140.219					
3.	4.		140.219					
3.	6.		7.01095					
3.	7.		1.					
3.	10.		.2804380					
9.	12.		-202.5	137.174681	256.000	197.2472		
6.		1.						
GENERATE 1ST ORDER MATRIX ELEMENTS								
1.		82.		.250.	180.	-.09		
10.								

```
P(GEV/C)= -.1822751
ES(INCHES)=   .280
BR(KG-IN)= -2.3937E+C5
E(GEV)= -.0900000
MATRIX INITIALIZED TO A UNIT MATRIX

  J  S(INCH) X(INCH)  Y(INCH)  R(INCH)   AZ(DEG)    INDEX     B(GAUSS)   DPHI(DEG)
  1    0.    28.046  -77.055   82.0000   250.0000  -1.529E-01 -22298.0   180.0000
 26    7.511 34.931  -76.897   84.4592   245.5654  -7.5.4E-C1 -22035.3   138.2582
 51   14.22  40.565  -80.883   90.4850   243.3647  -4.4.7E+CC -18835.6   106.9987
 76   21.033 43.374  -87.243   97.4299   243.5654  -7.637E+CC -11768.0    75.6025
101   28.044 43.828  -94.219  103.9134   245.0536  -9.624E+CC  -6799.4    61.8493
126   35.055 42.861 -101.156  109.8617   247.0370  -1.322E+01  -3674.0    55.2477
151   42.066 41.138 -107.950  115.5229   249.1391  -1.944E+01  -1639.6    53.0314
176   49.077 39.078 -114.651  121.1281   251.1788  -5.779E+01   -389.8    55.4664
201   56.088 36.931 -121.326  126.8218   253.0700  -2.998E+01    141.5    55.2534
226   63.099 34.801 -128.005  132.6518   254.7904   9.217F+CC    143.5    57.2402
251   70.109 32.699 -134.694  138.6059   256.3545   1.651E+C1     83.9    58.9952
276   77.120 30.614 -141.387  144.6638   257.7827   6.C71E+C1     23.0    60.5136
301   84.131 28.534 -148.083  150.8C65   259.C935   2.463E+C1      5.2    61.8432
326   91.142 26.455 -154.778  157.0228   260.3006  -1.866E+02      0.     63.0540
351   98.153 24.376 -161.474  163.3C35   261.4154   0.              0.    64.1688
376  105.164 22.298 -168.170  169.6414   262.4472   0.              0.    65.2906
401  112.175 20.219 -174.865  176.0304   263.4044   0.              0.    66.1578
426  119.186 18.140 -181.561  182.4651   264.2943   0.              0.    67.0477
451  126.197 16.062 -188.257  188.9408   265.1234   0.              0.    67.8768
476  133.208 13.983 -194.953  195.4534   265.8974   0.              0.    68.6509
501  140.219 11.904 -201.648  201.9994   266.6214   0.              0.    69.3748


---------FCRWARD MATRIX---------          ---------REVERSE MATRIX---------
 -14.25347    .02188   0.      -2.60426      .11574   -.02188   0.       -.17729
-121.08877   -.11574   0.     -21.87644  121.08877  -14.25347   0.        3.53129
    0.        0.      -1.20170    .05394     0.        0.        .16905    -.05394
    0.        0.     -22.30461    .16905     0.        0.      22.30461  -1.20170
     RACIAL DETM=  1.00001    VERTICAL DETM=  1.00000

 J  S(INCH) X(INCH) Y(INCH)  R(INCH)  AZ(DEG)   INDEX    B(GAUSS) DPHI(DEG)
MESON WHEEL LOCATION INTERCEPTS ( 12.000,-201.341),(-202.500, 11.640)
DISTANCE FROM WHEEL CENTER DX,DY=   -.360    1.159

WHEEL ANGLE =   17.247 DEGREES                 LINE INTERSECTIONS =   -.00080     .00001
```

FIG. 88. Partial printout showing the orbit trajectory plotted in Fig. 86 as well as the forward and reverse matrices. Note the similarity of the values with those produced by program SOTRM (see Fig. 85).

210   V. Integration of Equations of Motion

The first-order transfer matrix at the meson wheel center is calculated by this program and verified by the program TRANSPORT. The paraxial ray-trace should pass through the center of the wheel.

The data cards required, their proper sequence, and their associated functions, are shown in full detail in Table 45. The card following the title card describes the minimum, maximum radii at which the field extends (RMIN, RMAX), and the radius interval (DR). The next 25 cards give the radial median plane magnetic field in order of increasing radius followed by a blank card. The next six code 3.0 cards initialize the starting parameter values while the code 9.0 card describes the meson wheel. The code 1.0 card, following the comment card, describes the particle input. The code 10.0 card terminates the job.

The output from the problem just described consists of the coordinates of the calculated orbit, the transfer matrices at the location specified, and in this specific case, information pertaining to the meson wheel coordinates and angle (Fig. 88). In this figure, the forward and reverse matrices have been verified by running the same problem with program SOTRM, previously described (see Fig. 85 for comparison). The coordinates of the paraxial ray pass very close to the specified center of the meson wheel, that is, by $DX = -0.360$ inches and $DY = 1.159$ inches. This trajectory has been plotted in the schematic representation of the problem (see Fig. 66). The coordinates of the paraxial ray solved with program SOTRM produced similar displacements, $DX = -0.515$ inches and $DY = 0.984$ inches.

TRAJECTORY has proved to be a very valuable program for the calculation of particle trajectories in different magnetic field geometries. The ability to calculate orbits accurately, with relative ease of data preparation and speed of execution, makes it a desirable program for this type of calculation.

32. Program GOC3D

a. Instruction

This is a general and versatile program for calculating particle orbits based on the procedures of Gordon and Welton [70]. The code has been subsequently revised and modified by Hopp [71], Shaw [72] and Paul [73]. The version described herein is based on Paul's revision and is perhaps the most general and adequately documented version. Some of the outstanding characteristics of this code are:

1. Flexibility in selecting magnetic field input geometries, e.g., the main magnetic field array may be: radial (one-dimensional), median plane

polar (two-dimensional R, θ), non-median plane (two-dimensional r, z), or three-dimensional with input of B_r, B_z, B_θ.
2. Capability of ray tracing and tracking of phase-space.
3. Determination of equilibrium orbit properties.
4. Flexibility in selecting various types of output (CalComp and printer).

GOC3D is written in Fortran and is operational on the CDC 6600/7600 computer system. There are two versions available: one with 3000 points in the main perturbation field arrays requiring $60K_8$ memory locations, and one with 30000 points in the perturbation field map requiring $157K_8$ memory locations. A typical problem will execute in less than one minute.

b. Program Description

GOC3D is a general purpose orbit code with applications in many diverse areas of accelerator design. Specifically, it has been used in the improvement studies of the 184-inch cyclotron at Lawrence Berkeley Laboratory, in the study of resonances for the electron ring accelerator project at LBL, and at the Lawrence Livermore Laboratory high current storage ring. The flexibility of GOC3D is attributed partly to the variety of models of magnetic fields that it can accept (two- and three-dimensional), the variety of tasks it can perform, and the flexible output it produces.

Basically the program numerically integrates the equations of motion to determine the trajectory of a particle, given the particle's energy, its position and direction (E, r, z, p_r, p_z). The equations to be integrated are derived from the Lorentz equation

$$\dot{\mathbf{p}} = e/c(\mathbf{V} \times \mathbf{B}) \tag{32.1}$$

and the initial conditions, $r, p_r, p_z, z, \theta, F$. The magnetic fields are static, although betatron acceleration can be simulated by using the appropriate scaling options. The total energy is conserved, therefore the total momentum is

$$p^2 = 1/c^2(E^2 - E_0^2), \tag{32.2}$$

$$p_\theta = (p^2 - p_r^2 - p_z^2)^{1/2}, \tag{32.3}$$

$$\dot{p}_z = e/c(V_r B_\theta - V_\theta B_r), \tag{32.4}$$

$$\dot{p}_r = e/c(V_\theta B_z - V_z B_\theta) + mr\dot{\theta}^2, \tag{32.5}$$

$$\dot{p}_\theta = e/c(V_z B_r - V_r B_z) - 2m\dot{r}\dot{\theta}. \tag{32.6}$$

The "dot" above p and θ denotes differentiation with time. Changing to azimuth as the independent variable and eliminating time by $d/dt = (p_\theta/mr)d/d\theta$,

$$p_z' = e/c(rp_r B_\theta p_\theta^{-1} - rB_r), \tag{32.7}$$

$$p_r' = e/c(rB_z - rp_z B_\theta p_\theta^{-1}) + p_\theta, \tag{32.8}$$

$$t' = E/c^2 r p_\theta^{-1}, \tag{32.9}$$

$$r' = p_r p_\theta^{-1} r, \tag{32.10}$$

$$z' = p_z p_\theta^{-1} r, \tag{32.11}$$

where the prime (') denotes differentiation with respect to azimuth, $d/d\theta$, and the following parameters are defined as

$p_z(\theta)$ = axial component of the momentum

$p_r(\theta)$ = radial component of the momentum

$r(\theta)$ = radius for the orbit to be calculated

$z(\theta)$ = axial displacement of the particle

t = the time

E = total relativistic energy.

The units for the variables in the program are given in cyclotron units as follows:

1. Magnetic field—units of B_0, which can be chosen as the central field value of the cyclotron
2. Length—units of $m_0 c^2/eB_0$
3. Velocity—units of velocity of light, c, so that velocity and beta are identical
4. Momentum—units of $p_0 = m_0 c$
5. Time—units of $1/\omega_0$, where ω_0 is the cyclotron frequency eB_0/p_0
6. Energy—units of $m_0 c^2$, the rest energy of the particle

Once the magnetic field is specified, Eqs. (32.7) through (32.11) are integrated simultaneously to obtain the desired orbit information. The magnetic field at any point along the path is numerically determined by a Langrangian interpolating polynomial within a uniform mesh. Table 46 lists the main subroutines and functions of GOC3D.

32. Program GOC3D

Table 46

GOC3D Subroutines and Functions

Name	Description
GOC	Main routine, reads input data and acts as a control program to initiate calls to the rest of the subroutines
BREAD	Calls the appropriate magnetic input routine
BTIME	Calculates the time-dependent magnetic field scale factor and the betatron energy gain simulating a time-varying field
B3DIN	Reads in the 3-dimensional magnetic field array
BRZIN	Reads in the 2-dimensional R, Z field array
BTRIN	Reads in the 2-dimensional R, Θ field array
FIELD	Evaluates the magnetic field at orbit point for each field-type present
FOURIER	Synthesizes field (from a sum of sine and cosine functions) with a given symmetric radial profile
INTERP	Performs 2-dimensional Lagrangian interpolation
LAGRAN	Performs 1-dimensional Lagrangian interpolation for irregularly spaced points
MATRIX	Calculates betatron frequencies
MPLANE	Calculates median-plane field expansion
ORBIT	Main integration routine using Runge-Kutta method
ORBEQ	Calculates equilibrium orbit for a given energy
PSASEP	Calls ORBIT for a grid of r and p_r values in order to generate a phase-plot about a given R and PR or about a calculated equilibrium orbit
PLOT, PPLT, PLOTQU	Plotting routines
SUMMARY	Prints particle summary table

c. Preparation of Input Data

The magnetic field is supplied to the program either on magnetic tape or cards. Data points are given in terms of the magnetic field values in an array of points on a uniform grid. The type of grid (two- or three-dimensional) to be used is controlled by the user through input cards to be described later. The rest of the input consists of cards punched in the format described below. The flow of information is controlled by code digits punched on cards which

V. Integration of Equations of Motion

specify the appropriate function to be performed. Interpretation of these code digits along with the parameters occupying the remaining fields of the data cards follows:

1. Code digit 1.0 (particle input)—allows for the input of particles. The quantities associated with this code are

      1.0   N   E   THETA   LAB   IRMAXT   JTHETO   ORB   (8F10.0)

where N = number of particles (maximum 200), E = kinetic energy of the particles in MeV, THETA = revolution azimuth, LAB = label, IRMAXT = maximum number of revolutions, JTHETO = starting azimuth in degrees, ORB = output frequency. A code 1.0 card is always followed by N cards specifying the initial conditions of the input particles; the quantities associated with these cards are

      R   PR   Z   PZ   E   THETA   IR   LAB   (4E15.8,F10.3,F3.0,F2.0,F5.0)

where R and Z are the coordinates of the orbit to be calculated in CU (cyclotron units), PR and PZ are the radial and axial components of momentum in CU, THETA is the starting angle, IR is the revolution counter and LAB = label if different from those on the 1. cards.

2. Code digit 3.0 (matrix calculation)—allows the perturbation of the initial orbit to produce a matrix at TMA1 $\neq$ 0, and allows its use in connection with code digit 14.0. The quantities associated with this code are

      3.0   DR   DRP   DZ   DPZ   TMA1   TMA2   TMA3   (8F10.0)

where DR, DRP, DZ, and DPZ are the radial or vertical inputs for phase plot, TMA1, TMA2, and TMA3 are the angles where matrix is to be calculated. If DZ, DPZ, TMA1, TMA2, and TMA3 all equal zero, see code digit 14.0.

3. Code digit 4.0 (magnetic field input)—allows for the specification of the type of magnetic field input, the parameters necessary are

      4.0   J   REST   BUNIT   DTHETA   KFLDTYP   KTAPE   (8F10.0)

where J = code digit (see below), REST = rest energy (or code digit for J = 7), BUNIT = scaling factor, DTHETA = integration step, KFLDTYP = field type code, and KTAPE = magnetic tape number. The magnetic field is the appropriate combination of fields from three field arrays as follows:

    a. J = 1, radial array of 120 points maximum, field is entered in ascending order of radius.

b. J = 2, $B_z(r, z)$ and $B_r(r, z)$ are maximum of 1500 points each; field is recorded with B_z in ascending order of r then z, followed by B_r in ascending order of r then z.

c. J = 3, $B_z(r, \theta)$, maximum of 3000 points on the median plane; field points are entered in ascending r, then θ.

d. J = 4, three-dimensional field $B_z(r, \theta, z)$, $B_r(r, \theta, z)$, and $B_\theta(r, \theta, z)$; maximum of 11,000 points each, field points are entered in asending order of r, θ for each component at each z.

e. J = 5, perturbation field, same as J = 4, the allowed combinations of field types are:

$$1, \quad 2, \quad 3, \quad 4$$
$$1+3, \quad 1+5, \quad 1+3+5$$
$$2+5, \quad 3+5$$

f. J = 7, allows for the scaling of the field as follows:

If REST = 2.0, field to be scaled is of type J = 2.

If REST = 3.0, field to be scaled is of type J = 3.

If KTAPE = 4 or 8, magnetic field is read from tape 4 or tape 8 in BCD (8F10.0).

If KTAPE = −4 or −8, magnetic field is read from tape 4 or tape 8 in binary mode.

If KFLDTYP = −1, the program expects profile definition.

If KFLDTYP = 0, the program reads an array of points.

If KFLDTYP = N, the program expects N Fourier coefficients to synthesize the field.

The following examples will further demonstrate the usage of code 4.0:

Example 1. Read radial field only,

1st card 4.0 1.0 RESTM BUNIT DTHETAO
2nd card   RMIN RMAX DR

Magnetic field cards [$B(R)$ radius (R) from R_{min} to R_{max} in steps of DR]

Example 2. Read $B_z(r, z)$, $B_r(r, z)$,

1st card 4.0 2.0 RESTM BUNIT DTHETAO KFLDTYP
2nd card   RMIN RMAX DR ZMIN ZMAX DZ

216   V. Integration of Equations of Motion

Magnetic field cards
$$\begin{bmatrix} BZ(RMIN \cdots RMAX,0) \\ BZ(RMIN \cdots RMAX,DZ) \\ \vdots \\ BZ(RMIN \cdots RMAX,ZMAX) \\ BR(RMIN \cdots RMAX,0) \\ BR(RMIN \cdots RMAX,DR) \\ \vdots \\ BR(RMIN \cdots RMAX,ZMAX) \end{bmatrix}$$

Example 3. Read median plane field $B_z(r, \theta)$,

1st card	4.0	3.0	RESTM	BUNIT	DTHETAO	KFLDTYP	
2nd card	RMIN	RMAX	DR	TMIN	TMAX	DTHETA	NSEC

Magnetic field cards
$$\begin{bmatrix} BZ(RMIN \cdots RMAX,TMIN) \\ \vdots \\ BZ(RMIN \cdots RMAX,TMAX) \end{bmatrix}$$

where NSEC = number of sectors the magnetic field represents in a revolution.

Example 4. Read three-dimensional field or perturbation field,

1st card or	4.0	4.0	RESTM	BUNIT	DTHETAO	KFLDTYP	
	4.0	5.0	RESTM	BUNIT	DTHETAO	KFLDTYP	
2nd card	RMIN	RMAX	DR	TMIN	TMAX	DTHETA	NSEC
3rd card	ZMIN	ZMAX	DZ	INTYPE	CALCTYPE		

Magnetic field cards
$$\begin{bmatrix} BZ(R,THETA,Z) \cdots \\ BR(R,THETA,Z) \cdots \\ B\theta(R,THETA,Z) \cdots \end{bmatrix}$$

If INTYPE = 0, enter only BZ(R,THETA,0), i.e., median plane only.

4. Code digit 5.0 (Time varying fields)—allows for entering data for time varying fields. The following parameters are:

5.0   JSW   TIME1   TIME2   DTIME

$$\begin{bmatrix} B(T) \cdots \\ E(T) \cdots \end{bmatrix}$$

where

JSW = 1, for synchronous oscillations in which case B(T) is the time, and E(T) is the frequency.

JSW = 2, the main field is simulating betatron acceleration by scaling the field B(T) and E(T) is the energy at time T; T = TIME1, TIME1+DTIME+···, TIME2.

5. Code digit 6.0 (Acceleration)—GOC3D provides for the adding of energy to the orbit at specified azimuthal positions in the magnetic field. The following parameters are required:

6.0 ERF TRF1 TRF2 IACCEL   (8F10.0)

where ERF = energy gain/passage by the accelerating electrode. TRF1, TRF2 = azimuths of accelerating electrode. IACCEL = type of acceleration with the following options: if = 1, hard edge; if = 2, phase dependent; if = 3, sinusoidal; if = 4, betatron approximation.

6. Code digit 9.0 (Parameter input)—used to input various parameters.

9.0 J X   (8F10.0)

J = 1, X = initial time of starting of particles.
J = 2, X = initial phase angle.
J = 3, X = initial reference energy.
J = 4, X = initial reference vertical amplitude.
J = 5, X = output only every X revolution.
J = 6, X = average R,N,DR,DZ and quadratic every X revolution.
J = 7, X = reference radius for radial quadratic.
J = 8, X = NEQDN.   Number of equilibrium orbits to be calculated with energy less than the starting energy.
J = 9, X = NEQUP.   Number of equilibrium orbits to be calculated with energy greater than the starting energy.
J = 10, X = DE.   Energy interval for J = 8, J = 9.
J = 11, X = EQOUT.   Output intermediate equilibrium orbits.
J = 12, X = THETAEQ. Starting azimuth for equilibrium calculation.
J = 13, X = IORBEQ. DTHETA for print in equilibrium orbit calculation.
J = 14, X = JTRYEQ. Maximum number of tries for equilibrium orbit before fail.
J = 15, X = IRMAXT. Maximum number of revolutions to be tracked.
J = 16, X = DIRECTION. If X = −1, track particles backwards; if = +1, tracks particle, forward.
J = 17, X = REQTX.
J = 18, X = PREQTX.
⎡ If the equilibrium orbits cannot be found, the user should try starting conditions on a grid given by REQTX and PREQTX (12 points). ⎤

218 V. Integration of Equations of Motion

7. Code digit 11.0 (Tracking)—parameters needed are:

   11.0   NREV   THOUT   ORB   K1   K2   DK

where

NREV = number of revolutions, if NREV < 0 track backward.
THOUT = starting point of revolution counter (NREV).
ORB = output frequency.
K1, K2, DK are parameters to track particle K1 to particle K2 in DK steps.

8. Code digit 12.0 (Calculation of equilibrium orbits)—allows for the calculation of equilibrium orbits at energy E and initial radius (RID) and at a guessed momentum (PRID).

   12.N   EEQ   RID   PRID   TSTART   TEND   ORB   JEQOUT

If $N \neq 0$, i.e., 12.1 or 12.2, etc., read the following card,

   NEQDN   NEQUP   DEEQ   JUNK   DPREQ   EPSEQ   JTRYEQ

where the parameters in the former card are: EEQ = energy for which equilibrium orbit is to be calculated; RID and PRID = starting points for equilibrium orbit; TSTART = beginning angle of orbit; TEND = ending angle of equilibrium orbit; ORB = output frequency; and JEQOUT = extra output for equilibrium orbit.

9. Code digit 13.0 (Input/output options)—parameters are:

   13.0   J   X

where J = 1, print radial field; J = 2, print B_z and B_r field arrays; J = 3, print midplane field array; J = 4, print three-dimensional field array; J = 5, print perturbations field array; J = 6, particle summary written on magnetic tape; J = 9, orbit output from subroutine TABLE on tape; J = 12, print orbits used to calculate matrix.

10. Code digit 14.0 (Phase plots)—parameters necessary to initiate this option are:

   14.0   R1   R2   DR   PR1   PR2   DPR   JRMAX·SLANT

which define a regular grid, where DR = radial interval; DPR = momentum interval; JRMAX = cycle at which phase space to be initiated. This code digit (14.0) allows the user to specify a two-dimensional mesh of R, PR, Z, PZ points to be run, if used with a 2.0   P   PR   0.0   0.0   0.0   0.0 card. Similarly, a mesh of Z, PZ points may be run, if used with a 2.0   Z   PZ   0.0   0.0   0.0

0.0 card. The R, PR points are considered displacements from the equilibrium orbit if the code digit 14.0 card is followed by a code 12.0 card; otherwise, R, PR are actual radial phase points in cyclotron units.

Example

        12.0 EEQ PRID TSTART TEND ORB VEQOUT

        14.0 R1 R2 DR PR1 PR2 DPR IRMAX

11. Code digit 15.0 (Unit conversion)—allows changing units; the parameters are:

        15.0 J DIM(J) UNIT(J)

where DIM(J) is the alphameric name of the unit, such as inch or centimeter and UNIT(J) is the conversion factor to convert the data input values into the standard values. The value and significance of parameter J is shown as follows:

J	Quality	Standard unit	Comment
0	—		
1	R, Z	Cyclotron units	Particle input
2	PR, PZ	Cyclotron units	Particle input
3	E	Mega electron Volts	Energy input
4	P	Cyclotron units	Momentum output unit
5	T	Cyclotron units	
6	R, Z	Cyclotron units	Phase space output
7	PR, PZ	Cyclotron units	Phase space output
8	RIN, ZIN	centimeters	
9	—	centimeters	Magnetic field distance
10	B	gauss	Magnetic field unit

12. Code digit 17.0 (Title information)—this code allows the reading of N title cards. The parameters needed are:

        17.0 N

13. Code digit 18.0 (Output magnetic field)—this code causes the magnetic field to be calculated and printed on the specified three-dimensional grid. Parameters needed for this option are:

1st card   18.0 R1 R2 R3

2nd card   THETA1 THETA2 DTHETA

3rd card   ZMIN ZMAX DZ

where R, THETA, and Z are the coordinates of the three-dimensional grid.

220   V. Integration of Equations of Motion

d. Illustrative Example

In this example, GOC3D is used to investigate betatron amplitude growth upon traversing resonances during the compression cycle of the electron ring accelerator at Lawrence Berkeley Laboratory. The study of such resonances is useful in predicting betatron amplitude growth and the type of magnetic field perturbation causing large growth for a specific compressor geometry and coil-energizing sequence. To perform such an experiment with GOC3D, one must simulate closely the prevailing magnetic field, follow particles through it, and observe the betatron amplitude.

Table 47 shows a listing of the data necessary to accomplish the tasks stated above. Interpretation of the data follows the instructions given earlier in the preparation of input data, however, the remarks that follow will clear some of the questions that may arise during input preparation.

1. The first card (17.0) allows the reading of the next three title cards.
2. The five following unit conversion cards (15.0) allow the use of units other than the standard program units.
3. The next four (9.0) code cards allow for the input of the parameters shown.
4. Next, the three (13.0) code cards allow the choice of the input/output options.
5. The (4.0) card specifies the type of magnetic tape input. In this example, the magnetic field is specified on an azimuthally asymmetric field ($J = 4$), for approximately the time at which a particular resonance is crossed. The remaining parameters in this card have been described previously.
6. The next two cards are associated with (4.0) code card and describe the region coordinates of the median plane magnetic field, i.e., RMIN = 8.0, RMAX = 20.0, DR = 0.5, TMIN = -10.0, TMAX = 370.0, DTHETA = 5.0, ZMIN = ZMAX = 0.0, and DZ = 1.0. In the second of the two cards, the last two zeros indicate that only median plane field is specified (INTYPE = 0) and that the calculation is not three-dimensional (CALCTYPE = 0.0).
7. Next, follows the $B_z(r, \theta)$ field perturbations in ascending R from RMIN to RMAX in steps of DR.
8. Next, follows the input of the azimuthally symmetric non-midplane fields followed by the B_z and B_r field values on and off the median plane.
9. Once these fields have been read the program transfers control to code digit 9.0 which directs the code to read the initial time of starting particles as 100 microseconds. This card is followed by a code digit of 5.0 which instructs the program to accept data for time varying fields. In this 5.0 card, the JSW = 2.0 informs the program that the main field is varied by scaling

32. Program GOC3D 221

Table 47

GOC3D Sample Input Data

```
                                    Card Column
  1         10        20        30        40        50        60        70        80

 17.0        3.0
TEST RUN FOR BETATRON AMPLITUDE GROWTH.
E = 4.1 MEV, N = 0.5, T = 100 SEC
C 3.5 NSEC/REV  500 REV = 1.75 USEC
 15.0        3.0
 15.0        1.0       CM        .035200391389
 15.0        5.0       USEC      .0059540878785
 15.0       11.0       USEC     1000000.0
 15.0        0.0
  9.0        1.0       100.0
  9.0        5.0       1.0
  9.0        6.0       10.0
  9.0        7.0       14.5
 13.0       16.0       1.0
 13.0       17.0       1.0
 13.0       11.0
  4.0        4.0       0.511     60.0      1.0       0.0
  8.0       20.0       0.5      -10.0    370.0       5.0
  0.0        0.0       1.0        0.0      0.0
           -3.8629   -3.4119   -2.1921    -.4788    1.2530    2.6360    3.6119    4.3477
            5.0292    5.7580    6.5726    7.4932    8.5353    9.7026   10.9791   12.3313
            :         (Total 1925 B_z median plane field perturbations)        :
            .                                                                  .
            5.3680    6.3082    7.2697    8.2933    9.3938   10.5610   11.7642   12.9592
           14.0963   15.1274   16.0102   16.7097   17.2008   17.4712   17.5305   17.4236
           17.2462
  4.0        2.0       0.511     60.0      2.0
  0.0       20.0       0.5        0.0      5.0       0.5
         1264.1716 1247.2120 1230.9644 1215.6467 1201.2834 1187.6843 1174.4694 1161.1387
         1147.1716 1132.1305 1115.7437 1097.9497 1078.8979 1058.9100 1038.4153 1017.8782
          997.7317  978.3277  959.9100  942.6062  926.4363  911.3302  897.1495  883.7094
            :         (Total 275 B_z symmetric on/off median plane main fields) :
            .                                                                   .
         1355.6811 1395.2439 1387.3399 1332.9611 1237.5697 1124.1581 1025.9790  957.3667
          914.3211  889.0781  875.4222  869.0516  867.0804  867.5717  869.2115  871.0995
          872.6167        .         .         .         .         .         .         .
            0.        0.        0.        0.        0.        0.        0.        0.
            0.        0.        0.        0.        0.        0.        0.        0.
            :         (Total 275 B_r symmetric on/off median plane main fields) :
            .                                                                   .
           18.7722  -84.7624 -194.8461 -294.5254 -362.9905 -381.0962 -354.3469 -307.0143
         -257.0594 -212.1557 -174.6692 -144.8082 -121.9827 -105.3631  -94.1008  -87.4092
          -84.5915        .         .         .         .         .         .         .
  9.0        1.0       100.0
  5.0        2.0       100.0     115.0     5.0
  1.0        1.040     1.078     1.115
  1.0        1.024     1.047     1.069
 17.0        1.0
N=.50 2 CM RADIAL AMPLITUDE
  9.0       24.0       3.0
 18.0       10.0       20.0      1.0
  0.0       45.0       5.0
  0.0        0.0       1.0
  1.0        1.0       4.13
 16.2                  0.         .1        0.
 11.         2.        0.        1.        1.        1.        1.
 19.0        3.0
 11.0     2000.0       0.0       0.0       1.0       1.0       1.0
 19.0       -3.0
 20.0
```

B(T) and E(T), from TIME1 = 100.0 to TIME2 = 115.0 in steps of DTIM = 5.0. The next card gives the values of B(T) at the specified time intervals; followed by another card giving the values of E(T) at the specified time intervals. This variation is at the same rate as they change during compression.

10. The (17.0) card provides for entering titles or other identifying information.

In the same manner the remaining data appearing in Table 47 may be identified. Once these data have been submitted to the CDC 6600/7600 computer system, the partial printout appearing in Fig. 89 is produced. The quantities appearing in this figure are self explanatory, e.g., IR = revolution counter, R(CU) = radius in cyclotron units (CU), PR = radial momentum, etc.

Figure 90 shows the betatron amplitude growth in traversing the $2\nu_r - 2\nu_z = 0$ ($n = 0.50$) resonance. The particle is followed for only a few microseconds during which this resonance is traversed. The variation of the normalized gradient index n with the number of revolutions is shown in the upper portion of Fig. 90. The time for each revolution is about 3.5 microseconds. The initial and final radial and vertical betatron amplitudes are 1.5 cm and 0.1 cm, respectively.

Interpretation of the theoretical implications of this resonance crossing will be found in the work of Laslett and Perkins [74]. Here, it is sufficient to say that GOC3D has provided valuable information in simulating resonance crossings, and determining which of the resonances are of importance for a particular type of compressor.

```
DATA =    1.0000    1.0000    4.1300   -0.        -0.        -0.        -0.        -0.
             1   1.620000000000E+01   0.                    1.000000000000E-01  0.                -0.        0   -0   -0
DATA =   11.0000    2.0000    0.         1.0000    1.0000    1.0000    1.0000

K= 1     E=    4.1300  P= 9.0269711  GAMMA= 9.0821918  BETA=  .9939199                 KLABLE=  0-     1  KASENO-K=    1-  1
------------------------------------------------------------------------------------------------------------------------------

MAGNETIC RIDGITY= 1.53865373280101E+04  GAUSS-CM
 IR THETA   R(CU)        PR(CU)       Z(CU)        PZ(CU)        E(MEV)   P(CU)    DEG   R(CM)    Z(CM)     N     B(I,R)   R AVE     R AMP
  0   .57024634050180         0.                             3.52003913889999E-03          0.                               ZMAX      ZAMP
  0  .57024634 0.          3.52004E-03 0.                4.130000  9.02697    -0.    16.200    .100
  0  .57016778-3.56462E-02 3.51456E-03-2.48337E-03     4.130000  9.02697    4.00   16.198    .100     .6056    1003.42
  0  .56993171-7.14743E-02 3.49817E-03-4.94360E-03     4.130001  9.02697    8.00   16.191    .099     .6037    1003.78
  0  .56953747-1.07477E-01 3.47103E-03-7.36688E-03     4.130002  9.02697   12.00   16.180    .099     .5981    1004.47
  0  .56898486-1.43530E-01 3.43333E-03-9.75414E-03     4.130002  9.02698   16.00   16.164    .098     .5930    1005.31
  0  .56827442-1.79480E-01 3.38524E-03-1.21077E-02     4.130003  9.02698   20.00   16.144    .096     .5902    1006.20
  0  .56740751-2.15165E-01 3.32694E-03-1.44279E-02     4.130004  9.02698   24.00   16.119    .095     .5897    1007.11
  0  .56638634-2.50416E-01 3.25864E-03-1.67118E-02     4.130005  9.02698   28.00   16.090    .093     .5906    1008.05
  0  .56521399-2.85053E-01 3.18056E-03-1.89541E-02     4.130005  9.02698   32.00   16.057    .090     .5922    1008.97
  0  .56389445-3.18897E-01 3.09294E-03-2.11479E-02     4.130006  9.02698   36.00   16.020    .088     .5941    1009.88
  0  .56243248-3.51798E-01 2.99609E-03-2.32862E-02     4.130007  9.02698   40.00   15.978    .085     .5959    1010.79
  0  .56083350-3.83650E-01 2.89032E-03-2.53623E-02     4.130007  9.02699   44.00   15.933    .082     .5983    1011.75
                   .                                              .
                   .                                              .
                   .                                              .

  0  .46600434 3.82921E-01-3.26591E-03 1.87101E-02     4.130049  9.02707  296.00   13.239   -.093     .4092    1117.01
  0  .46743057 4.06325E-01-3.19515E-03 2.04532E-02     4.130049  9.02707  300.00   13.279   -.091     .4130    1116.62
  0  .46894352 4.28219E-01-3.11789E-03 2.21710E-02     4.130050  9.02707  304.00   13.322   -.089     .4170    1115.94
  0  .47053871 4.48709E-01-3.03418E-03 2.38579E-02     4.130051  9.02707  308.00   13.367   -.086     .4209    1114.88
  0  .47221201 4.67894E-01-2.94406E-03 2.55079E-02     4.130051  9.02707  312.00   13.415   -.084     .4246    1113.44
  0  .47395955 4.85833E-01-2.84765E-03 2.71155E-02     4.130052  9.02707  316.00   13.465   -.081     .4279    1111.68
  0  .47577759 5.02548E-01-2.74503E-03 2.86761E-02     4.130052  9.02707  320.00   13.516   -.078     .4310    1109.68
  0  .47766236 5.18036E-01-2.63633E-03 3.01859E-02     4.130053  9.02708  324.00   13.570   -.075     .4348    1107.54
  0  .47961004 5.32298E-01-2.52168E-03 3.16424E-02     4.130054  9.02708  328.00   13.625   -.072     .4384    1105.27
  0  .48161678 5.45344E-01-2.40123E-03 3.30428E-02     4.130054  9.02708  332.00   13.682   -.068     .4424    1102.37
  0  .48367871 5.57180E-01-2.27513E-03 3.43843E-02     4.130055  9.02708  336.00   13.741   -.065     .4469    1100.33
  0  .48579186 5.67778E-01-2.14356E-03 3.56636E-02     4.130056  9.02708  340.00   13.801   -.061     .4519    1097.71
  0  .48795202 5.77065E-01-2.00669E-03 3.68769E-02     4.130056  9.02708  344.00   13.862   -.057     .4572    1095.10
  0  .49015454 5.84933E-01-1.86473E-03 3.80207E-02     4.130057  9.02708  348.00   13.925   -.053     .4627    1092.57
  0  .49239424 5.91269E-01-1.71790E-03 3.90910E-02     4.130057  9.02708  352.00   13.988   -.049     .4685    1090.20
  0  .49466550 5.95995E-01-1.56645E-03 4.00833E-02     4.130058  9.02709  356.00   14.053   -.045     .4747    1087.94
```

FIG. 89. Partial output showing orbit information.

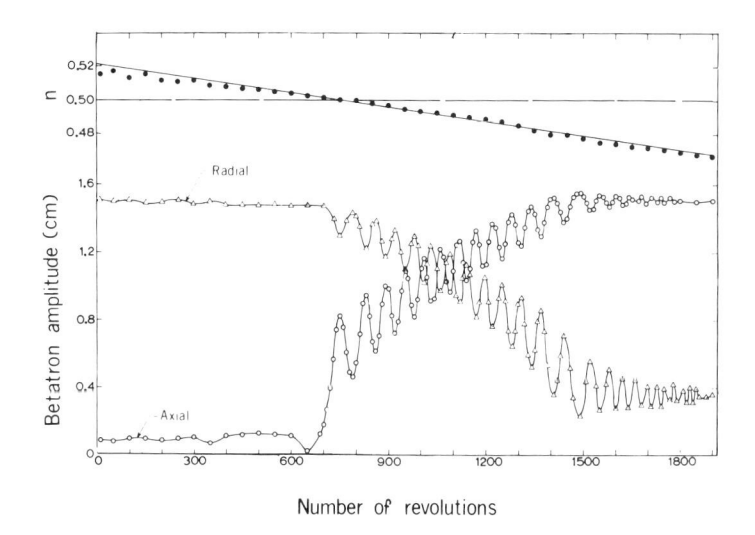

FIG. 90. Radial and axial betatron amplitudes versus time (number of revolutions) as the $\eta = 0.5$ resonance is traversed by a particle in the computer calculation. The initial radial and axial betatron amplitudes are 1.5 cm and 0.1 cm respectively. The upper graph shows how η is varying during this time (from Laslett and Perkins [74]).

33. Program MAGOP

a. Introduction

This program by Brady [75] calculates the orbit of a charged particle in the median plane of a symmetric two-dimensional magnetic field and determines the first-order horizontal and vertical transfer matrices associated with the orbit. MAGOP is written in Fortran for the CDC 6600/7600 computer system; it occupies $40K_0$ memory locations, and takes only a small fraction of a minute to compute a transfer matrix.

b. Program Description

MAGOP solves the vector equation of a charged particle moving in a static magnetic field defined by

$$d^2\mathbf{r}/ds^2 = q/p \, d\mathbf{r}/ds \times \mathbf{B}, \tag{33.1}$$

where $\mathbf{r}$ is the position vector of the particle, s is the arc length, q the particle charge, p the momentum, and $\mathbf{B}$ is the magnetic field. The equations of motion, of those particles near the central orbit in the horizontal and vertical direction,

224   V. Integration of Equations of Motion

are taken to be

$$d^2\eta/ds^2 = -[q/p\, dB_z/d\eta + (q/pB_z)^2]\eta + q/pB_z\, dp/p, \qquad (33.2)$$

and

$$d^2\sigma/ds^2 = q/p\, dB_z/d\eta\ \sigma, \qquad (33.3)$$

where η is the displacement of the deviating orbit in the median plane, measured positively in the direction of the vector $\mathbf{k} \times d\mathbf{v}/ds$; $p + dp$ is the momentum of the deviating orbit in the median plane, and σ is the displacement of the deviating orbit in the vertical direction measured from the median plane.

The program calculates one of two sets of transfer matrices, depending upon the sign chosen for the step size, ds. For a positive step size, s, the forward transfer matrices corresponding to transformations from the initial to the final point are calculated. For a negative step size, $-s$, the reverse transfer matrices corresponding to the transformations from the final to the initial point are calculated. The horizontal matrices operate on a vector (η, $d\eta/ds$, dp/p), and the vertical matrices on the vector (σ, $d\sigma/ds$). The B_z component of the magnetic field on the median plane, at a point (x, y), is calculated by making at least-square fit of a second-degree polynomial to the field values at a set of nine points lying on a rectangular grid and chosen so that (x, y) lies near the center of the grid.

c. Preparation of Input Data

MAGOP input is controlled by subroutines TABLE and MARAP. TABLE reads the median-plane magnetic field from TAPE21 in format 10F7.0. These values must lie in a rectangular grid and must be recorded in a right-hand coordinate system. Once the magnetic field is read, the program expects the following quantities (format 6F10.5) in the following order:

XMAX, XMIN—maximum and minimum x-coordinates of the field grid (inches)

YMAX, YMIN—maximum and minimum y-coordinates of the field grid (inches)

DX, DY—grid spacing in the x and y direction (inches).

The magnetic field values are stored in the two-dimensional array BTBL(I, J) where I = 1, 2, ... (XMAX-XMIN)/DX+1 and J = 1, 2, ... (YMAX-YMIN)/DY+1. Although this version of MAGOP reads the field values from a magnetic tape containing Bevatron fringe-field measurements, the user may substitute his own field.

Subroutine MARAP reads the parameter values needed to compute any orbit and associated transfer matrices. It may read different card types, numbered 0, 1, 2, 3, and 4, that are punched, one per card, in column 60.

Type 1, columns 1–10, initial x-coordinate of particle (inches)
               11–20, initial y-coordinate of particle (inches)
               21–30, initial particle direction, θ (degrees), θ is measured clockwise from the positive y-axis
               31–40, particle momentum, p (GeV/c)
               41–50, particle charge, q (units of elementary charge)
               51–59, blank
                  60, integer 1 (type)

Type 2, columns 1–10, arc-length step size, DS, of central orbit
               11–59, blank
                  60, integer 2 (type)

Type 3, columns 1–10, maximum number of integration steps of central orbit
               11–59, blank
                  60, integer 3 (type)

Type 4, columns 1–10, upper limit of integration in x direction (inches)
               11–20, lower limit of integration in x direction (inches)
               21–30, upper limit of integration in y direction (inches)
               31–40, lower limit of integration in y direction (inches)
               41–50, blank
                  60, integer 4 (type).

All quantities entered on cards of type 1, 2, 3, and 4, are floating-point quantities unless specified otherwise. Calculation begins when the program encounters a type 1 card, while types 2, 3, or 4, allow subsequent cards to be read. When the calculation for a given set of initial values has ended, the program calls subroutine MARAP to read another card, and this process continues until a type 0 card is encountered.

d. An Illustrative Example

The example given will demonstrate the data preparation procedures and describe the output produced by MAGOP. The data appearing in Table 48 corresponds to the data necessary to obtain a family of trajectories in the fringe field of the Bevatron at the Lawrence Berkeley Laboratory. The magnetic field is already stored on magnetic tape (TAPE21), therefore, the only data needed are shown in Table 48 which shows the parameters necessary to define the rectangle encompassing the fringe field and to establish the grid of points

226   V. Integration of Equations of Motion

Table 48

Magop Sample Input Data

			Card Column			
1	11	21	31	41	51	61
36.0	-160.0	100.0	-38.0	2.0	2.0	
2.0						2
200.0						3
-89.67	7.33	68.43	1.8	-1.0		1
-89.67	7.33	68.43	3.0	-1.0		1
-89.67	7.33	68.43	4.2	-1.0		1

```
MAGOP INITIAL VALUES.
              X =     -89.67000 INCHES
              Y =       7.33000 INCHES
          THETA =      68.43000 DEGREES
       MOMENTUM =       1.80000 GEV/C
         CHARGE =      -1.00000 ELEMENTARY CHARGES
      STEP SIZE =       2.00000 INCHES
   MAX NO. STEPS =          200
     UPPER X LIM =     36.00000 INCHES
     LOWER X LIM =   -160.00000 INCHES
     UPPER Y LIM =    100.00000 INCHES
     LOWER Y LIM =    -38.00000 INCHES

    X         Y       FIELD    ARC LENGTH    THETA      HORIZONTAL TRANSFER MATRIX           VERTICAL TRANSFER MATRIX
 (INCHES) (INCHES)  (GAUSS)    (INCHES)    (DEGREES)    (INCHES, RADIANS, FRACTION)            (INCHES, RADIANS)

  -89.67    7.33    15157.4      0.         68.43     10.0000E-01  0.           0.            10.0000E-01  0.
                                                       0.         10.0000E-01  0.             0.          10.0000E-01

  -87.81    8.08    15099.9      2.00       67.70     99.9884E-02 19.9992E-01-12.8091E-03     10.0003E-01 20.0002E-01
                                                     -11.5900E-05 99.9884E-02-12.8007E-03     33.9689E-06 10.0003E-01

  -85.97    8.85    15025.4      4.00       66.97     99.9537E-02 39.9939E-01-51.1633E-03     10.0014E-01 40.0018E-01
                                                     -23.0573E-05 99.9540E-02-25.5422E-03     67.4593E-06 10.0013E-01

  -84.13    9.64    14948.4      6.00       66.24     99.8959E-02 59.9792E-01-11.4936E-02     10.0031E-01 60.0062E-01
                                                     -34.9933E-05 99.8941E-02-38.2183E-03     10.6451E-05 10.0033E-01

  -82.31   10.46    14871.9      8.00       65.52     99.8131E-02 79.9497E-01-20.3989E-02     10.0057E-01 80.0158E-01
                                                     -49.1385E-05 99.7974E-02-50.8160E-03     16.8459E-05 10.0074E-01

  -80.49   11.30    14608.7     10.00       64.80     99.6976E-02 99.8956E-01-31.8144E-02     10.0100E-01 10.0037E+00
                                                     -59.0894E-05 99.7003E-02-63.3382E-03     18.9533E-05 10.0101E-01
                                                 ⋮                                                 ⋮

   27.40   93.52    -395.0     146.00       50.94    -66.8174E-03 98.7555E+00-30.2359E+00     21.4925E-01 19.7017E+01
                                                     -10.4288E-03 44.7767E-02-20.9350E-02     12.7379E-03 16.3294E-01

   28.95   94.78    -378.4     148.00       50.96    -87.6749E-03 99.6516E+00-30.6545E+00     21.7472E-01 20.0282E+01
                                                     -10.4289E-03 44.7866E-02-20.9048E-02     12.7356E-03 16.3273E-01

   30.51   96.04    -365.1     150.00       50.98    -10.8533E-02 10.0547E+01-31.0723E+00     22.0018E-01 20.3547E+01
                                                     -10.4288E-03 44.7724E-02-20.8691E-02     12.7366E-03 16.33CCE-01

   32.06   97.30    -350.9     152.00       51.00    -12.9390E-02 10.1442E+01-31.4892E+00     22.2567E-01 20.6814E+01
                                                     -10.4288E-03 44.7791E-02-20.8408E-02     12.7370E-03 16.3286E-01

   33.61   98.56    -337.1     154.00       51.02    -15.0248E-02 10.2339E+01-31.9059E+00     22.5114E-01 21.0079E+01
                                                     -10.4288E-03 44.7749E-02-20.8105E-02     12.7378E-03 16.3253E-01
```

FIG. 91. Partial output showing orbit information and transfer matrices calculated by program MAGOP.

(the first card in Table 48). The second and third cards describe the central-orbit arc-length step size and the maximum number of integration steps, respectively. The next three cards specify the particles to be traced at the three momenta specified (1.8, 3.0, and 4.2 GeV/c). The last card (zero in column 60) signals termination of execution.

The output obtained from this simple run, shown in Fig. 91, is straightforward. The first part lists some of the input quantities while the rest of the output shows trajectory coordinates, the magnetic field encountered, as well as the horizontal and vertical matrices at all positions along the trajectory path. A plot of the three trajectories is shown in Fig. 92.

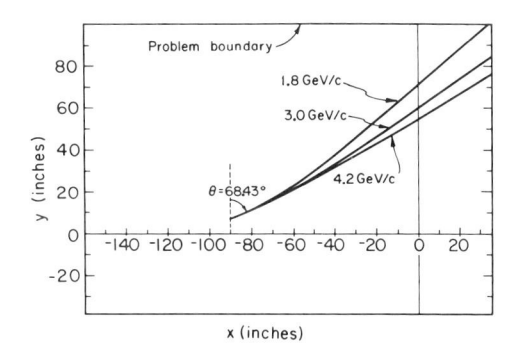

FIG. 92. Family of trajectories in the fringe field of the Bevatron at Lawrence Berkeley Laboratory produced by program MAGOP.

34. Program MAFCOIII

a. Introduction

This program by Perkins and Sackett [76] utilizes MAFCO, previously described (see Brown and Perkins [44]), in order to accommodate for time-varying fields and to calculate particle trajectories in a field generated by MAFCO. The code is written in Fortran and is operational on the CDC 6600/7600 computer systems. MAFCOIII occupies $70K_8$ words of memory and a typical run executes in less than one minute. A special interactive version of this code exists, making use of modern on-line communication media.

b. Program Description

Program MAFCOIII is a combination of MAFCO and ZAM [77]. MAFCO performs the calculation of the magnetic and/or electric fields resulting from the coil configuration specified; while ZAM performs the step-by-step solution of the system of first-order differential equations by fourth-order Adams-Moulton predictor-corrector method, to obtain the particle trajectories desired. The program is extremely flexible, as evidenced by the generalized coil geometry that MAFCO accepts (see 16. Program MAFCO).

c. Preparation of Input

Input preparation follows the same pattern as that outlined for program MAFCO; however, since time-varying currents are allowed, the data structure has been modified. In the description that follows, the user must refer to program MAFCO for symbols not explained herein.

The first card in the data deck contains an integer indicating the number of problems to be solved (right-adjusted, in columns 1–5).

The following group of cards is repeated for each problem: the first card contains the alphanumeric problem description in columns 1–80. The second card in format (10I5) contains in right-adjusted columns:

- 1–5 number of loops
- 6–10 number of arcs
- 11–15 number of straight lines
- 16–20 number of general current-element groups
- 21–25 number of point charges
- 26–30 number of constant electric fields in polar coordinates
- 31–35 number of electric fields in rectangular coordinate system
- 36–40 number of general time-varying currents and charges
- 45–50 geometry input tape (omit if input is via cards).

This second card calls any group or groups of cards as for loops, arcs, straight lines, etc. depending upon the problem geometry to be solved.

1. *Circular loop cards* (format 7F10.3,2I5)—one card for each loop with the following parameters:

$$X, Y, Z, A, ALPHA, BETA, I, GTVC, G.$$

GTVC is the general time-varying current: if GTVC = 0, the current is d.c.; G is the current/charge flag: if G = 0, use current only; if G = 1, use charge only. If G = −1, current and charge are indicated and one card will always follow with the parameter VOC (velocity of beam/velocity of light) with the format E20.6.

2. *Circular-arc cards* (format 7F10.3,2I5 and 2F10.3)—two cards for each arc are required: the first card contains

$$X, Y, Z, A, ALPHA, BETA, I, GTVC, G,$$

and the second card contains,
$$\theta_1, \theta_2.$$

3. *Straight-line cards* (format 7F10.3,2I5)—one card for each straight line is required with the following parameters:
$$X_1, Y_1, Z_1, X_2, Y_2, Z_2, I_{12}, GTVC, G.$$

I_{12} = current from 1 to 2.

4. *General current-element cards* (format 4F10.3,3I5 and 3F10.3)—one group of cards for each element is required, the first card contains:
$$X_1, Y_1, Z_1, I, GTVC, G.$$

The second card, or rather the NPTS succeeding points, follow with the values of
$$X, Y, Z.$$

5. *Point-charge cards* (format 3F10.3,E20.6,I5)—one card per point charge with the following parameters:
$$X, Y, Z, Q, GTVC.$$

where X, Y, Z are the coordinates of the point charge and Q is the charge in statcoulombs (electrostatic units).

6. *Constant-polar electric-field cards* (format 6F10.3 and 3F10.3)—two cards are required for each field: the first card contains the following parameters,
$$R_{min}, R_{max}, \theta_{min}, \theta_{max}, Z_{min}, Z_{max},$$

while the second card contains,
$$E_R, F_\theta, F_Z$$

7. *Constant-rectangular electric-field cards* (format 6F10.1 and 3F10.1)—two cards are required for each field: the first card contains the following parameters,
$$X_{min}, X_{max}, Y_{min}, Y_{max}, Z_{min}, Z_{max},$$

while the second card contains,
$$E_x, E_y, E_z.$$

8. *General time-varying current cards* (format I5,2F10.3 and 8F10.3)—two cards are required: the first card contains the following parameters,
$$NPOINT, T, I,$$

where NPOINT is the number of points, T is the time at the first point, and I is the current at the first point. The second card or rather NPOINT succeeding cards contain the parameters,

$$\text{T, I.}$$

These two parameters are arranged so that four pairs are punched per card, and there must be a minimum of five points per table.

9. *Boundary card* (format 6F10.3)—one card containing the following parameters,

$$R_{min}, R_{max}, \theta_{min}, \theta_{max}, Z_{min}, Z_{max}.$$

10. *Particle trajectory cards* (format 6F10.1,I5 and 6F10.1)—two cards are required for each trajectory. The first card has the following parameters,

$$\text{R}, \theta, \text{Z}, t_{min}, dt, t_{max}, \text{NOREV,}$$

where R, θ, and Z are the coordinates of the particle at the start of the trajectory, t_{min} is the starting time, dt is the printout time step, and t_{max} is the ending time for calculations. The second card has the following parameters,

$$w, V_x, V_y, V_z, m, q,$$

where V_x, V_y, and V_z, are the particle's starting velocity in the x, y, and z directions and w is the particles kinetic energy. If $w \neq 0$, V_x, V_y, and V_z are the particles normalized velocities; when $w = 0$, then V_x, V_y, and V_z are the particle's velocity components. If mass $(m) = 0$, the program uses m_e. If charge $(q) = 0$, the program uses $-e$.

d. An Illustrative Example

Perhaps the best way to show the generality of MAFCOIII is through an example, the data for which appear in Table 49. Here, a compressor coil of an electron ring accelerator is simulated by 48 circular loops and ten 100-degree arcs. Next, the data describe the time-varying currents of which there are six groups: the first three groups consist of five points each, the fourth group consists of thirteen points, and the last group of five points. The time-varying currents for the inflector coil were obtained from experimental data and the format conforms with the instructions just previously given. The last five cards pertain to the problem boundary (the first card of the last five) and the four particle trajectories to be traced.

The results of these calculations are shown in Fig. 93; the partial output is self-explanatory with all the quantities properly identified. In Fig. 94, the

Table 49

MAFCOIII Sample Input Data

			Card Column				
5	10	15	22	33	45	55	64	75

COMPRESSOR 4		COILS 1 + 2 + INFLECTOR						
48	10			6	2			
0.	0.	-25.11	32.7	0.	0.	1000.	1
0.	0.	-25.11	33.6	0.	0.	1000.	1
0.	0.	-25.11	31.8	0.	0.	1000.	1
0.	0.	-25.11	30.9	0.	0.	1000.	1
0.	0.	-24.19	32.7	0.	0.	1000.	1
0.	0.	-24.19	33.6	0.	0.	1000.	1
0.	0.	-24.19	31.8	0.	0.	1000.	1
0.	0.	-24.19	30.9	0.	0.	1000.	1
0.	0.	-23.27	32.7	0.	0.	1000.	1
0.	0.	-23.27	33.6	0.	0.	1000.	1
0.	0.	-23.27	31.8	0.	0.	1000.	1
0.	0.	-23.27	30.9	0.	0.	1000.	1
0.	0.	25.11	32.7	0.	0.	1000.	1
0.	0.	25.11	33.6	0.	0.	1000.	1
0.	0.	25.11	31.8	0.	0.	1000.	1
⋮	⋮	(Total of 48 circular loop entries)		⋮			
0.	0.	-1.022	22.7	0.	0.	-250.	4
90.	270.						
0.	0.	1.022	22.7	0.	0.	-250.	4
90.	270.						
0.	0.	-3.066	22.6	0.	0.	-250.	4
90.	270.						
⋮	⋮	(Total of 10 circular arc entries)		⋮			
5	0.	0.						
73.	2.8		113.	4.	200.	8.	10000.	0.
5	0.	0.						
60.		0.	134.	.16	220.	4.	10000.	0.
5	0.	0.						
73.	2.8		113.	4.	200.	8.	10000.	0.
13	0.	0.						
104.970	0.	105.	.50	105.005	.46	105.010	.40
105.015	.29	105.020	.17	105.025	.08	105.030	-.02
105.035	-.10	105.040	-.14	105.045	-.18	1000.	-.20
5	0.	0.						
73.	2.8		113.	4.	200.	8.	10000.	0.
5	0.	0.						
60.	0.	134.	.16	220.	4.	10000.	0.
0.	22.7	0.	360.	-5.	5.		
21.1	0.	0.	104.95	.00005	104.965	2	
3.72	0.	1.	0.				
21.1	0.	0.	105.015	.00005	105.030	2	
3.72	0.	1.	0.				

R-THETA trajectory of the electron is plotted for four different strengths of the field that have the indicated strengths of the inflector coil: (a) zero current, (b) I_0 amperes, (c) $2I_0$ amperes, and (d) $3I_0$ amperes. The dashed rectangle indicates the location of the injector pipe (septum). The sine-wave shaped curve shows that with no inflector field the electron strikes the injector pipe on the third turn. With I_0 amperes in the inflector coil, the electron just barely

232   V. Integration of Equations of Motion

```
PROBLEM NUMBER     1  ****    COMPRESSOR 4  COILS 1 + 2 + INFLECTOR
CIRCULAR LOOPS
           X           Y           Z           A        ALPHA      BETA       I/Q         I/Q-TABLE
    1   0.000000    0.000000   -25.110000   32.700000   0.000000   0.000000  1000.000000      1
    2   0.000000    0.000000   -25.110000   33.600000   0.000000   0.000000  1000.000000      1
    3   0.000000    0.000000   -25.110000   31.800000   0.000000   0.000000  1000.000000      1
                               (Total 48 circular loops)
   46   0.000000    0.000000   -12.180000   17.420000   0.000000   0.000000  1000.000000      2
   47   0.000000    0.000000   -13.150000   17.420000   0.000000   0.000000  1000.000000      2
   48   0.000000    0.000000   -14.120000   17.420000   0.000000   0.000000  1000.000000      2

CIRCULAR ARCS
           X           Y           Z           A        ALPHA      BETA       I/Q         I/Q-TABLE    PHI1        PHI2
    1   0.000000    0.000000   -1.022000   22.700000   0.000000   0.000000  -250.000000      4       90.000000   270.000000
    2   0.000000    0.000000    1.022000   22.700000   0.000000   0.000000  -250.000000      4       90.000000   270.000000
                               (Total 10 circular arcs)
    9   0.000000    0.000000   -5.100000   14.500000   0.000000   0.000000   166.700000      4       90.000000   270.000000
   10   0.000000    0.000000    5.100000   14.500000   0.000000   0.000000   166.700000      4       90.000000   270.000000

GENERAL TIME VARYING CURRENT AND/OR CHARGE    1 ***    5 POINTS
              1          2          3          4          5          6          7          8          9         10
  0  T    0.      7.3000E+01  1.1300E+02  2.0000E+02  1.0000E+04
     I(T) 0.      2.8000E+00  4.0000E+00  8.0000E+00  0.

GENERAL TIME VARYING CURRENT AND/OR CHARGE    2 ***    5 POINTS
              1          2          3          4          5          6          7          8          9         10
  0  T    0.      6.0000E+01  1.3400E+02  2.2000E+02  1.0000E+04
     I(T) 0.      0.          1.6000E-01  4.0000E+00  0.

GENERAL TIME VARYING CURRENT AND/OR CHARGE    3 ***    5 POINTS
GENERAL TIME VARYING CURRENT AND/OR CHARGE    4 ***   13 POINTS
GENERAL TIME VARYING CURRENT AND/OR CHARGE    5 ***    5 POINTS
GENERAL TIME VARYING CURRENT AND/OR CHARGE    6 ***    5 POINTS
              1          2          3          4          5          6          7          8          9         10
  0  T    0.      6.0000E+01  1.3400E+02  2.2000E+02  1.0000E+04
     I(T) 0.      0.          1.6000E-01  4.0000E+00  0.

TRAJECTORY NUMBER     1

BOUNDARY CONDITIONS
RMIN=   0.000   RMAX=   22.700   THMIN=   0.000   THMAX=   360.000   ZMIN=   -5.000   ZMAX=   5.000
REST MASS= 9.10850E-28 GM.   CHARGE= -4.80280E-10 ESU   TIME STEP= 5.000000E-05 MUSEC.   TMAX= 1.04965E+02 MUSEC.

     GYRAD           BR            BTH           BZ            PO           GAMMA
  2.00135E+01   -7.58504E-12    0.         7.00017E+02   4.20002E+00   8.28006E+00

     T              R          THETA          Z             W            VX            VY             VZ              B             E
  1.04950E+02   2.11000E+01    0.            0.         3.72000E+00   0.         2.97598E+10    0.          7.00017E+02   7.40240E-01

     T              R          THETA          Z             W            VX            VY             VZ              B             E
  1.04950E+02   2.10972E+01  4.04072E+00  -4.16052E-16   3.72000E+00  -2.21067E+09  2.96776E+10  -1.49212E-05   7.00091E+02   7.40179E-01
  1.04950E+02   2.10886E+01  8.08248E+00  -1.43050E-15   3.72000E+00  -4.40959E+09  2.94313E+10  -2.57763E-05   7.00312E+02   7.39995E-01
  1.04950E+02   2.10744E+01  1.21263E+01  -2.93804E-15   3.72000E+00  -6.58504E+09  2.90222E+10  -3.50934E-05   7.00681E+02   7.39689E-01
  1.04950E+02   2.10545E+01  1.61732E+01  -5.25347E-15   3.72000E+00  -8.72539E+09  2.84520E+10  -5.13167E-05   7.01196E+02   7.39260E-01

  1.04959E+02   1.70253E+01  1.71715E+02  -3.03249E-13   3.72024E+00  -6.95898E+09  -2.89348E+10   2.24633E-04   7.95513E+02   6.38266E-01
  1.04959E+02   1.71629E+01  1.76681E+02  -2.91694E-13   3.72024E+00  -4.49690E+09  -2.94182E+10   2.25016E-04   7.92637E+02   6.42206E-01
  1.04959E+02   1.73037E+01  1.78394E+02  -2.80867E-13   3.72024E+00  -2.01207E+09  -2.96918E+10   2.19235E-04   7.89667E+02   6.46172E-01

  N(R) =  .5771    NU(R) =  .6503    N(Z) =  .4601    NU(Z) =  .6783    TCYC =  3.78273E-03
  BR = -1.26047E-12    BTH =  4.41229E-14    BZ =  7.89667E+02
```

FIG. 93. Partial output produced by MAFCOIII; the upper half of the printout lists the input data corresponding to the compressor coil. The lower part contains information relating to the first orbit.

missed the injection pipe, but with $2I_0$, the electron misses with a good margin (which is necessary for a beam of finite width). With $3I_0$ in the inflector coil, the beam strikes the injector pipe on the first turn. Results of this type have been proven useful in understanding the inflector-coil trapping resulting for various parameters of the system.

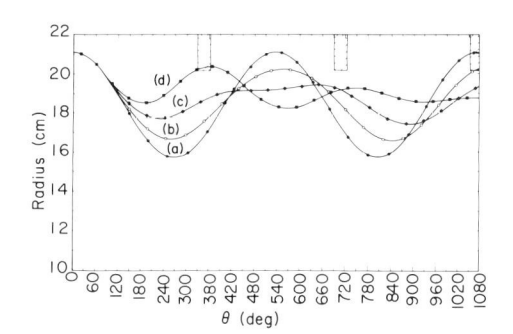

FIG. 94. R-THETA trajectories for various field strengths.

35. Program CYDE

a. Introduction

The computer programs described in this section under the collective name of CYDE have evolved from theoretical considerations devised by Arnette *et al.* [78] and Garren [79]. These codes have been used in connection with the development of the 88-inch cyclotron [80] at the Lawrence Berkeley Laboratory, California. Even though these programs have been adapted to this particular cyclotron, their use is not restricted to this type of machine.

Originally, CYDE was written for the IBM 704 computer and later was adapted to the IBM 7094 in a "chain form" to accomodate large memory requirements. The Fortran version described here has been documented by Close [81], revised and made operational on the CDC 6600/7600 computer systems. It requires approximately $66K_8$ words of memory and a typical computer run takes less than one minute on the CDC 6600.

The overlays of CYDE are now organized in a manner allowing maximum flexibility and input standardization. Restrictions have been removed thus permitting its use for various types of cyclotrons. Some of the capabilities of CYDE are summarized as follows:

1. Trim-coil currents may be predicted to give a least square fit to a desired magnetic field shape.

2. Trim-coil currents may be calculated which will minimize phase shift and gradient fluctuations.

3. Calculations are possible for magnetic fields produced by various combinations of trim-coil currents.

4. A calculated phase-history is possible for previously run and predicted beams.

5. Isochronous equilibrium orbit properties can be calculated.

The computer programs assembled to carry out these calculations will be described separately; their roles in the overall organization of the system will be identified, and some examples of their application will be given.

b. Program Organization

A simplified block diagram of CYDE appears in Fig. 95. Each box represents an overlay in the program, and the arrows indicate the present form of interconnections between overlays. Only a minimum of programming effort is required to alter the sequence of overlays or the addition of a new overlay. For example, an electrostatic-deflector computer code could be incorporated as a new overlay as indicated by the dotted lines.

CYDE may be used in two separate modes: the "fit mode" and the "run mode". In the fit mode, the program uses the least-square fitting procedure of the Michigan State University TRIMCO code, Berg [82] and Berg et al. [83] to calculate the optimum currents in the 17 trim coils that will produce an average magnetic field as close as possible to the required magnetic field at each radius of the acceleration region. The input includes: the required field, BREQDBI (expressed as a ratio to the isochronous field); the fitting weight for each radius and the maximum and minimum currents allowed in each trim coil. The output includes: the above input, the difference between the required and the resulting fitting fields, the phase history (adjusted for minimum phase slip over most of the acceleration region), and the calculated trim-coil and main-coil currents.

In the run mode, CYDE calculates the magnetic field and orbit properties from a given set of trim-coil and main coil currents included in the input. The output includes: the magnetic field produced by the trim and main coils; the ratio of this field to the isochronous field; and the phase history, adjusted as in the previous mode.

The output for either fit or run mode includes the equilibrium orbit properties for any of the isochronous magnetic fields, as selected by input parameter (IDR) (see i. Preparation of Input). It also includes the trim-coil effects at each radius. The number of input cards is the same for either mode.

In each overlay in the following paragraphs, particular emphasis will be made on the functions that are fulfilled, rather than on the mathematical considerations leading to the development of these routines. References cited at certain points will refer the reader to the appropriate literature describing the mathematical development.

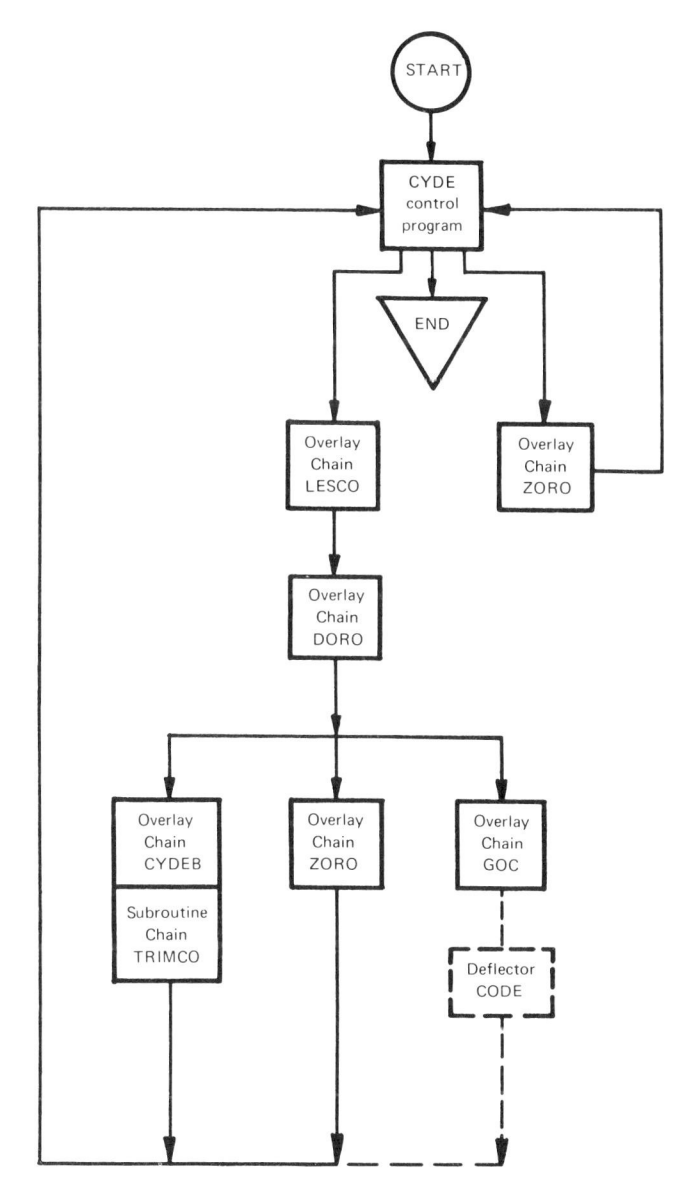

FIG. 95. Overlay structure of program CYDE.

c. Control Program

As the name implies, this control program establishes the proper operating conditions for the program as a whole. The input data is read in, checked, and printed; various sentinels and flags are set; and arrays initialized, generally, providing the controls needed to direct the "overlays" through the proper paths leading to a successful solution.

d. LESCO

This routine searches a magnetic tape which contains the main coil magnetic field (as a continuous function of current) in terms of the least-square coefficients applicable to the cyclotron under study [84]. Thus, given a current level, I, LESCO searches the magnetic tape and finds the appropriate coefficients corresponding to the current range containing I. The code then constructs the polynomial, interpolates for the base magnetic field $B(I, R, \theta)$, and prepares a new magnetic field tape to be used in later parts of the program (see DORO). This field is stored in a BCD (binary coded decimal) form with the format (20X,F10.1,20X,F10.1,F10.3) in the following order:

(degrees)	R(inches)	$B(R,\theta)$
0	0	$B(0,0)$
3	0	$B(0,3)$
6	0	$B(0,6)$
⋮	⋮	⋮
120	0	$B(0,120)$
0	1	$B(1,0)$
3	1	$B(1,3)$
⋮	⋮	⋮
120	67	$B(67,120)$

The magnetic field requirements described in this routine are those of the 88-inch cyclotron at the Lawrence Berkeley Laboratory (see Nelson [84]). However, measured field values for other accelerators may be incorporated, provided the user creates a magnetic tape in the fashion described above, in which case the LESCO overlay is bypassed completely.

e. DORO

This overlay evaluates certain orbit properties of cyclotron magnetic fields and prepares input required by other codes by implementing the following twelve subroutines:

1. READB—The radii and field values are read in from input tape IG. Second, the field values at the beginning and closing points of the sector are compared.

If these values are not identical they are matched by averaging and the difference is spread smoothly throughout the sector.

2. BODBO—If desired, a finer θ-mesh is interpolated and azimuthal derivatives are calculated.

3. ISOCHR—The synchronous radial profile is calculated using the options set by the user.

4. SYNC1B The azimuthal average of the field and the flutter squared at each radius is computed.

5. ISO2—An analytic approximation to the isochronous field is calculated from information obtained from SYNC1B.

6. ORO—The equilibrium orbit code (see Arnette *et al.* [78]).

7. TR—The time factors that were computed by ORO as a function of β, are converted here to be a function of the radius.

8. SYNC2B—β and β_s (isochronous field) are calculated using time factors obtained from ORO. The field error $\Delta B = B_{av} - B_s$ is obtained, and at each radius the field values are corrected for isochronism. Fields and derivatives are also normalized.

9. DFSQ—The derivatives of the flutter squared are calculated.

10. ABBCYD—Certain parameters are written on a scratch tape (Tape 11) for use in overlay CYDEB.

11. TAP2—Two binary tapes are written for use in a general orbit code [85, 86] and a deflector code such as CYBOUT [87, 87a, 88].

12. FOURCO—A Fourier analysis routine is provided for use with subroutine DPHI.

f. ZORO

This overlay is used to provide a calling program for subroutine ORO. ZORO allows the user to rotate the angle of output for ORO, thus furnishing orbit properties at different angles. ZORO also allows the user to adjust the radial step size or to start at various energy levels. The input required and the output provided by this overlay will be described in later sections.

g. CYDEB

CYDEB provides for trim-coil effects, dB/dI, calculates various derivatives, and prepares input for the subroutine TRIMCO. By use of the subroutine DOR88, CYDEB calculates a field determined by the trim-coil currents, which is used to build isochronous fields from currents generated by TRIMCO or to investigate empirically-determined fields.

h. TRIMCO

This subroutine of CYDEB provides for a fitting process based on specifying a desired average magnetic field vs. radius (see Berg and co-workers [82, 83]). Various field shapes are tried on a "model beam" and one is selected that gives the best beam intensity and quality. The field shape is varied in the 0- to 12-inch radius region (the "center bump"), and in the 35- to 40-inch region ("edge fall-off"). It is synchronous from 18- to 35-inches. The center bump adjusts the magnetic focusing and phase history, and the edge fall-off determines the resonance crossings (vertical blow-up) and phase-slip at extraction.

TRIMCO adjusts the trim-coil currents to produce approximate isochronism. This subroutine has been modified to reject coil currents that exceed the physical limitations by reducing the excessive current to the allowable maximum. This fitting process is repeated until all the remaining trim coils are within the allowable maximum.

This subroutine was written at Michigan State University and has been incorporated into CYDE by Clark [89] and Brady [90] of the Lawrence Berkeley Laboratory.

i. Preparation of Input Data

There are two magnetic input tapes required: one designated IG and the other, IN. Tape IG is read by the LESCO overlay chain and contains the polynomial coefficients used in constructing a main-coil magnetic field. (The format of this tape is described by Berg [82, p. 3].) The field in question pertains to the 88-inch cyclotron at Lawrence Berkeley Laboratory; however, other fields may be incorporated either by computing their polynomial coefficients from magnet measurements and creating a magnetic tape designated IG, or by bypassing LESCO completely, creating the magnetic field $B(R, \theta)$ on a tape, and using this tape directly.

Tape IN is read by CYDEB and contains values for dB/dI for the main coil and 17 trim coils at five different current levels. (The format of this tape is described by Close [81, p. 17].)

While previous versions of CYDE required data pertaining to each individual overlay of the program, the present version has combined all required input, and the proper calling sequence is determined by program flags set or reset by the user. Thus, the flexibility of the program to calculate various properties of a cyclotron is in the hands of the user, who, by proper input manipulation, creates the necessary conditions required to compute the desired quantities.

The description of the parameters needed is shown in some detail in Table 50. The user must be very careful to maintain the same order of the parameters shown and to punch the exact number of parameters in each card with the specified format when preparing his input data. The various combinations of input, producing alternate modes of operation, will become more obvious as the user gains more experience with running this program. The illustrative example in the following section, will also provide additional help in interpreting and using Table 50.

j. Illustrative Example

The application of CYDE to the solution of problems relating to cyclotron design can be demonstrated by comparing the calculated phase history of a 120-MeV alpha-particle beam with that measured in the 88-inch cyclotron at the Lawrence Berkeley Laboratory. This sample problem also provides a very sensitive test of the accuracy of the code, and simultaneously demonstrates the manner by which data are prepared for the successful operation of program CYDE.

Table 51 shows the card input sequence and the data necessary for this sample problem. The first card is the OPTION card signifying whether this is the first run (1) or a continuation run (2). The next three cards represent alphanumeric information identifying the run under consideration. Since the code expects these 3 cards, blank cards must be substituted if the problem title information has been described by fewer than 3 cards. The remaining input cards are specified in the same order as listed in Table 50; for example, the next card describes the desired energy (EF), the mass of the particle (ATWT), etc.

The output is controlled by various flags appearing in the input data. For example, the last entry in the fifth card, (IRF = −1), tells the code to print the run output; while the last entry of the sixth card (IPR = 1), directs the program to print the isochronous field table and trim coil effects. In a similar manner, the user may avail himself of the output options described in Table 50.

Figure 96 shows the equilibrium orbit table obtained from the isochronous field. The quantities shows in Fig. 97 are the phase-history computations.

1. R(I). Radius.
2. BISOC(I). The isochronous average magnetic field (gauss).
3. BIRON(I). The main coil average magnetic field (gauss).
4. DBRUN(I). The difference BRUN(I) − BIRON(I).
5. BRUN(I). The run mode, average magnetic field calculated from the trim-coil currents (RUNCUR).
6. BRUN(I)-BI(I). Here, BI(I) = BISOC(I).
7. BRUN(I)/BI(I). Ratios of "run" field to isochronous field.
8. SPHIR(I). The sine of phase angle.

Table 50

CYDE Input Parameter Specifications

Card sequence	Parameter name	Format	Description
1	MODE	I1	MODE = 1, the first run ≧ 2, a continuation run = 0, terminates the job
2,3,4	TITLE	10A8	Comment cards which pertain to the case being run (required for first run only)
5	MESSAGE	10A8	An additional comment card required in first and all continuation runs

6	EF,ATWT,ATNO,C,
	VD,HARM,IDR,IRF
		2(F5.0,F10.0),
		2F5.0,2I5
			EF - Desired energy in MeV units at radius RF
			ATWT - Mass of particle in units of proton mass
			ATNO - Particle charge in units of elementary charge
			C. - Main coil current in amperes
			VD - Dee voltage (kV)
			HARM - Harmonic number
			IDR = -1, equilibrium orbits (run) on field created from run currents
			       = 0, equilibrium orbits (run) on isochronous fields
			       = +1, equilibrium orbits (run) on field created from calculated currents
			IRF = -1, print run-mode output; if = 1, print fit-mode output
			       = 2, print run and fit modes output

7	RF,PF,BF,FREQ,IPR	4F10.5,I5	RF - Reference radius (inches) for determining the synchronous field PF - Desired momentum in units of $m_o c$ at reference radius RF BF - Desired magnetic field in gauss at radius RF FREQ - Desired frequency in megahertz IPR = 0, no print; if ≠ 0, print isochronous field table (trim-coil effects)
8	IOP1,IOP2,IOP4, IOP5,IOP6,IOP7, IOP9,IOP12,IOP13, IOP14	10I3	IOP1 - Number of ORO iterations IOP2 = 0, print last ORO iteration only; if ≠ 0, print each ORO iteration IOP(1) times IOP4 = 0, call LESCO after DORO; if = 1, call CYDEB after DORO IOP5 = 0, run ORO on base field; if = 1, run ORO on isochronous field IOP6 = 1, write Tape 5 and Tape 15 (assuming IOP7 = 1) = 0, do not write IOP7 = 1, write Tape 21 for ZORO; if = 0, do not write IOP9 = 0, base field is written on Tape 5; if = 1, isochronous field is written on Tape 5 IOP12 = 1, Fourier analysis of the field; if = 0, no Fourier analysis required IOP13 - Number of harmonics desired in the analysis IOP14 - Number of radii for which the field is analyzed
9	IG,IT,IL,NSECT	4I5	IG - Logical unit number of file containing main coil field $B(R,\theta)$ IT - Logical unit number of file containing coil field coefficients IL - If = 0, no print; if ≠ 0, print magnetic field calculations from coefficients NSECT - Number of field sectors

Table 50 (cont.)

Card sequence	Parameter name	Format	Description
10	SCALE,RMIN,RMAX, DR,DTHETM,DTHETO, EPSB	7F5.1	SCALE - Makes full scale dimension of model dimensions RMIN - Minimum radius (inches) of the magnetic field RMAX - Maximum radius (inches) of the magnetic field DR - Radial interval (inches) between input grid points of the magnetic field DTHETM - Azimuthal interval (degrees) between magnetic field input grid points DTHETO - Azimuthal interval (degrees) between magnetic field output grid points EPSB - Difference allowed between the synchronous field and the preassigned value (gauss) at the preset radio frequency
11	IX,IN,ILL,WGTM	3I5,F10.0	IX - If = 1, call TRIMCO from CYDEB; if = 0, do not call TRIMCO IN - Logical unit number of the file containing trim coil effects DB/DI ILL - If = 0, interpolate for the trim coil effects; if $\neq$ 0, use the measurements from Tape IN WGTM - Weight of main field relative to the required field that will allow trim coils to sit in an intermediate B_{eff} for their gauss/ampere calculation (WGTM = 0.60 is a good value for 88-inch fields)
12 . . (NL entries)	BREQDBI	10F6.4	Ratio of the required field to the isochronous field as a function of radius NL entries are required (NL = RMAX - RMIN / DR +1)
13	NMN(I)	7I5	The parameter NMN has 7 entries: NMN(1) = 0, call EXIT; if $\neq$ 0, call TRIMCO NMN(2) = 0, read maximum current for each coil; if $\neq$ 0, NMN(2) is the maximum current (amperes) NMN(3) = 0, read minimum current for each coil; if $\neq$ 0, NMN(3) is the minimum current (amperes) NMN(4) = 0, set RHO(I) = 1.0; if $\neq$ 0, read RHO(I) from cards NMN(5) = 0, read trim coil set; if $\neq$ 0, calculate trim coil field using air coil approximation NMN(6) = 0, maximum and minimum currents of each coil limited; if $\neq$ 0, there are no current limits NMN(7) not presently used
14	UNITCR,UNITM, NCOIL,L,LSKIP, NPLUG,SPHIO	2F10.0,4I5, F10.3	UNITCR - Unit current on each trim coil (amperes) UNITM - Unit current on the main coil (amperes) NCOIL - Total number of coils, including main and trim L - Number of radii used for field fitting LSKIP - Number of radii to be skipped NPLUG - If = 0, center plug is not counted on a trim coil, if $\neq$ 0, center plug is counted on a trim coil SPHIO - The sine of phase at R = 0
15 . . (L entries)	RHO(I)	18F4.0	Least-square weights for each radial point to be fitted; L values are read only if NMN(4) $\neq$ 0; otherwise, bypass
16 . . (NCOIL entries)	AM(I)	18F4.0	Maximum current value for each trim coil; NCOIL values are read only if NMN(2) = 0; otherwise, bypass
17 . . (NCOIL entries)	AS(I)	18F4.0	Minimum current value for each trim coil; NCOIL values are read only if NMN(3) = 0; otherwise, bypass
18 . . (NCOIL entries)	RUNCUR	10F6.0	Trim coil currents used on the cyclotron run (amperes); NCOIL values must be read

Table 51

CYDE Sample Input Data

```
                                   Card Column
1         10        20        30        40        50        60        70        80

1
120 MEV ALPHA,  PHASE HIST. COMP. CYDE CALC. AND BEAM MEAS.,  WGTM=.6,  C=2074 A.
   RUN DATA                MAIN C=2073.8 A,  FREQ.=12.0519 MHZ,  VD= 65 KV.
   FIT DATA   -   NOT USED
120.0    3.9724   2.0    2074.0        65.0  1.0        -1    -1EF,ATWT,ATNO,C,VD,HARM,IDR,IRF
         38.7    0.0           0.0            0.0            1       RF, PF, BF, FREQ, IPR
      3  0  1  1  0  0  0  0  0  0         IOP1,2,4,5,6,7,9,12,13,14
           4     1     0     3                              IG,, IT, IL, NSECT
1.0   0.0   66.0  1.0    3.0    3.0    1.0     SCALE,RMIN,RMAX,DR,DTHETM,DTHETO,EPSB
      1   16        0   0.6                      IX, IN, ILL, WGTM
1.035 1.032 1.026 1.016 1.007 1.000  .997  .995  .995  .996 BREQDBI(0-9)-I-50A
 .997  .998  .999  .999 1.000 1.000 1.000 1.000 1.000 1.000 BREQDBI(10-19)-I-50A
1.000 1.000 1.000 1.000 1.000 1.000 1.000 1.000 1.000 1.000 BREQDBI(20-29)-I-50A
1.000 1.000 1.000 1.00031.00061.00081.0006 .9997 .998  .992 BREQDBI(30-39)-I-50A
 .97   .93   .87   .80   .72   .63   .55   .47   .41   .35 BREQDBI(40-49)-I-50A
 .30   .25   .20   .16   .13   .11   .10   .09   .08   .07 BREQDBI(50-59)-I-50A
 .06   .05   .04   .04   .03   .02   .02   .01   .01   .0  BREQDBI(60-69)-I-50A
   0     0     0     1     0     0     0              NMN(1-7)
100.0      10.0            18     51      0      0  0.35   UNITCR,UNITM,N,L,LSKIP,NPLUG,
   0   1   2   3   4   5   6   7   8   9  10  11  12  13  14  15  16  17W(0-17)
  18  19  20  21  22  23  24  25  26  27  28  29  30  31  32  33  34  35W(18-35)
  36  37  38  39   0   0                                              W(36-41)
 750  750  750  750  .01  500  .01  500  .01  700  .011000  .01200020002000  .01 200AM(1-18)
  20   20   20   20    0   20    0   20    0   20    0   20   20   20    0   0AS(1-18)
-444 -341 -340   80 -175  160 -450  180 -397 -332RUNCUR(1-10)120ALPHA
-469 -569 -658 -1358 -1919 -1889 -1889    0         RUNCUR(11-18)120 A
```

Figure 98 shows a plot of the sine of the beam phase relative to the radio-frequency voltage on the dee for a 120-MeV alpha-particle beam in the 88-inch cyclotron at the Lawrence Berkeley Laboratory. This curve was taken near the maximum magnetic field of the cyclotron, and is a good test of the code. A zero phase means that the beam crosses a dee gap at the maximum of the

ZRO CODE RUN ON ISOCHRONOUS FIELDS

EQUILIBRIUM ORBIT TABLE

	BETA	P (MZERO C)	E (MC**2)	TIME	NUZ	NUR	KE (MEV)	R (INCH)	PR (MZERO C)	RAVE (INCH)	AMP (INCH)
1	.0064	.006405	1.0000	1.00000	91.000	1.000	.076	1.000	.00000	1.000	.000
2	.0128	.012810	1.0001	1.00000	91.001	1.000	.306	2.001	.00003	2.000	.001
3	.0192	.019217	1.0002	.99999	91.000	1.001	.688	3.003	.00013	3.000	.007
4	.0256	.025627	1.0003	1.00001	.040	1.002	1.224	4.008	.00037	4.000	.021
5	.0320	.032039	1.0005	1.00001	.076	1.003	1.912	5.016	.00079	4.999	.043
6	.0384	.038456	1.0007	1.00000	.101	1.005	2.755	6.026	.00136	5.998	.075
7	.0448	.044877	1.0010	1.00000	.116	1.007	3.751	7.038	.00202	6.996	.111
8	.0512	.051304	1.0013	1.00000	.124	1.009	4.902	8.051	.00273	7.994	.149
9	.0576	.057737	1.0017	1.00000	.132	1.011	6.207	9.065	.00345	8.991	.189
10	.0640	.064178	1.0021	1.00000	.137	1.012	7.667	10.080	.00417	9.988	.228
11	.0705	.070626	1.0025	1.00000	.141	1.013	9.284	11.097	.00487	10.985	.267
12	.0769	.077083	1.0030	1.00000	.143	1.014	11.056	12.117	.00552	11.982	.304
13	.0833	.083550	1.0035	1.00000	.145	1.014	12.986	13.140	.00612	12.979	.341
14	.0897	.090027	1.0040	1.00000	.147	1.015	15.073	14.167	.00664	13.976	.379
15	.0961	.096516	1.0046	1.00000	.147	1.016	17.318	15.197	.00708	14.973	.416
16	.1025	.103016	1.0053	1.00000	.147	1.017	19.724	16.230	.00742	15.971	.451
17	.1089	.109529	1.0060	.99999	.149	1.017	22.289	17.266	.00764	16.968	.485
18	.1153	.116057	1.0067	1.00000	.150	1.018	25.015	18.306	.00776	17.965	.516
19	.1217	.122599	1.0075	1.00000	.151	1.019	27.904	19.348	.00775	18.963	.546
20	.1281	.129156	1.0083	1.00000	.151	1.020	30.957	20.392	.00764	19.961	.575
21	.1345	.135730	1.0092	1.00000	.153	1.021	34.173	21.437	.00743	20.958	.610
22	.1409	.142321	1.0101	1.00000	.153	1.022	37.556	22.483	.00711	21.956	.641
23	.1473	.148931	1.0110	.99999	.153	1.023	41.106	23.528	.00669	22.954	.670
24	.1537	.155559	1.0120	1.00000	.154	1.024	44.824	24.573	.00618	23.952	.695
25	.1601	.162208	1.0131	1.00000	.155	1.026	48.712	25.615	.00557	24.951	.719
26	.1665	.168878	1.0142	1.00000	.157	1.027	52.772	26.656	.00486	25.949	.749
27	.1729	.175569	1.0153	1.00000	.159	1.028	57.005	27.694	.00406	26.948	.775
28	.1793	.182284	1.0165	1.00000	.161	1.029	61.412	28.728	.00317	27.947	.795
29	.1857	.189023	1.0177	.99999	.163	1.031	65.997	29.757	.00219	28.945	.811
30	.1921	.195786	1.0190	.99999	.168	1.032	70.759	30.780	.00113	29.945	.836
31	.1985	.202576	1.0203	1.00000	.171	1.033	75.702	31.796	-.00003	30.945	.854
32	.2049	.209392	1.0217	1.00001	.177	1.035	80.828	32.804	-.00127	31.945	.863
33	.2114	.216237	1.0231	1.00001	.184	1.036	86.138	33.802	-.00260	32.946	.872
34	.2178	.223111	1.0246	1.00001	.187	1.038	91.634	34.789	-.00399	33.947	.883
35	.2242	.230015	1.0261	.99999	.187	1.041	97.320	35.763	-.00542	34.946	.882
36	.2306	.236950	1.0277	.99999	.185	1.042	103.197	36.723	-.00685	35.945	.877
37	.2370	.243918	1.0293	.99989	.187	1.041	109.267	37.675	-.00824	36.947	.876
38	.2434	.250920	1.0310	1.00005	.180	1.037	115.535	38.625	-.00955	37.955	.864
39	.2498	.257956	1.0327	1.00015	.098	1.042	122.001	39.571	-.01071	38.961	.850
40	.2562	.265029	1.0345	1.00001	91.030	1.043	128.670	40.505	-.01149	39.959	.835
41	.2626	.272139	1.0364	1.00001	.133	1.035	135.543	41.437	-.01149	40.965	.781
42	.2690	.279287	1.0383	1.00026	.241	1.042	142.625	42.370	-.01047	41.985	.678
43	.2754	.286476	1.0402	1.00019	.253	1.065	149.917	43.285	-.00866	42.992	.540
44	.2818	.293706	1.0422	.99993	.219	1.066	157.424	44.196	-.00657	43.988	.398
45	.2882	.300978	1.0443	.99986	.087	1.058	165.149	45.128	-.00463	44.990	.273
46	.2946	.308295	1.0464	.99997	91.090	1.052	173.096	46.082	-.00310	45.995	.178
47	.3010	.315657	1.0486	.99997	91.168	1.051	181.267	47.051	-.00201	46.998	.112

FIG. 96. Equilibrium orbit table; calculated from the isochronous field, the quantities listed are self-explanatory.

sine wave dee voltage. The phase was measured with a current probe connected to a sampling oscilloscope. At a sin ϕ of ± 1, the beam stops accelerating. The sin ϕ computed by CYDE appears in Fig. 97 as SPHIR. The values appearing under that column were adjusted by an additive constant for an arbitrary starting phase, and by another additive constant to account for the resonance of the frequency and the magnetic field. The agreement is good between calculated and measured phase histories and shows the code can be used to predict sin ϕ within about ± 0.1. This means the fit mode of CYDE can be used reliably to calculate trim coils for the whole range of particles and energies.

120 MEV ALPHA, PHASE HIST. COMP. CYDE CALC. AND BEAM MEAS., WGTM=.6, C=2074 A
RUN DATA - 10-1-71, MAIN C=2073.8 A, FREQ.=12.0519 MHZ, VD= 65 KV.
FIT DATA - NOT USED
UNITS= GAUSS, AMPS.

R	BISOC	BIRON	DBRUN	BRUN	BRUN-BI	BRUN/BI	SPHIR
(IN)	(GAUSS)	(GAUSS)	(GAUSS)	(GAUSS)	(GAUSS)		
0	15673.5	17180.6	-1148.0	16032.6	359.1	1.02291	.350
1	15673.7	17154.4	-1148.9	16005.5	331.8	1.02117	.307
2	15674.6	17077.6	-1148.1	15929.5	254.9	1.01626	.199
3	15675.1	16965.2	-1145.6	15819.7	144.5	1.00922	.075
4	15672.7	16853.3	-1136.8	15716.5	43.8	1.00280	-.006
5	15666.3	16769.5	-1119.7	15649.8	-16.4	.99895	-.021
6	15658.0	16714.2	-1092.8	15621.5	-36.6	.99766	.015
7	15651.3	16675.1	-1061.3	15613.8	-37.5	.99760	.074
8	15647.4	16646.3	-1028.4	15617.9	-29.5	.99812	.136
9	15646.2	16626.3	-996.1	15630.2	-15.9	.99898	.184
10	15647.6	16614.5	-967.4	15647.1	-.4	.99997	.203
11	15651.3	16608.9	-942.8	15666.1	14.8	1.00095	.185
12	15656.9	16608.1	-926.6	15681.5	24.6	1.00157	.129
13	15664.0	16611.4	-921.0	15690.5	26.4	1.00169	.050
14	15672.6	16618.6	-922.4	15696.2	23.6	1.00151	-.034
15	15682.7	16628.7	-928.1	15700.6	17.9	1.00114	-.108
16	15693.6	16641.1	-933.1	15708.0	14.5	1.00092	-.170
17	15705.2	16655.5	-938.5	15717.0	11.9	1.00075	-.223
18	15717.2	16670.6	-947.0	15723.6	6.3	1.00040	-.263
19	15730.0	16685.9	-958.5	15727.4	-2.7	.99983	-.271
20	15743.6	16700.5	-968.5	15732.0	-11.6	.99926	-.237
21	15758.1	16714.9	-970.5	15744.5	-13.6	.99914	-.173
22	15773.6	16729.2	-963.4	15765.8	-7.8	.99951	-.117
23	15790.5	16743.5	-952.2	15791.2	.8	1.00005	-.098
24	15808.4	16757.6	-946.7	15810.9	2.5	1.00016	-.107
25	15827.2	16770.5	-948.0	15822.5	-4.6	.99971	-.101
26	15846.7	16781.3	-949.1	15832.3	-14.4	.99909	-.041
27	15867.2	16789.1	-940.3	15848.8	-18.4	.99884	.065
28	15888.8	16792.9	-919.2	15873.8	-15.0	.99905	.177
29	15912.2	16792.4	-888.2	15904.2	-8.0	.99949	.257
30	15937.2	16787.0	-851.9	15935.1	-2.0	.99987	.293
31	15963.7	16776.8	-810.4	15966.5	2.8	1.00018	.290
32	15991.3	16761.4	-764.0	15997.4	6.1	1.00038	.256
33	16020.7	16739.7	-710.5	16029.2	8.5	1.00053	.199
34	16050.8	16710.0	-649.5	16060.5	9.7	1.00060	.125
35	16082.7	16669.9	-575.2	16094.7	12.0	1.00075	.035
36	16115.9	16612.8	-482.1	16130.7	14.8	1.00092	-.079
37	16147.4	16525.4	-362.5	16162.8	15.4	1.00095	-.212
38	16173.7	16383.0	-216.2	16166.8	-6.9	.99957	-.250
39	16203.8	16142.3	-50.5	16091.8	-112.0	.99309	-.298
40	16261.2	15738.6	117.4	15856.1	-405.1	.97509	2.730
41	16340.4	15090.8	270.3	15361.1	-979.4	.94006	9.374
42	16400.8	14151.5	385.9	14537.5	-1863.3	.88639	23.302
43	16424.6	12953.2	449.0	13402.2	-3022.5	.81598	47.781
44	16421.4	11615.8	462.4	12078.2	-4343.2	.73551	85.560
45	16416.4	10251.3	442.3	10693.5	-5722.9	.65139	138.394
46	16424.1	8951.2	403.3	9354.5	-7069.6	.56956	207.014
47	16445.3	7760.3	364.8	8125.1	-8320.1	.49407	291.272
48	16474.8	6700.1	326.3	7026.4	-9448.4	.42649	390.468
49	16508.9	5763.9	291.8	6055.6	-10453.2	.36681	503.677
50	16545.3	4942.2	259.9	5202.1	-11343.2	.31442	629.941

FIG. 97. Phase history printout.

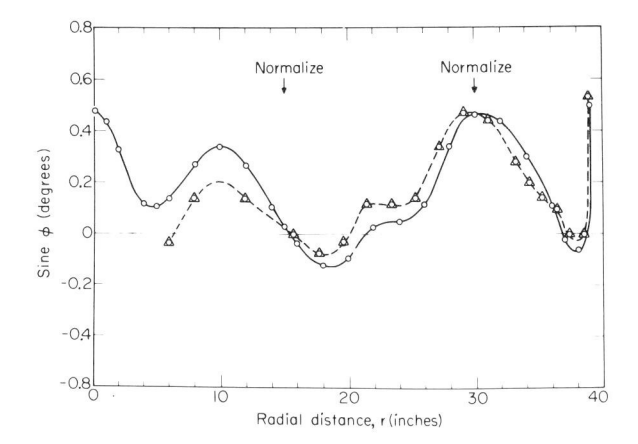

FIG. 98. Comparison of measured and computed phase history for a 120-MeV alpha-particle beam in the 88-inch cyclotron at the Lawrence Berkeley Laboratory. (-△-) measured values, (June 26, 1968); (-○-) CYDE-computed values.

k. General Remarks

Program CYDE has proved to be extremely useful in predicting trim-coil settings of new beams and is being utilized at the Lawrence Berkeley Laboratory in the following fashion.

The fit mode is used to calculate trim-coil settings for a commonly used experimental beam. Estimated required shapes are used to give a "center bump" at 0–12 inches and an "edge fall-off" of 35- to 40-inches radius. These currents are then used in a cyclotron run.

The inner and outer trim coils are optimized to effect the best external beam intensity and quality. The inner trim coils control the center bump which adjusts the central magnetic vertical focusing. The outer coils control the edge fall-off, which determines the resonance crossings (vertical blow-up) and phase slip near extraction. The resulting trim-coil settings of this optimized "model beam" are used in the CYDE run mode to get the ratio of the magnetic field to the isochronous field, BRUN(I)/BI(I).

These values are then used as input in the CYDE fit mode as BREQDBI(I) to calculate trim and main coil currents for other beams having a similar number ($N = E/qV_D$) of particle revolutions during acceleration. For example, the model beam of 50 MeV α-particles with 50 kV on the dee, has the same number of revolutions as 60 MeV α-particles or 30 MeV protons with 60 kV on the dee. The goodness of the fit is given by the output BF(I) — BREQ(I).

In using the fit mode, we use the minimum number of trim coils necessary to give a good fit and to obtain well-determined solutions which vary smoothly with energy. The other trim coils are turned off (I = 0.01 ampere). The lower and upper limits of the coils are set according to power supply characteristics. The main-coil current, C, is estimated, and the variation from this value is treated as trim coil 18 in the program.

The phase history is calculated for all cases. The parameter WGTM is used to adjust the effective field (BEFF) in which the trim coils act, so that it lies between the main-coil field and the final main-coil + trim-coil field. This results in an agreement of ± 0.1 in $\sin \phi$ (6 degrees) between measured and calculated phase histories for a 30-MeV or a 120-MeV α beam. The best agreement was found for WTGM $= 0.6$. In the previous version of this code, the agreement was ± 0.7 (± 40 degrees) for a 120-MeV α beam when the effective field was assumed to be the isochronous field. A further refinement, which might make the phase history agreement even better than ± 0.1, would be to include the change in hill-valley difference as the trim coils are turned on. The present version assumes that the trim coils add an equal amount to each azimuthal position at a given radius.

Nearly all of the heavy-ion beams have been run with these predicted trim-coil settings; examples are 160-MeV N^{+4} and 250-MeV N^{+5}, based on the 80-MeV α model beam.

36. Program PINWHEEL

a. Introduction

This program, written in Fortran IV, is used for tracking orbits of charged particles in a combined electric and magnetic field. It was originally written at Michigan State University, by Reiser and Kopf [91], and subsequently modified by Close [92] (see Brady [90]).

The code is operational on the CDC 6600/7600 computer systems with approximately 55K_8 memory locations required to run a job in less than one minute on the CDC 6600 computer system.

b. Program Description

The problem considered by this code may be defined by the following equations of motion of a charged particle in a combined electric and magnetic field

$$dp_x/dt = qE_x + qv_y B_z, \tag{36.1}$$

$$dp_y/dt = qE_y - qv_x B_z, \tag{36.2}$$

$$dx/dt = v_x, \tag{36.3}$$

$$dy/dt = v_y, \tag{36.4}$$

where E is the electric field, B is the magnetic field, p is the particle's charge momentum, and v is the velocity.

It should be noted that this code restricts itself to two-dimensional problems with $E = (E_x, E_y)$ and $B = B_z$. The parameters E, B, q, and v are treated in a dimensionless form by using velocity in units of c, momentum in units of $m_0 c$, and distance in cyclotron units. Specifically $V_x = v_x/c$, $V_y = v_y/c$, $P_x = p_x/m_0 c$, $P_y = p_y/m_0 c$, and $X = x/a$, $Y = y/a$, where $a = c/\omega_0$. The independent variable is the electrical time defined as $\tau = \omega_0 t$, where $\omega_0 = e_0 B_0/m_0$ is the cyclotron frequency with B_0 denoting the central isochronous magnetic field.

The equations of motion in dimensionless form then, are

$$(dP/d\tau)_x = \partial U/\partial X \cos(N\tau) + V_y B, \tag{36.5}$$

$$(dP/d\tau)_y = \partial U/\partial Y \cos(N\tau) - V_x B, \tag{36.6}$$

$$dX/d\tau = V_x, \tag{36.7}$$

$$dY/d\tau = V_y, \tag{36.8}$$

$$P^2 = P_x^2 + P_y^2, \tag{36.9}$$

$$V_x = P_x(1 + P^2)^{-1/2}, \tag{36.10}$$

$$V_y = P_y(1 + P^2)^{-1/2}, \tag{36.11}$$

$$E_k = E_0[(1 + P^2)^{1/2} - 1]. \tag{36.12}$$

where E_k is the kinetic energy. The potential function U is defined as

$$U = eV_0/m_0 c^2 f(X, Y), \tag{36.13}$$

where $f(X, Y)$ represents the measured electric potential distribution carried on a two-dimensional grid of data points normalized to 1, and V_0 is the maximum dee voltage. The parameter N defines the ratio of the electrical frequency, ω_e, to the cyclotron frequency, ω_0 and determines the mode of cyclotron operation. $B = B_z$ is the axial component of the magnetic field evaluated on the median plane.

Runge-Kutta integration is used to solve this first-order system of equations. The integration step size is presently fixed at $\Delta\tau = (7.50/N) \times (\pi/180)$ radians.

Program PINWHEEL can easily be divided into three routines: control, plotting, and integration. The control section is responsible for reading the input data, determining problem requirements and setting appropriate "flags" to route the program through the correct channels. The plotting section provides for the utilization of the routines necessary to perform CalComp plotting. The integration section uses the routines necessary to perform the actual integration and print the results. A simplified flow chart of this program appears in Fig. 99.

248 V. Integration of Equations of Motion

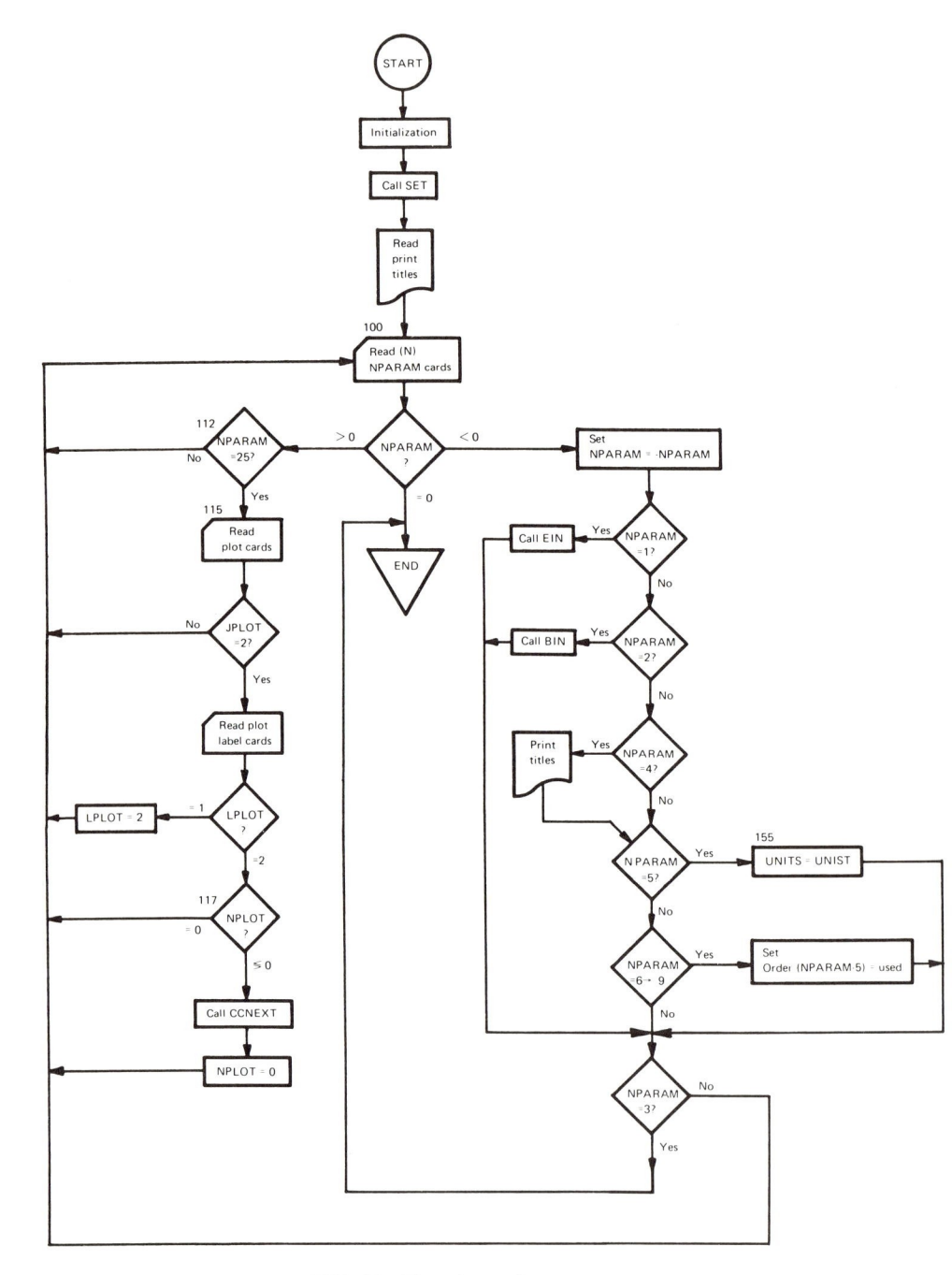

FIG. 99. Flow chart of program PINWHEEL.

c. Input Organization

There are five different types of data cards used for input in PINWHEEL; the format and required program variables are as follows.

1. Type A Card (format 8A10), provides for problem identification. The two cards required are the Title card which is printed once at the beginning of the job, and the message card which is printed when NPARAM = −4 (see Table 52).

Table 52

Program PINWHEEL Card B parameter List (Format I3)

NPARAM	Description
0	Stops program
-1	Calls subroutine EIN that inputs onto the electric field grid the desired values of the potential function $f(X,Y)$. This is to be input as $V(i,j)$, where $i = X_i$, $j = Y_i$ are the grid lines indexed from the lower left-hand corner (x-min,y-min)
-2	Calls subroutine BIN that is used to verify and store the magnetic field, B/B_0, in the form of Fourier series, $$B = B_0 \left[B(r) + \sum_{n=3,6,9} H_n \cos(n\phi) + \sum_{n=3,6,9} G_n \sin(n\phi) \right]$$ subroutine BIN reads the $B(r)$, H_n, and G_n coefficients
-3	Enters main program that initiates calculation
-4	Prints Type A message card
-5	Prints output headings in centimeters; otherwise, output headings in inches
6	Prints coordinates of orbit center points
-7	Used in relation with the field orientation (-1 is left-handed, +1 is right-handed)
-8	Assumes standard isochronous B field, that is, $B/B_U = 1.0$
-9	Prints results in polar coordinate form
0 < NPARAM < 25	Reads NPARAM Type C cards
25	Reads Type D card that will start and stop plotting, and may or may not initiate a read of Card E

2. Type B card, (format I3), is needed to furnish information relating to various options provided by the program (see Table 52).

3. Type C Card, (format I3,F10.5), defines the parameters needed to properly describe the problem under consideration. A list of all possible parameters and prevailing standard values is provided (see Table 53.)

Table 53

PINWHEEL Card C Parameter List (Format I3,F10.5)

NPARAM	Description	Standard value
	FIELD PARAMETERS	
1	The number of rows in the electric field grid row, implies constant y	67.0
2	The number of columns in the electric field grid column, implies constant x	67.0
3	The grid spacing, D (inches)	0.25
4,5	The x minimum and y minimum (inches)	-8.25
6	Angle between E and B field grids (degrees)	0.0
7	The frequency (megahertz/second) - determines cyclotron unit $a = c/\omega_0$	12.3
	OUTPUT PARAMETERS	
8	The number of Runge-Kutta steps between printouts. When plotting, this corresponds to the location of graph identification marks	4.0
9	The maximum number of printouts desired	50.0
10	This scales graphical output; if PARAM(10) = 2.0, plot full scale; if = 1.0, plot half scale	2.0
11	Magnetic field scaling, e.g., 0.5 implies field used is one-half the input field	1.0
	PARTICLE AND OPERATION PARAMETERS	
12	The charge state	1.0
13	The rest energy (million-electron-volts)	1876.05
14	Harmonic mode of operation	1.0
15	The dee voltage (kilovolts), corresponds to maximum E value. For a (+) charge state use negative kilovolts; for a (-) charge state use positive kilovolts	-70.0
	INITIAL CONDITIONS	
16	The x (inches) from center point (x,y) = (0,0)	0.0
17	The y (inches) from center point (x,y) = (0,0)	0.0
18	The starting kinetic energy (electron volts)	0.0
19	The starting angle (degrees) that the velocity vector makes with respect to the y axis, measured positively in a clockwise direction	0.0
20	The starting time in electrical degrees, t = 0 starts zero phase, i.e., at peak voltage. Note, voltage varies as a cosine function	0.0

4. Type D Card, (Format 4X,I1,I5,7A10), is used to furnish information relating to graph plotting and has the following parameters:

   a. JPLOT—Flag indicating the initiation of plotting, if JPLOT = 2 results will be plotted. If JPLOT ≠ 2 plotting is bypassed.

   b. NPTS—Number of points to be plotted.

Table 54

PINWHEEL Sample Input Data

```
                              Card Column
   1        10        20        30        40        50        60        70        80

PINWHEEL -  CYCLOTRON CENTER REGION PROGRAM
30 MEV ALPHA BEAM,  RUN B FIELD,  88-8 E FIELD,   E0=9 KV,    VD=30 KV
 -6
 18                                                         READ 18 CARD C PARAMS
  1 100.0
  2 100.0
  3 0.10                                                    GRID SPACING - CYC SC
  4 -5.0                                                    XMIN - CYC SCALE
  5 -5.0                                                    YMIN - CYC SCALE
  6  0.0
  7  6.1311                                                 FREQ
  8  1.0
  9 720.0                                                   TOTAL NO. OF STEPS
 10  1.0                                                    PLOT SCALE
 12  2.0
 13 3727.3                                                  REST ENERGY
 15 -30.0                                                               DEE VOLTS
 16  0.0
 17  0.0
 18 9000.0                                                  E INITIAL
 19 180.0
 20 -210                                                                PHI INITIAL
 -2
       1.039     1.042     1.039     1.031     1.019     1.008     1.000    M2
        .995      .992      .993      .992      .992                        M3
                                                                            M 4
                                                                            M17
                                                                            M18
                                                                            M19
                                                                             .
                                                                             .
                                                                             .
                                                                            M26
                                                                            M27
                                                                            M28
 -1
 -0.003 -0.003 -0.002 -0.002 -0.002 -0.002 -0.002 -0.003 -0.003 -0.002 88-8    1
 -0.002 -0.002 -0.002 -0.002 -0.001 -0.001 -0.001 -0.001 -0.001 -0.    88-8    2
 -0.    -0.    -0.    -0.     0.     0.001  0.001  0.001  0.001  0.001 88-8    3
  0.001  0.001  0.001  0.001  0.002  0.002  0.002  0.002  0.002  0.003 88-8    4
                                                                             .
                                                                             .
                                                                             .
 -0.989 -0.989 -0.989 -0.989 -0.989 -0.989 -0.989 -0.989 -0.989 -0.989 88-8  997
 -0.989 -0.989 -0.989 -0.989 -0.989 -0.989 -0.989 -0.989 -0.989 -0.989 88-8  998
 -0.989 -0.989 -0.989 -0.989 -0.989 -0.989 -0.989 -0.989 -0.989 -0.989 88-8  999
 -0.989 -0.989 -0.989 -0.989 -0.989 -0.989 -0.989 -0.989 -0.989 -0.989 00 8 1000
 25
     2  120  30 ALPHA,  88-8 E FIELD, RUN B FIELD, E0=9 KV,   VD=30 KV
PHI= -210            PHI= -180        PHI= -150         PHI= -120
 -3
 -4
  2
  7  6.1311
 20 -180
 -3
 -4
  2
  7  6.1311
 20 -150
 -3
 -4
  2
  7  6.1311
 20 -120
 -3
  0
```

252   V. Integration of Equations of Motion

```
PINWHEEL - CYCLOTRON CENTER REGION PROGRAM

30 MEV ALPHA BEAM, RUN B FIELD, 88-8 E FIELD,  E0=9 KV,  VD=30 KV

    1     N. ROW          100.000
    2     N. COL          100.000
    3     GRD SP             .100
    4     X MIN.          -5.000
    5     Y MIN.          -5.000
    6     ANG*EB           0.000
    7     FREQ.            6.131
    8     RK/PR            1.000
    9     RK TOT         720.000
   10     PL SCL           1.000
   11     B SCL.           1.000
   12     CHARGE           2.000
   13     E REST        3727.300
   14     HARM.            1.000
   15     D VOLT         -30.000
   16     X INIT           0.000
   17     Y INIT           0.000
   18     ENERGY        9000.000
   19     ANG IN         180.000
   20     T INIT        -210.000
   -6     C.POIN    YES
   -7     HAND.R    NO
   -8     ISO.MG    NO
   -9     POLAR     NO
```

X INCHES	Y INCHES	PX	PY	REV.	TIME (DEGREES)	ENERGY (MEV)	EX	EY	B
0.000	0.000	.000	-.673	1	-210.0	.009000	.001000	.003000	1.04200
.003	-.059	.063	-.669	1	-205.0	.008974	.000805	-.000611	1.041963
.011	-.117	.123	-.662	1	-200.0	.009001	-.000961	.000135	1.041909
.024	-.174	.182	-.648	1	-195.0	.008999	-.000426	-.000469	1.041838
.043	-.230	.240	-.630	1	-190.0	.009028	.000067	-.001083	1.041747
1.295	1.759	-1.902	1.373	1	355.0	.109250	.000336	-.003413	1.022285
1.124	1.871	-2.017	1.198	2	.0	.109252	-.000107	.001170	1.021093
.943	1.968	-2.116	1.014	2	5.0	.109292	.000323	-.000475	1.020497
.755	2.048	-2.198	.822	2	10.0	.109304	-.000000	-.000000	1.020547

ORBIT CENTER POINTS

X	Y	T	REV
.646	.000	-210.000	1
.242	.063	-43.926	1
-.138	-.102	36.736	1
.029	-.135	123.953	1
.291	-.060	206.112	1
.180	-.043	303.278	1
-.049	-.087	28.861	2
.030	-.094	115.155	2
.179	-.057	200.842	2
.139	-.043	294.117	2

FIG. 100. Partial printout obtained by program PINWHEEL showing orbit information.

c. PPLOT—Alphanumeric information pertaining to the identification of the plot.

5. Type E Card, (format 8A10), furnishes a source to enter arbitrary information used to identify the CalComp plots in the following way: Array PLBS(I) is reserved to store a maximum of four identifiers for four graphs which may be overlaid. Only four orbits per graph are permitted, however, additional graphs can be plotted on a new sheet of CalComp paper using the stored values of card D and PLBS(I), starting with PLBS(1). Thus, card D furnishes a graph title and card E furnishes identification of the overlaid graphs.

d. An Illustrative Example

To utilize the input instructions just described, consider a center-region study for the 88-inch cyclotron at Lawrence Berkeley Laboratory (see Clark [89]). In this example we will trace orbits of 30-MeV α-particles with different

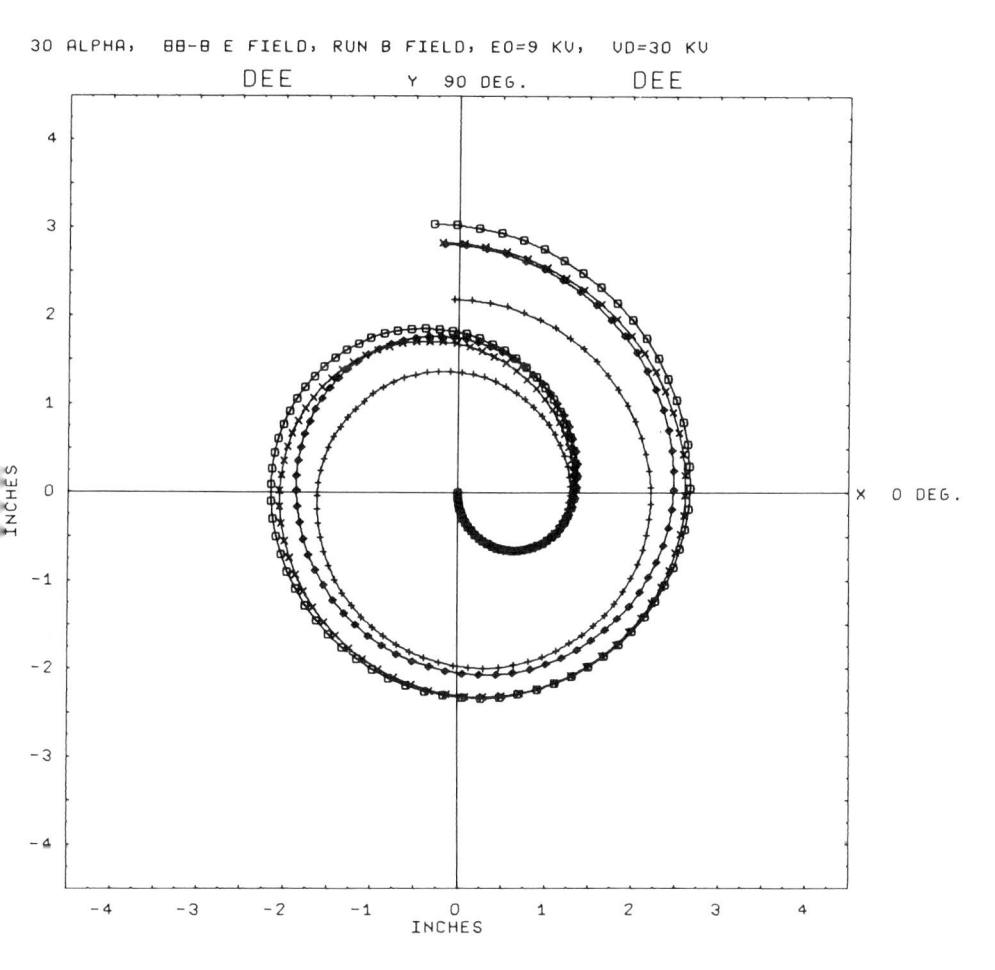

FIG. 101. CalComp plot of cyclotron trajectories for various starting phase angles. (+) phi = $-$ 210, ($\times$) phi = $-$ 180, ($\square$) phi = $-$ 150, ($\diamondsuit$) phi = $-$ 120.

starting phases (see Table 54). The first two cards (type A) are used for the alphanumeric description of the problem. The third card, Type B card, indicating the 18 Type C parameter cards that are to be read. In most cases, control is returned to a Type B card, which instructs the code as to the next operation. The exception is with a Type B, -3 card, which begins execution before returning control to the next Type B card. After the 18 Type C parameter cards have been read, a Type B, -2 card directs the program to accept the Fourier coefficients of the magnetic field data. In a similar fashion, the reader may interpret the rest of the data by consulting Tables 52 and 53. When all the options in the data have been executed, a Type B, 0 card will terminate the program.

e. Output Description

PINWHEEL provides for printer output in either polar or Cartesian coordinates and CalComp plots. A partial output (see Fig. 100) shows various orbit parameters computed from 5 to 10 revolutions for the illustrative example, and a CalComp plot of particle orbits for $1\frac{1}{4}$ revolutions (120 points). The different trajectories in this plot are for different starting phases relative to the radio frequency (see Fig. 101). The beam starts at the center of the cyclotron, as injected from an external ion-source through the pole axis, and is accelerated at each crossing of the x axis when it enters or leaves the dee.

37. Program SYNCH

a. Introduction

SYNCH is a computer program by Garren, Kenney, and Eusebio [93] that calculates linear properties of transverse orbit motion in synchrotrons or storage rings.

The program calculates transfer matrices for beam elements (drift spaces, bending magnets, quadrupoles, etc.) and combinations of beam elements. Of particular importance are the transfer matrices that correspond to the effect of one focusing period or cell on the orbits. The betatron functions [94] and the off-momentum functions are obtained from these matrices for use in computing transverse beam dimensions. This process is carried out for one-cell matrices that start (and end) at each constituent element of the cell in order to get beam envelopes for every azimuth. The closed orbit errors caused by magnet misalignments may also be obtained.

Some important additional features of SYNCH may be noted: the cell parameters required to produce desired orbit properties are calculated along with the straight section insertions.

The linear action of beam elements on particular orbits defined by the transfer matrices can be augmented by nonlinear mappings to describe sextupoles, octupoles, etc. These mappings are effected by Fortran subroutines compiled by the user with the SYNCH program.

Individual particles can be tracked repetitively through the linear and nonlinear elements of the synchrotron ring to generate phase plot information.

Off momentum closed orbits, betatron functions and other properties of orbits with nonlinear elements present can be calculated.

Program SYNCH has been written in Fortran for the CDC 6600/7600 computer systems and requires $121 K_8$ memory locations for the complete program. Since it is a multi-purpose code, shorter versions can be compiled for certain applications by deleting unneeded subroutines. Another version of SYNCH has been written to run on IBM 360 series computers. The original version of SYNCH, which has not been updated, was written for the IBM 7094 and is still available. A typical run on the CDC 7600 takes a few seconds.

b. Program Organization

SYNCH has been organized in a most flexible way. The user builds the synchrotron structure by a series of modular input instructions that correspond to simple verbal statements descriptive of the accelerator. These instructions comprise a special-purpose language.

Each SYNCH instruction contains three components: a name, an operation, and a set of data. The name is an arbitrary set of characters which can be used as data in subsequent instructions to refer to the element defined by the instruction. The operation is a particular set of characters that characterizes the nature of the elements described and of the calculations to be performed. The data completes the specification of the element and provides input parameters for the operation. The instruction, together with quantities calculated, such as matrices, are stored for use by subsequent instructions.

c. Summary of Selected SYNCH Instructions

The instructions listed in Table 55 are representative of the intrinsic potentialities of SYNCH. Below, a few of the instructions will be described in detail. For a more complete treatment on SYNCH, the reader is referred to Garren, et al. [93].

1. *Operations defining beam elements and combinations of elements.* These instructions, when the program is in the standard mode, generate the horizontal transfer matrix that expresses the change in the vector $(x, dx/ds, \Delta p/p)$ caused by the corresponding element, and the vertical matrix that expresses the change in the vector $(y, dy/ds, \Delta p/p)$.

Table 55

SYNCH Input Instructions

Element	Description
MAG,MAGS	Defines magnet with or without field errors or misalignments
DRF	Calculates matrices for drift space length, l (meter)
CRD	Reads in matrices from cards
SHF	Shift matrix definition
EQU,INV,REF,**	Operations on single matrices: EQU equates matrices; INV inverts matrices; REF allows matrix reflection; ** raises matrix to a power
MMM,CYA,CYB,FXPT	Operations on two or more matrices: MMM defines matrix multiplication; CYA calculates cycled matrix product, saves cycled matrices; CYB calculates matrix product; FXPT determines closed orbit, betatron functions with nonlinear transformations
WMA,WBE	Output instructions: WMA prints transfer matrices; WBE prints betatron functions
CELL,ALTC,CELC, CFD,DCFD	AGS repeat lengths (cells): CELL calculates matrices for FDDF symmetric cell; ALTC same as CELL and varies input parameters of CELL to obtain correct phase advance; CELC calculates unequal straight section lengths; CFD calculates unequal F and D magnet lengths; DCFD designs CDF cell and adjusts parameters to obtain μ_x and μ_y
STR2,STPI,STR4,STRN	Straight sections: STR2 designs Collins straight section with specified phase advance, μ_x, for insertion in specified cell; STPI designs π-type straight section (DFDF or FDFD); STR4 calculates 4-element antisymmetric straight section; STRN allows for more than four quadrupoles to be used in the straight section
INCR	Increment floating-point number
RUN,FIN,STOP,BMIS, EMIS,REM	Computer control instructions: RUN begins execution of problem; FIN allows the user to process sequencial problems; STOP ends entire calculation cycle; BMIS begins misalignment mode; EMIS ends misalignment mode; REM allows remarks printed on output
MAP,TRK	Particle tracking: MAP calculates nonlinear transformation (user-written subroutines TRK tracks particles through beam system

When the program is in the misalignment mode, the third component of the vectors is to be thought of as unity. When the horizontal, vertical, and closed orbits do not decouple, the program constructs a single 7×7 matrix that acts on the vector $(x, dx/ds, y, dy/ds, \Delta s, \Delta p/p, 1)$.

No printed output is generated by these instructions.

MAG Bending magnet, or quadrupole. Input: length, l; gradient dB/dR; magnetic rigidity, $B\varrho$, field, B_0, entrance and exit angles. Alternatively, the input quantities 2, 3, 4 can all be divided by B_0, to represent $K = (1/B_0)(dB/dR)$, ϱ, 1.

MAGS Magnet with misalignment or field error. Input: name of magnet to be misaligned, transverse displacements of magnet ends, field error. Figure 102 shows the geometry of the central trajectory in a bending magnet. The angles e_1 and e_2 are positive for the case illustrated.

In Table 56 some examples of instruction MAG are given. The labels *BB* or *Q*1 represent an arbitrary name chosen by the user. Any

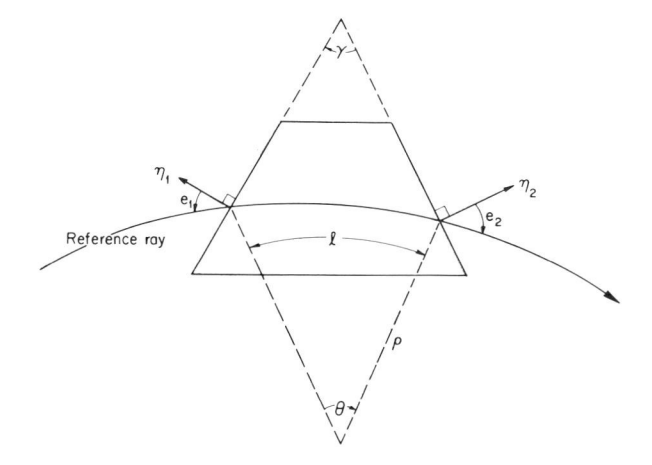

FIG. 102. Bending magnet showing notation and sign convention.

reference to this name by other instructions will fetch the transfer matrix generated by *BB* or *Q*1 to be used by that instruction.

DRF Drift space. *Input*: length.
CRD Arbitrary transfer matrices. Input: matrix elements.
INV Inverse matrix. *Input*: name of element whose matrix is to be inverted.
MMM Beam line. *Input*: the ordered sequence (in beam direction) of named, previously defined beam elements. Calculates the product of the matrices of the constituent elements.
** Repeated element. Raises matrix to a power.
ROTZ Rotation. Calculates matrix (7×7) corresponding to rotation about the beam axis.

2. *Operations involving calculation of betatron functions and closed orbits.* The betatron functions describe the beam envelope of a matched beam in a

Table 56

Examples of Instruction MAG

2-6	8-12	13-15	17-19	21	31	41	51	61	71
Element	OP	K	L	Data.....					
				l	k	ρ	ω	γ	δ
BB	MAG			2.35	0.0	12.71648	1.0	0.0	0.0
Q1	MAG			0.5	-10.512	50.03454	0.0	0.0	0.0

synchrotron. If the area of the beam in the x, dx/ds phase plane is $\pi\varepsilon_x$, then the beam ellipse is given by

$$\varepsilon_x = \beta x'^2 + 2\alpha x x' + \left(\frac{1+\alpha^2}{\beta}\right)x^2 \qquad (37.1)$$

where $x' = dx/ds$. The betatron functions β, α are functions of orbit azimuthal position s. They are calculated by SYNCH from the 2×2 transfer matrix for one period of the lattice, which can be written

$$M(s/s+L) = \begin{pmatrix} \cos\mu + \alpha(s)\sin\mu & \beta(s)\sin\mu \\ \dfrac{-1+\alpha(s)^2}{\beta(s)}\sin\mu & \cos\mu - \alpha(s)\sin\mu \end{pmatrix} \qquad (37.2)$$

where L is the length of the period. The third betatron function μ is the phase advance of the betatron oscillations through the period. Horizontal and vertical betatron functions are calculated from the corresponding 2×2 matrices.

The closed orbit displacement vector

$$H(s)(\Delta p/p) = (\eta(s)(\Delta p/p), \eta'(s)(\Delta p/p), \Delta p/p)$$

is calculated from the horizontal 3×3 matrix by solving the equation

$$M(s)H(s) = H(s). \qquad (37.3)$$

WBE Write betatron functions of matrices corresponding to previously defined elements using formulas (1) and (2).

CYA Compute betatron functions at the beginning of each element of a lattice period. Input: beam line as in MMM, $A_1, A_2, \ldots, A_n$.

    Calculation: a) $M_1 = A_n A_{n-1} \cdots A_2 A_1$ matrix product.
                b) Compute and print β, α, μ, η, η' from M_1 using (37.1) and (37.2).
                c) $M_2 = A_1 M_1 A_1^{-1}$.
                d) Compute and print betatron functions from M_2.
                e) $M_3 = A_2 M_2 A_2^{-1}$
                etc.

The betatron functions computed from the cycled matrix M_{k+1} give the beam envelope at the end of element A_k, for any k.

CYB Same as CYA except beam line is that of a previous MMM.

3. *Operations relating to design of lattice cells.* Synchrotrons are made up of a number of identical sequences of magnets and drift spaces called cells,

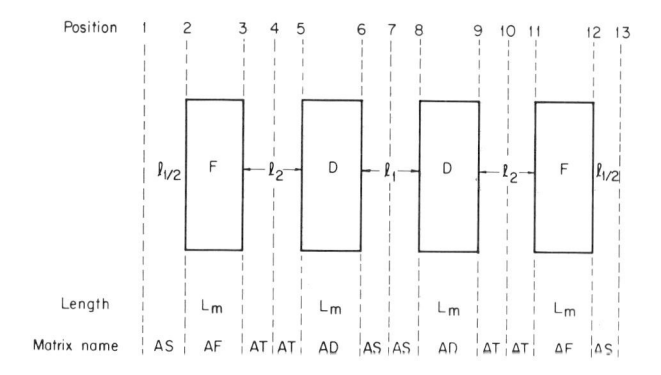

FIG. 103. FDDF symmetric cell showing notation.

so the focusing forces are periodic and repeat every cell length. Figure 103 shows a diagram of such a cell (FDDF) with its defining quantities. The following instructions are useful in designing the most common types of cell.

CELL Construction of FDDF type cell. Constructs matrices for elements of a cell with structure

$$L_1/2 \ F \ L_2 \ D \ L_1 \ D \ L_2 \ F \ L_1/2$$

where F and D are equal length focussing and defocussing magnets with equal and opposite gradient and L_1, L_2 are drift spaces (see Table 57 for an example).

ALTC Design FDDF cell with specified horizontal phase advance μ_x, by varying any of the input parameters (e.g. length or gradient).

DCFD Design cell with specified values of μ_x and μ_y by letting the lengths of the F and D magnets be unequal.

Table 57

Data Structure for Instruction CELL

2-6	8-12	13-15	17-19	21	31	41	51	61	71
Element	OP	K	L	Data.....					
A	CELL			L_m ω_x	k	ρ	ℓ_1	ℓ_2	e
C	CELL			6.06 0.0	4.5	556.0	4.57	0.6	0.0

260   V. Integration of Equations of Motion

4. *Matching insertions and replacement structures.* Besides the repetitive cells, synchrotrons sometimes include special structures including an extra long drift space. If the transfer matrix of an insertion has betatron functions β, α identical to that of the cell at the insertion point, the beam envelope in the normal cells is unchanged. The following operations calculate parameters for certain kinds of insertions.

- STR2   Calculates parameters for an insertion of structure $l\,QD\,L\,L\,QF\,l$ to be inserted between F and D magnets of a previously calculated CELL.
- STR4   Calculates parameters for a structure of the form

$$l\,QF\,a\,\overline{QD}\,L\,L\,\overline{QF}\,a\,QD\,l$$

  that replaces part of a normal cell.
- STRN   Similar to STR4 but the outer quads QF, QD are replaced by previously defined composite structures.

5. *Design problems with more general structures;* SYNCH *subroutines.* In case cells or insertions are to be designed of a structure not included in the special-purpose instructions of 3 and 4, one can make use of the SUB and FIT instructions to achieve definite betatron functions.

- SUB   An instruction that marks subsequent instructions as constituting a SYNCH subroutine.
- END   Marks the termination of a SYNCH subroutine.
- FIT   Varies certain parameters of instructions included in a previously named SYNCH subroutine to fit any two betatron functions of an MMM-defined matrix in that subroutine.

SYNCH subroutines can be nested to any depth.

If one desires to perform calculations repetitively over a parameter range, the following instructions are used with SUB and END:

- REPL   Replace a particular floating point input parameter.
- INCR   Increment a particular floating point input parameter.
- CALL   Execute a named subroutine a specified number of times.
- MESH   Call a subroutine repetetively varying any set of included parameters over a mesh of values.

6. *Facilities for nonlinear transformations.* To effect any transformation of the particle's orbit vector $V = (x, x', y, y', \Delta s, \Delta p/p, 1)$ not included in the SYNCH repertoire, there is provision for compiling user-written subroutines with SYNCH that carry out the desired transformation of the vector. These SUB-routines are named MAP1, ..., MAP9 and contain in their arguments

symbols for the vector and input parameters that define the mapping. These are transmitted to the subroutine by the SYNCH instruction MAPk:

    A   MAPk P1   P2 $\cdots$.

When A is used in a subsequent SYNCH instruction, the user's routines MAPk is executed with arguments P1, P2, ... Two instructions that use these mappings follow:

    TRK   Tracks vectors through a beam line containing linear and/or nonlinear transformations.

    FXPT   Searches for the vector that is unchanged by passage through a beam line. Calculates complete closed orbit and perturbed betatron functions about closed orbit.

7. *Use of* TRANSPORT *with* SYNCH. It is not possible to fit as many constraints with SYNCH as with TRANSPORT. To design very complicated insertions of types not provided in SYNCH it is sometimes expedient to use TRANSPORT as follows. First SYNCH is used to design the normal lattice cells, and provide betatron functions at the point where a long straight section is to be inserted. From these functions we obtain beam ellipse data for TRANSPORT

$$x = \sqrt{\varepsilon_x \beta_x}, \qquad x' = \sqrt{\varepsilon_x \gamma_x}, \qquad y = \sqrt{\varepsilon_y \beta_y}, \qquad y' = \sqrt{\varepsilon_y \gamma_y} \qquad (37.4)$$

where ε_x and ε_y are the horizontal and vertical emittances assumed for the beam and $\gamma = (1 + \alpha^2)/\beta$. One can then design the insertion with TRANSPORT. It will be matched to the cells, provided that at either end the beam ellipse σ-matrix is constrained to agree with Eq. (37.4). Finally the cells and the insertion can be linked together in another SYNCH run to obtain betatron functions through the entire synchrotron ring.

e. Illustrative Example

The combination of SYNCH and TRANSPORT was instrumental in the design of the magnetic lattice for the Stanford Positron-Electron Storage Ring (SPEAR) [95].

A schematic block diagram of the lattice structure is shown in Fig. 104 and the corresponding SYNCH input data in Table 58.

Input data always begins with a RUN instruction. It is followed by nine MAG instructions, each of which refers to a different quadrupole type. The other two MAG's define bending magnets. Next are six drift spaces, defined by DRF instructions, which complete the specification of all the elementary beam elements used in the SPEAR lattice.

262   V. Integration of Equations of Motion

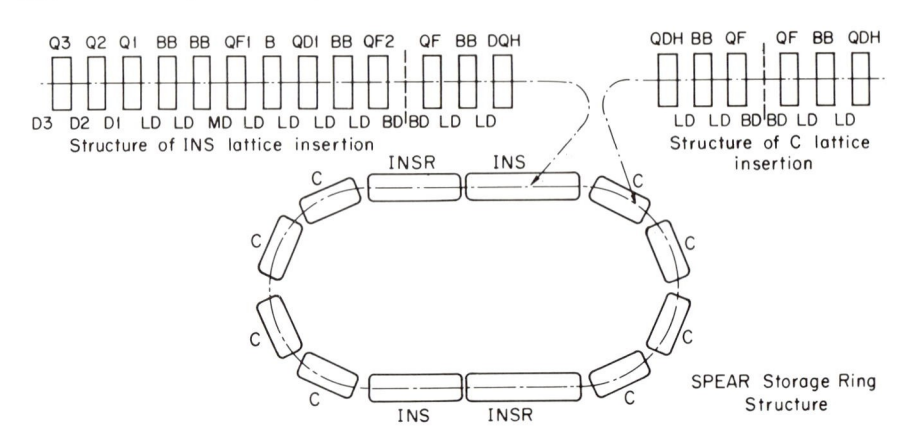

FIG. 104. SPEAR Storage Ring structure showing computer notation used in program SYNCH.

A normal lattice cell, C, is defined in the next instruction, MMM, which is made up of an ordered sequence of twelve previously defined magnets and drifts. This operation generates the horizontal and vertical transfer matrices of C. The next element INS, also an MMM operation, defines half of an insertion

Table 58

SYNCH Data for Illustrative Example

2-6	8-12	13-15	17-19	21	31	41	51	61	71
Element	OP	K	L	Data.....					
AL1	RUN			SPEAR LATTICE					
QDH	MAG			0.25	−32.753	50.03454	0.0	0.0	0.0
QD	MAG			0.50	−32.753	50.03454	0.0	0.0	0.0
QD1	MAG			0.5	−3.59	50.03454	0.0	0.0	0.
QF	MAG			0.5	23.494	50.03454	0.0	0.0	0.0
QF1	MAG			0.5	13.071	50.03454	0.0	0.0	0.0
QF2	MAG			0.5	−12.153	50.03454	0.0	0.0	0.0
Q1	MAG			0.5	−10.512	50.03454	0.0	0.0	0.0
Q2	MAG			1.25	16.819	50.03454	0.0	0.0	0.0
Q3	MAG			1.0	−30.845	50.03454	0.0	0.0	0.0
BB	MAG			2.35	0.0	12.71648	1.0	0.0	0.0
B	MAG			1.175	0.0	12.71648	1.0	0.0	0.0
LD	DRF			0.625					
BD	DRF			1.5					
MD	DRF			2.83682					
D1	DRF			5.4699					
D2	DRF			0.75					
D3	DRF			2.5					
C	MMM	12		QDH LD	BB LD	QF BD	BD QF	LD BB	LD QDH
INS	MMM	27		D3 Q3	D2 Q2	D1 Q1	LD BB	LD BB	MD CF1
				LD B	LD QD1	LD BB	LD QF2	BD BD	QF LD
				BB LD	QDH				
INSR	REF			INS					
SR	MMM	12		INS C	C C	C	INSR INS C	C C	C INSR
	CYB	12		SR	C				
	FIN								
	STOP								

from the beam interaction point to a cell. The other half of the insertion, INSR, which is the mirror reflection of INS, is defined by the REF operation. The entire storage ring lattice is defined in the instruction named SR, which is composed of cells C and the two halves of the insertion, INS and INSR. The following CYB instruction causes calculation and printing of the orbit functions at the boundaries between the constituent composite blocks that make up SR.

The next instruction, CYB, instructs the program to cycle the previously-defined matrix product (SR) without saving the cycled matrices. Specifically, CYB causes the program to calculate κ' (in this case $\kappa' = 12$) of the successive cycled products of SR, a product of κ ($> \kappa'$) matrices defined by the preceeding instruction. The output from CYB is dictated from the value of the integer in the data field. Since this value is zero, the betatron functions of the 12 cycled products are printed.

The output produced by SYNCH (see Fig. 105) varies according to the output instructions used. For the example mentioned here the quantities listed in the

```
           SYNCH RUN AL1            SPEAR LATTICE
================================================================================
***  QDH    MAG          //  0.25     -32.753   50.03454  0.0      0.0      0.0
***  QD     MAG          //  0.50     -32.753   50.03454  0.0      0.0      0.0
***  QD1    MAG          //  0.5       -3.59    50.03454  0.0      0.0      0.
***  QF     MAG          //  0.5       23.494   50.03454  0.0      0.0      0.0
***  QF1    MAG          //  0.5       13.071   50.03454  0.0      0.0      0.0
***  QF2    MAG          //  0.5      -12.153   50.03454  0.0      0.0      0.0
***  Q1     MAG          //  0.5      -10.512   50.03454  0.0      0.0      0.0
***  Q2     MAG          //  1.25      16.819   50.03454  0.0      0.0      0.0
***  Q3     MAG          //  1.0      -30.845   50.03454  0.0      0.0      0.0
***  BB     MAG          //  2.35      0.0      12.71648  1.0      0.0      0.0
***  B      MAG          //  1.175     0.0      12.71648  1.0      0.0      0.0
***  LD     DRF          //  0.625
***  BD     DRF          //  1.5
***  MD     DRF          //  2.83682
***  D1     DRF          //  5.4699
***  D2     DRF          //  0.75
***  D3     DRF          //  2.5
***  C      MMM    12    //  QDH  LD  BB  LD  QF  BD  BD  QF  LD  BB  LD  QDH
***  INS    MMM    27    //  D3   Q3  D2  Q2  D1  Q1  LD  BB  LD  BB  MD  QF1
  *                      //  LD   B   LD  QD1 LD  BB  LD  QF2 BD  BD  QF  LD
  *                      //  BB   LD  QDH
***  INSR   REF          //  INS
***  SR     MMM    12    //  INS  C   C   C   C   INSR INS C   C   C   C  INSR
***         CYB    12    //  SR   0
--------------------------------------------------------------------------------
         SR
POS        L      PSIX      BETAX   ALPHAX     XEQ      DXEQ      PSIY     BETAY   ALPHAY     YEQ   DYEQ
 0         0.     0.        1.2504  -.00000   4.99932  -.00000    0.       .0499   -.00000    0.    0.
 1 INS    35.13172 .49200    .9714  -.00083    .45426   .00019    .69439 29.8191  -.00163    0.    0.
 2 C      46.83172 .90233    .9703  -.00096    .45452  -.00011   1.00324 29.7790   .00110    0.    0.
 3 C      58.53172 1.31297   .9697  -.00000    .45414   .00000   1.31239 29.8635   .00000    0.    0.
 4 C      70.23172 1.72261   .9703   .00096    .45452   .00011   1.62155 29.7790  -.00110    0.    0.

 5 C      81.93172 2.13314   .9714   .00083    .45426  -.00019   1.93040 29.8191   .00163    0.    0.
 6 INSR  117.06344 2.62514  1.2504  -.00000   4.99932  -.00000   2.62478   .0499  -.00000    0.    0.
 7 INS   152.19516 3.11714   .9714  -.00083    .45426   .00019   3.31917 29.8191  -.00163    0.    0.
 8 C     163.89516 3.52747   .9703  -.00096    .45452  -.00011   3.62802 29.7790   .00110    0.    0.
 9 C     175.59516 3.93771   .9697  -.00000    .45414   .00000   3.93717 29.8635   .00000    0.    0.

10 C     187.29516 4.34795   .9703   .00096    .45452   .00011   4.24633 29.7790  -.00110    0.    0.
11 C     198.99516 4.75828   .9714   .00083    .45426  -.00019   4.55518 29.8191   .00163    0.    0.
12 INSR  234.12688 5.25028  1.2504  -.00000   4.99932  -.00000   5.24957   .0499  -.00000    0.    0.

R =    37.26245                          THETA =  6.28318528        QX =   5.25028098         QY =   5.24956578
MAXIMA   BETAX( 1)=  1.25042001  XEQ( 1)=   4.99931832   BETAY( 3)= 29.86353185
                                                                        TRANSITION GAMMA =   4.70925752
--------------------------------------------------------------------------------
***       FIN     0    0 //    LFILE =   508,  LINF =   4128,  UNUSED STORAGE =   3620.
================================================================================
END OF SYNCH RUN AL1
```

FIG. 105. SYNCH output showing computed orbit functions at various points along the storage ring structure.

264   V. Integration of Equations of Motion

output include closed orbit information (XEQ, DXEQ, YEQ, DYEQ), the path length 5, the cumulative phase advance to the beginning of the i-th element (PSIX, PSIY), as well as the betatron function (BETAX, BETAY).

The FIN card signals the end of the run, and the STOP card signals the end of the entire problem.

38. Program AGS

a. Introduction

Written at CERN, by Keil and Strolin [96], this program is used to analyze the behavior of alternating gradient synchrotrons and aid the experimenter in arriving at an optimal design. The high cost of constructing such instruments makes it most desirable to determine a priori the optimum lattice structure before the construction phase.

Program AGS computes the transformation matrices of the elements that make up the synchrotron, computes the betatron and closed orbit functions, the coordinates of the equilibrium orbit, and other pertinent quantities. It has been written in Fortran for the CDC 6600 computer system and a typical run is completed in less than one minute. Memory requirements depend on the number of elements that the program can handle. For maximum efficiency the program is overlaid and more than 1500 elements can be accommodated in less than $50K_8$ memory locations.

b. Program description

Program AGS is, in many respects, similar to program SYNCH, described previously. AGS is organized around a set of control cards, each of which directs the program to perform a specific function. This function may be to read data necessary to define a synchrotron cell or superperiod, to perform some input/output function, or to direct the program to some other task. The control cards in proper sequence are read one at a time, executed by the program, and the indicated operations are performed. The program checks the validity of the information read, and prints various error diagnostics to aid the user in case of input errors.

c. Illustrative example

Data for a simple FOFDOD structure appear in Table 59. The first control card (TE) instructs the program to read the next card and store it as a heading.

Table 59

AGS Sample Input Data

			Card Column			
1	5	10	20	30	40	50

```
TF
FOFDOD   STRUCTURE.           ILLUSTRATIVE EXAMPLE.
S3
    50
(
    1       3      5 F1
    2
    1       3      5 F2
    1       4      6 D1
    2
    1       4      6 D2
)
E
    0
    6
    12.5
    21.5
    30.031
    40.032
    5-0.06
    60.06
```

The third control card (s3) instructs the program to read and store the cell structure that follows. The next card (50) calls for 50 repetitions of the FOFDOD cell defined in the following bracketed 6 cards. The control card (E) signals the end of structure reading, and the beginning of reading and storing momentum error cards. In this case since the next card contains zero, no momentum cards are read. The next card indicates to the program that six parameter cards follow: the first parameter card indicates that the length of the magnets is 2.5 meters; the second card, indicates that the length of the straight sections is 1.5 meters; the third and fourth cards indicate that the angles are 0.031 and 0.032 radians respectively for the focusing and defocusing magnets while the last two cards show the normalized gradient parameters to be -0.06 and 0.06 (meters^{-2}) respectively for the focusing and defocusing magnets.

The output produced by this sample problem is shown in Fig. 106 and is mostly self-explanatory; lengths are printed in meters, angles in milliradians. (However, they are punched on cards in radians.) The gradient parameter

FOFDOD STRUCTURE. AGS ILLUSTRATIVE EXAMPLE.

CIRCUMFERENCE = 650.0000   50 SUPERPERIODS   WEDGE MAGNETS

NO		LENGTH (METERS)	DL/DX	D2L/DX2	ANGLE (MRAD)	1/(METERS)	K(V)	DK(V)/DX	D2K/DX2	X (METERS)	XX (RADIANS)	BETAX	BETAZ	ALPHAX	ALPHAZ	MUX/2PI	MUZ/2PI
1	F1	2.5000	0.000	0.000	31.00000	0.000	-0.0600000	0.0000000	0.0000000	1.63495	0.24657	13.61	13.75	-2.164	2.183	0.02288	0.03912
2		1.5000	0.000	0.000	0.00000	0.000	0.0000000	0.0000000	0.0000000	1.95320	0.00000	19.71	8.69	0.038	0.087	0.03501	0.06674
3	F2	2.5000	0.000	0.000	31.00000	0.000	0.0600000	0.0000000	0.0000000	1.95320	0.00000	19.71	8.69	-0.038	-0.087	0.05789	0.10586
4	D1	2.5000	0.000	0.000	32.00000	0.000	0.0600000	0.0000000	0.0000000	1.63495	-0.24657	13.61	13.75	2.164	-2.183	0.09745	0.12852
5		1.5000	0.000	0.000	0.00000	0.000	0.0000000	0.0000000	0.0000000	1.33599	0.00000	8.59	19.90	0.088	-0.038	0.12539	0.14053
6	D2	2.5000	0.000	0.000	32.00000	0.000	0.0600000	0.0000000	0.0000000	1.33599	0.00000	8.59	19.90	-0.088	-0.038	0.16494	0.16319

		BETAMAX(H)	COSMU(V)	Q(V)	BETAMAX(V)	XMAX(H)	GAMMA TR.
DP/P	COSMU(H)	Q(H)					
0.0000	0.50934	8.2472	19.7066	0.51880	8.1595	19.8981	1.9532	7.9381

FIG. 106. Typical printout produced by program AGS; in general the amount of output produced is controlled by IPRINT which is the third entry after the E-type control card (see Table 59).

k_r is defined as

$$k_r = 1/B\varrho(dB_z/d_x), \tag{38.1}$$

where $B\varrho$ is the magnetic rigidity of the particle. The momentum compaction functions (X,XX) are given in meters and radians respectively. The betatron functions are printed at the entrance of the element; while the accumulated phase advance (MUX,MUZ) is given at the exit of the element in radians/2π.

Part C | **LINEAR ACCELERATOR PROGRAMS**

39. Introduction

The computer programs described in this chapter deal with problems occuring in the design of linear accelerator-type cavities. These programs solve a wave-type equation in terms of an eigenfunction expansion (with the exception of program AZTEC); the eigenvalue is calculated from a variational principle or some other way [97]. The resulting equations are reduced to finite difference relations applied to a grid and the solutions are obtained by a relaxation procedure. The type of relaxation method used varies, perhaps, the most popular being the PSOR method (Point-Successive Overrelaxation). However, experimenters have reported that all methods produce similar results in similar convergence times provided the initial guesses are good.

Program AZTEC is perhaps more general in the sense that it solves the wave equation for the electromagnetic fields and self-impedance of an azimuthally bunched beam in a cylindrical structure of arbitrary axial geometry. This cavity may consist of many layers of different materials.

Programs of the type described in this chapter have proved their usefulness in the design of cavities as well as in instability studies for electron ring accelerators.

VI | Programs for Linac Cavities

40. Program CURE

a. Introduction

This program by MacRoberts and Rich [98] is used to calculate electromagnetic fields, frequency, and other pertinent quantities arising in the design of resonant cavities; it is a modified version of program LALA [99, 100, 101]. CURE is capable of handling cavities of any reasonable cylindrically symmetric geometry described by a combination of straight line segments and circular arcs superposed on a square mesh, the size of which depends on the memory capacity of the computer. The cavity dimensions must be an integral number of mesh spacings. If this can not be done, one must run several cases with modified cavity dimensions which are integral multiples of the mesh spacing and interpolate the results for the actual cavity dimensions.

Program CURE was developed at the Los Alamos Scientific Laboratory and has been used there extensively in the design of the linear accelerator cavities for the meson facility (see MacRoberts and Rich [98]). Because of the similarity of cavities to certain microwave power tubes, CURE may also be used successfully for their design. The program is written in Fortran and is operational on the CDC 6600/7600; it occupies $66K_8$ memory locations for a 0.25×0.25 mesh size and $110K_8$ memory locations for a 0.1×0.1 mesh size. A typical problem, such as described later, requires less than two minutes of central processor time (CDC 6600).

b. Program Description

The mathematical model used to simulate this particular type of problem has been adequately described by Hoyt [99] and Hoyt et al. [100].

It is sufficient here to say that CURE solves a wave equation of the form:

$$\nabla^2 \mathbf{H} = (1/c^2)(\partial^2 \mathbf{H}/\partial t^2), \qquad (40.1)$$

in which **H** is assumed to have only a θ-component given by

$$H_\theta = He^{j\omega t}. \qquad (40.2)$$

This condition applies for the lowest frequency mode of the cavity which is the case of interest.

Expressing Eq. (40.1) in cylindrical coordinates and setting $F = rH_\theta$ to simplify the computational model, we obtain

$$\partial^2 F/\partial r^2 - 1/r(\partial F/\partial r) + \partial^2 F/\partial z^2 + K^2 F = 0, \qquad (40.3)$$

where the eigenvalue K^2 is related to the frequency ω and the wavelength λ by

$$K^2 = \omega^2 \mu \varepsilon = (2\pi/\lambda)^2. \qquad (40.4)$$

The eigenvalue K^2 is computed by:

$$K^2 = \frac{\iint 1/r[(\partial F/\partial r)^2 + (\partial F/\partial z)^2]\, dr\, dz}{\iint F^2/r\, dr\, dz}. \qquad (40.5)$$

Equations (40.3) and (40.5) and the appropriate boundary conditions are replaced by their equivalent finite difference equations which are solved by a point-by-point relaxation scheme. The problem converges to the lowest eigenvalue using a double iterative process. This process involves first the guessing of a value for the eigenvalue to be used to compute F, which in turn is used to compute a new value for K^2. This process continues until the convergence criteria have been satisfied.

The overrelaxation parameter, ϱ, is set to 1.9 until the lowest eigenvalue is obtained. Then ϱ is reduced by 10% and this value is used until the F's have converged. The convergence is rapid and a solution is usually obtained in less than two minutes on the CDC 6600 computer.

This algorithm has been implemented in CURE by making use of the following subroutines:

1. MESH, MSHPNT, PRPLT, WL, and SUM generate the mesh of the described cavity, calculate and store quantities needed, print various quantities (depending on the appropriate value of the print control card-PC), plot on the generated mesh.

2. LAMCAL calculates the eigenvalue K^2.

272   VI. Programs for Linac Cavities

3. SDVR computes D^2F/DR^2 at $Z = 0$.
4. EPC calculates the electric and magnetic fields plus the surface lengths.
5. SOA computes the shunt impedance off the axis (ZT^2).
6. SKOT makes a plot of equal F values.
7. FLOAD provides for the initial loading of F.

c. Preparation of Input Data

Program CURE provides the experimenter with a flexible input structure, relative ease of data preparation and a variety of options from which to print a selective output. Data are organized utilizing the Fortran NAMELIST statement. Briefly, this statement allows the input and output of character strings, consisting of names and values, without a format specification. Each name within the NAMELIST statement consists of 1 to 7 characters and must be unique within the program unit in which each is used. Table 60 shows all the names reserved for the NAMELIST statement, their function within the program, and their default values.

The order of data cards that constitute the input deck is as follows

1st card	Problem title. All 80 columns may be used
2nd card	$NINPUT—Reserved word for reading the NAMELIST parameters ($ must be in column 2)
3rd–Nth card	NAMELIST parameters (selected from Table 1)
Nth + 1 card	* END TAPE (* in column 1, END TAPE begins from column 8).

In preparing the cards for the description of the boundary of the cavity under consideration, one should observe the following rules:

1. The boundary must be connected by a single continuous line.
2. Each line segment must be recorded by the beginning coordinate only. Each segment joins the segment following, except for the last segment which joins the first.
3. The segments are recorded so that the interior of the problem is always to the right.
4. The coordinates of the center (RC, ZC), and the radius (R) of circular arcs are entered at the beginning points of the arc.
5. The radius of the arc is positive if the arc is followed in a counterclockwise direction; negative if clockwise (with the problem interior to the right).
6. Enter the material for each line segment.

Table 60
NAMELIST Parameters for Program CURE

Parameter	Description	Standard value
RB(I)	Beginning r coordinate of each line segment ($1 \leq I \leq 20$)	(Specify)
ZB(I)	Beginning z coordinate of each line segment	(Specify)
RC(I)	The r coordinate center of circular arc	(Specify)
ZC(I)	The z coordinate center of circular arc	(Specify)
R	Radius of curvature of circular arc	0.0
METAL(I)	Material type: if METAL = 1, metallic surface; if METAL = 0, non-metallic	0
TYPE(I)	Boundary condition type: if TYPE = 0, Neumann boundary; if TYPE $\neq$ 0, Dirichlet boundary	0
H	Mesh size	0.1 cm
LOAD	Initial loading condition: if LOAD = 1, Bessel function; if = 2, linear bottom to top; if = 3, linear lower left to upper right with F = minimum(r,z)	1
CONVL	Eigenvalue convergence criterion: Abs($\lambda_n^2 - \lambda_0^2 / \lambda_n^2$) < CONVL	1.0E-5
CONVF	F convergence = maximum Abs($F_n - F_0 / F_n$) < CONVF	5.0E-3
MODE	π-mode or 2π-mode	3.14159
NOL	Number of line segments and circular arcs used to describe geometry of problem ($1 \leq NOL \leq 20$)	(Specify)
W	Energy gain of particle as it crosses cavity	1.25 MeV
NLINE	Number of lines of equal F to be plotted ($1 \leq NLINE \leq 200$)	10
FREQ	Initial frequency guess	$1.15E4/r_{max}$
CAV	If CAV = 1, use half-cavity; if CAV = 2, use full cavity	1
CONVV	Voltage convergence criterion: Abs($V_S - V_A$)/V_S < CONVV	0.05
SIG	Conductivity of all metalic surfaces (Mhos/meter)	5.91E7
RUNTIM	Maximum time allocated for convergence before output computation	270.0 sec
RSNT(I)	Values of r coordinate along which shunt impedance (ZT^2) is to be calculated off the axis	0.0
REH(I)	Values of r coordinate along which the electric and magnetic fields are to be calculated	0.0
FLOW	If FLOW = .TRUE., run flow-type problem, plot output values of equal F	.FALSE.
RELAX	Overrelaxation parameter ($1.0 \leq RELAX \leq 2$)	1.9
PC1,PC2, PC3,PC4	If PC1 = .TRUE., mesh printed as generated by subroutine MESH; if PC2 = .TRUE., line-by-line listing; if PC3 = .TRUE., an ordered listing of the areas of the boundary mesh will be made; if PC4 = .TRUE., print every point at end of each y cycle	.FALSE.
PC5,PC6	If PC5 = .TRUE., print eigenvalue λ^2, change $\lambda^2(d\lambda^2/\lambda^2)$ and maximum abs ($\Delta F/F$) for every cycle; if PC6 = .TRUE., print electric- and magnetic-field values at every boundary point	.FALSE.
PC7,PC8	If PC7 = .TRUE., print F at all mesh points when converged or time limit reached; if PC8 = .TRUE., print mesh point types	.FALSE.
PC9	If PC9 = .FALSE., plots will not be produced	.TRUE.

274   VI.   Programs for Linac Cavities

The illustrative examples that follow will demonstrate the manner by which data are prepared and the application of the simple rules outlined above.

d. Illustrative Examples

Consider the shaped linac cavity shown in Fig. 107. Since calculations for this cavity have been performed by program LALA and reported by Hoyt [101], it will be sufficient here to illustrate the manner by which data are prepared and compare some of the results obtained.

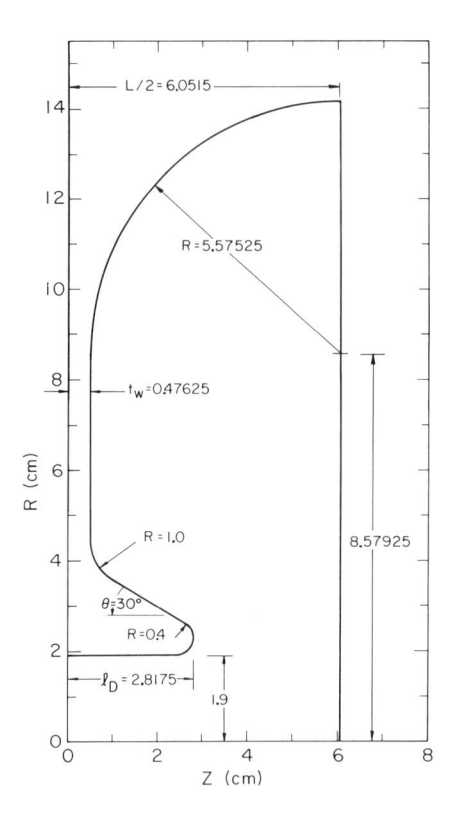

FIG. 107. Shaped 805-MHz proton linac cavity.

Table 61 shows the input data for this cavity. The geometry of the cavity is described in terms of the beginning coordinates of the straight-line segments (RB,ZB) and the circular arcs (RC, ZC). If one relates these coordinates to the physical dimensions of the cavity appearing in Fig. 107, he will observe that the rules outlined earlier prevail, and that these six cards (RB, ZB, RC, ZC, R, METAL) adequately describe the cavity under consideration. The rest of the cards describe various quantities and set various flags for selective output.

Table 61

CURE Input Data Linac Cavity

			Card Column				
2	10	20	30	40	50	60	70

```
805 MHZ PROTON LINAC CAVITY.TEST CASE.
$NINPUT
RB(1)=0.0,1.9,1.9,2.6464,3.584,4.45,8.57925,14.15175,0.0,
ZB(1)=0.0,0.0,2.4127,2.6127,0.97625,0.47625,0.47625,6.0515,6.0515,
RC(1)=0.0,0.0,2.3,0.0,4.45,0.0,8.57925,0.0,0.0,
ZC(1)=0.0,0.0,2.4127,0.0,1.47625,0.0,6.0515,0.0,0.0,
R(1)=0.0,0.0,0.4,0.0,-1.0,0.0,-5.57525,0.0,0.0,
METAL(1)=0,1,1,1,1,1,1,1,0,
H=0.20,NOL=9,RUNTIM=200.,CAV=1,
PC2=.T.,
PC3=.T.,
PC5=.T.,
PC6=.T.,
PC7=.T.,
$
*      END TAPE
```

Figures 108 and 109 show some of the output produced by CURE. Figure 108a shows an outline of the generated mesh; Fig. 108b shows a plot of the equipotential lines (lines of equal $\Gamma - rH_0$), while Fig. 108c shows some of the calculated quantities useful in accelerator cavity design. The upper half of Fig. 109 shows a plot of the electric field calculated on the axis of the cavity, while the lower half shows the magnetic field calculated by the program.

As it can be seen from Table 62, CURE compares very favorably with program JESSY and MESSYMESH (see Sections 41 and 42) for the linac cell whose geometry appears in Fig. 110. The resonant frequency of this cell as well as other pertinent quantities such as shunt impedance and transit time factor were computed by program CURE. A listing of the data necessary to describe this geometry is shown in Table 63.

The effect of mesh size on the accuracy of the solution, as well as the timing considerations of the program, are shown in Fig. 111.

(a)

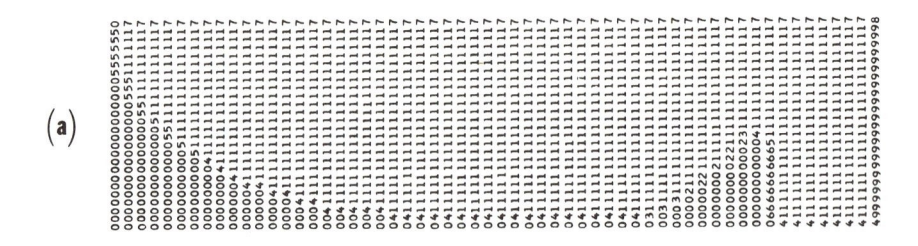

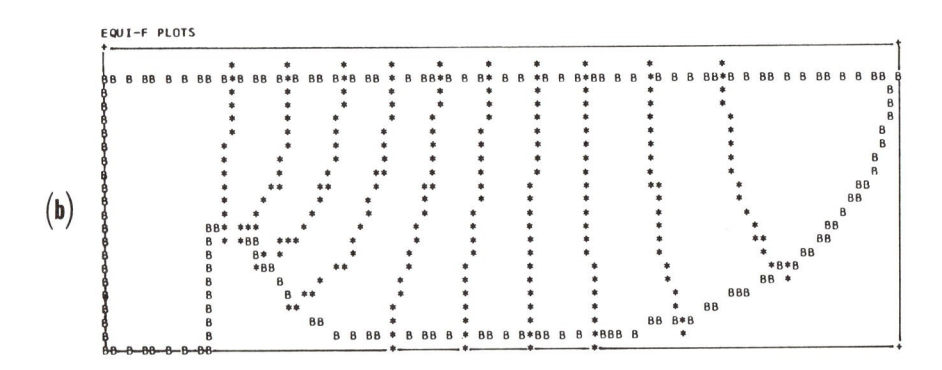

(b) EQUI-F PLOTS

(c)
```
OUTPUT 805 MHZ PROTON LINAC CAVITY.TEST CASE.

NORMALIZATION FACTOR = 38.96152E+01

CAVITY LENGTH = 12.10300E+00 CENTIMETERS

FREQUENCY =      804.19 MEGA-HERTZ

W = 12.50000E-01 MEV/METER

BETA (RELATIVE VELOCITY) = .649323

VOLTAGE  SURFACE INTEGRAL = 18.14130E+04 VOLTS   AXIAL INTEGRAL = 18.31892E+04 VOLTS

PARTICLE ENERGY = 29.54517E+01 MEV

TRANSIT TIME FACTOR  =  -83.33814E-02

ENERGY STORED IN CAVITY = 21.54904E-03 JOULES    POWER LOSS = 64.77716E+02 WATTS

MAXIMUM SURFACE ELECTRIC FIELD = 67.54300E+05 VOLTS/METER AT R = 25.00000E-01  Z = 27.38638E-01

SHUNT IMPEDANCE (Z) = 51.87518E+05 OHMS

Z/L =  42.86142E+06  Z/(LQ) =  25.49893E+02  ZT2 =   36.02859E+05  ZT2/L =  29.76831E+06  ZT2/(LQ) =  17.70964E+

Q FACTOR = 16.80910E+03

   SURFACE     POWER LOSS
      2        91.57358E-02
      3        22.94654E+00
      4        23.30490E+01
      5        20.37074E+01
      6        83.15072E+01
      7        26.49578E+02
      8        25.36015E+02
```

FIG. 108. (a) Computer generated problem outline indicating type of point for every mesh unit (b) computer generated equipotential plot ($F = rH$) and (c) typical output showing quantities computed.

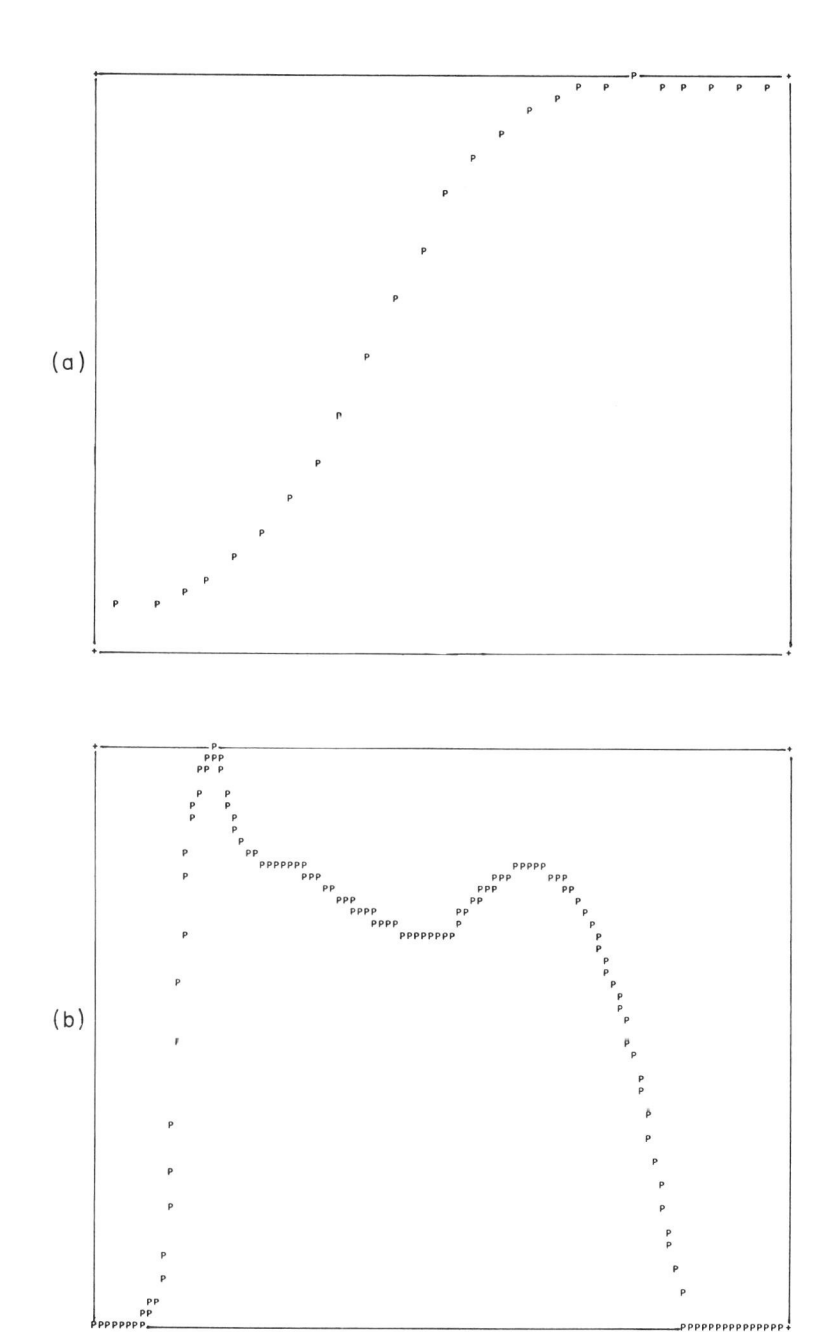

FIG. 109. A computer generated plot of (a) the electric field (volts/meter), axial electric field length = 6.55150E + 00, maximum field = 2.36423E + 06, and (b) the magnetic field (Ampere-turns/meter), magnetic field length = 4.12171E + 01, maximum field = 3.20773E + 03.

Table 62

Comparison of Measured and Computed Resonant Frequencies for Several Linac Cavities at MURA

Length (L/2) (centimeters)		3.350	4.039	4.717	5.565	6.304
Length (G/2) (centimeters)		0.947	1.128	1.293	1.633	1.864
Frequency (MHz)	Measured	202.47	200.70	198.81	200.57	198.94
	JESSY	203.07	201.64	199.59	200.24	199.23
	MESSYMESH	203.06	201.16	199.24	200.19	199.20
	CURE	202.80	201.24	199.19	200.33	199.49
Relative error (%)	JESSY	0.30	0.47	0.39	-0.17	0.14
	MESSYMESH	0.29	0.23	0.21	-0.18	0.13
	CURE	0.16	0.23	0.19	-0.12	0.28

Table 63

CURE Input Data for Linac Cell Geometry

		Card Column				
2	10	20	30	40	50	60

```
SAMPLE CAVITY FOR COMPARISON WITH PROGRAM JESSY.

$NINPUT

RB(1)=0.0,1.0,1.0,2.0,7.0,9.0,9.0,49.62,49.62,0.0,

ZB(1)=0.0,0.0,1.403,2.403,2.403,0.403,0.0,0.0,3.350,3.350,

RC(1)=0.0,0.0,2.0,0.0,7.0,0.0,0.0,0.0,0.0,0.0,

ZC(1)=0.0,0.0,1.403,0.0,0.403,0.0,0.0,0.0,0.0,0.0,

R(1)=0.0,0.0,1.0,0.0,2.0,0.0,0.0,0.0,0.0,0.0,

METAL(1)=0,1,1,1,1,1,1,0,0,0,

H=0.25,NOL=10,RUNTIM=200.,CAV=1,

PC2=.T.,

PC3=.T.,

PC5=.T.,

PC6=.T.,

PC7=.T.,

$

*       END TAPE
```

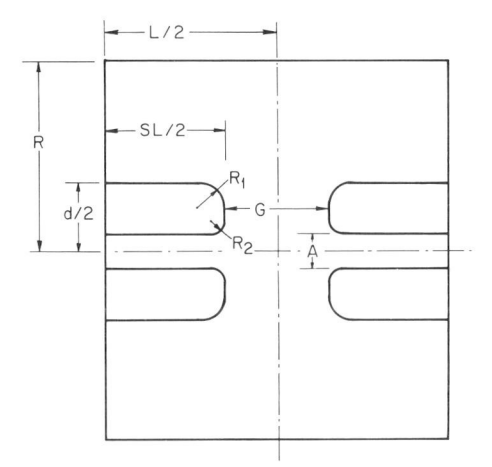

FIG. 110. Geometry of a linac cell.

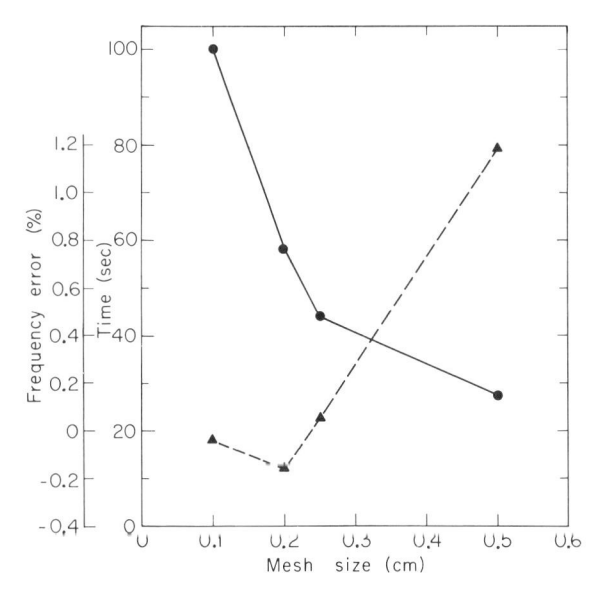

FIG. 111. Plot showing the dependence of mesh size on the accuracy and of the solution and the computer time needed. (●—●) time, (▲- -▲) error.

41. Program JESSY

This program, developed by Parzen [102] and Dahl et al. [103] considers the problem of computing the resonant frequency and the electromagnetic fields of linac cavities. It is essentially similar to program CURE (just described) in that it replaces the wave equation by a set of difference equations for each point on the rectangular mesh, which are solved by the point overrelaxation method.

The rate of convergence may be accelerated by:

1. Performing a run with a coarse mesh-interval and then using the results as the initial loading to a second run.

2. Improving the initial guess for F ($F = rH_0$) by the empirically-found expression,

$$F = r^2/(\bar{R} + d/2)^2(L/2)/\bar{R} + (r/R)^2[1 - 0.5(r/R)^2] \tag{41.1}$$

where

$$\bar{R} = [r^2 + (z - L/2)^2]^{1/2} \tag{41.2}$$

The parameters R, d, L are defined in Fig. 110.

3. Monitoring and modifying the overrelaxation parameter in a manner producing faster convergence.

A comparison of this program (see Dahl *et al.* [103]) with program CURE was made and the results appear in Table 62. Although program CURE used a 0.25×0.25 cm mesh, program JESSY, having an unequal mesh size, may accomodate smaller sizes in the z-direction thus effecting a higher accuracy. The results indicate an upper limit of about 0.5% in frequency.

42. Program MESSYMESH

This program, written in 1961, by Edwards [104, 105] and Austin *et al.* [106] was perhaps the first serious attempt to use a numerical model for the solution of the wave equation, as applied to problems relative to the design of resonant cavities. It was originally written in a combination of Fortran and machine language (SAP) subprograms appropriate to the IBM 704 computer system. Subsequently modified and revised, it is presently operational on the IBM 360 computer system.

The computational accuracy of this program, verified by other experimenters (see Dahl *et al.* [103]) is compared with JESSY and CURE also in Table 62.

The present status of this program is semi-operational, and perhaps obsolete, since programs like CURE and JESSY perform the same type of calculation faster, with a smaller program, and with comparable accuracy.

43. Program AZTEC

a. Introduction

This program by Sackett and Garren [107] calculates the fields due to an azimuthally bunched beam in cylindrically symmetric structure of arbitrary geometry. The computed fields are then used to calculate the self-impedance of the beam for stability studies. Any combination of dielectric, magnetic, and conducting materials is allowed. Material properties, however, are assumed to be isotropic and linear. Boundaries between different materials within the problem space may be arbitrary curves, as may the contour defining the beam region.

AZTEC has been written in Fortran and Compass for the CDC 7600 computer. The program requires $45K_8$ words of SCM (small core memory) to load; after loading, the program will automatically increase the field length to the size required for the mesh input. Running time depends upon the harmonic range input, the number of beam positions, and the number of mesh points. For a 2-mode problem with 15 beam positions, the run time is approximately fifty seconds in a 11×47 mesh. AZTEC is presently utilized at Lawrence Berkeley Laboratory to investigate the evidence of negative mass instabilities in the "compressor 4" of the Electron Ring Accelerator project [108]. The accuracy of the program has been verified with analytic solutions and the upper limit is approximately one percent.

b. Mathematical Development

The mathematical development as well as operational details of this program will be published by the authors (see Sackett and Garren [107]), briefly the equations governing this program are Maxwell's equations

$$\nabla \times \mathbf{E} = -\partial \mathbf{B}/\partial t, \quad \nabla \times \mathbf{H} = \mathbf{J} + \partial \mathbf{D}/\partial t, \tag{43.1}$$

conservation of charge

$$\nabla \cdot \mathbf{J} = -\partial \varrho/\partial t, \tag{43.2}$$

the constitutive relations

$$\mathbf{D} = \varepsilon_0 \varepsilon \mathbf{E}, \quad \mathbf{B} = \mu_0 \mu \mathbf{H}, \tag{43.3}$$

and Ohm's law

$$\mathbf{J} = \sigma \mathbf{E}. \tag{43.4}$$

To simplify the problem, we assume that all quantities vary as $e^{i(n_0\phi - \omega t)}$. We also assume that $\varepsilon = \mu = 1$, $\sigma = 0$, in the beam region, and that $J_r = J_z = 0$ everywhere. Using these assumptions and equations (43.1) thru (43.4) we obtain

$$\nabla \times \mathbf{E} = i\omega^* \mu \mathbf{H}^0 \tag{43.7}$$

$$\nabla \times \mathbf{H}^0 = \mathbf{J}^0 - i\omega^* \varepsilon' \mathbf{E} \tag{43.8}$$

where $\varepsilon' = \varepsilon + i\sigma/\varepsilon_0 \omega$, $\omega^* = \omega/c$, $\mathbf{H}^0 = Z_0 \mathbf{H}$, $\mathbf{J}^0 = Z_0 \mathbf{J}$, and $Z_0 = (\mu_0/\varepsilon_0)^{1/2} = 1/\varepsilon_0 c$. The exponential dependence has been removed so that $\mathbf{E}$ and $\mathbf{H}^0$ are complex. Equations (43.7) and (43.8) provide six complex equations for the six complex unknowns E_r, E_ϕ, E_z, H_r^0, H_ϕ^0, H_z^0. The r and z components of Eqs. (43.7) and (43.8) are algebraic in E_r, E_z, H_r^0, H_z^0. Solving for these quantities and substituting the resulting expressions, which involve derivatives of E_ϕ and H_ϕ^0, into the ϕ components of Eqs. (43.7) and (43.8) we have the following system of coupled complex equations for E_ϕ and H_ϕ^0

$$\frac{\partial}{\partial r}\left(\varkappa_1 \frac{\partial u}{\partial r}\right) + \frac{\partial}{\partial z}\left(\varkappa_1 \frac{\partial u}{\partial z}\right) - \frac{\partial}{\partial r}\left(\varkappa_3 \frac{\partial v}{\partial z}\right)$$
$$+ \frac{\partial}{\partial z}\left(\varkappa_3 \frac{\partial v}{\partial r}\right) - \omega^* \varepsilon' \frac{u}{r} = iJ_\phi^0, \tag{43.9}$$

$$\frac{\partial}{\partial r}\left(\varkappa_2 \frac{\partial v}{\partial r}\right) + \frac{\partial}{\partial z}\left(\varkappa_2 \frac{\partial v}{\partial z}\right) + \frac{\partial}{\partial r}\left(\varkappa_3 \frac{\partial u}{\partial z}\right)$$
$$- \frac{\partial}{\partial z}\left(\varkappa_3 \frac{\partial u}{\partial r}\right) - \omega^* \mu \frac{v}{r} = 0, \tag{43.10}$$

where $u \equiv rE_\phi$, $v \equiv rH_\phi^0$, $\varkappa \equiv (n_0^2 - \omega^{*2} r^2 \mu \varepsilon')^{-1}$, $\varkappa_1 \equiv \varkappa \omega^* \varepsilon' r$, $\varkappa_2 \equiv \varkappa \omega^* \mu r$, $\varkappa_3 \equiv \varkappa n_0$.

To obtain a numerical solution to Eqs. (43.9) and (43.10) we recast the problem as one in the calculus of variations. Define a functional

$$I(u) = \operatorname{Re} \int_A \left[\varkappa_1 \left(\frac{\partial u}{\partial r}\right)^2 + \varkappa_1 \left(\frac{\partial u}{\partial z}\right)^2 + \varkappa_2 \left(\frac{\partial v}{\partial r}\right)^2 \right.$$
$$+ \varkappa_2 \left(\frac{\partial v}{\partial z}\right)^2 - 2\varkappa_3 \frac{\partial u}{\partial r}\frac{\partial v}{\partial z} + 2\varkappa_3 \frac{\partial u}{\partial z}\frac{\partial v}{\partial r} + \frac{\omega^*}{r}(\varepsilon' u^2 + \mu v^2)$$
$$\left. + i2J_\phi^0 u \right] dr\, dz \tag{43.11}$$

If we can find u and v such that Eq. (43.11) is rendered stationary in the domain A, then Eqs. (43.9) and (43.10) will be satisfied in A along with the natural boundary conditions

$$\varkappa_1 \frac{\partial u}{\partial n} + \varkappa_3 \frac{\partial v}{\partial s} = 0, \qquad \varkappa_2 \frac{\partial v}{\partial n} - \varkappa_3 \frac{\partial u}{\partial s} = 0, \tag{43.12}$$

on the contour bounding A, and the natural interface conditions

$$\left[\varkappa_1 \frac{\partial u}{\partial n}\right] = -[\varkappa_3] \frac{\partial v}{\partial s}, \qquad \left[\varkappa_2 \frac{\partial v}{\partial n}\right] = [\varkappa_3] \frac{\partial u}{\partial s}, \qquad (43.13)$$

on all material interfaces within A. Note that Eq. (43.13), along with the assumed continuity of u and v, implies that we have continuity of parallel components of **E** and **H** and normal components of **B** and **D**, as required by Maxwell's equations. Application of the "finite element" method using Eq. (43.11) in a triangular mesh gives the approximating set of algebraic equations solved in AZTEC. Because of the unstable nature of problems of this type when an iterative method is used, AZTEC has been programmed to use direct (elimination) method which takes advantage of the symmetry and band structure of matrix equations.

c. An Illustrative Example

Program AZTEC is very flexible in the type of geometrical configurations that it can model as evidenced from the geometries shown in Fig. 112. As an example, the pillbox geometry is used and the results will be compared with those obtained by the computer programs mentioned by Entis *et al.* [108]. The quantity to be compared is the dimensionless complex number g_n defined by Neilsen *et al.* [109]:

$$E_n = -(ienN_n/2\pi r^2 \gamma^2)g_n, \qquad (43.14)$$

where $\gamma = (1 - \beta^2)$, $\beta = v/c$, and E_n is the azimuthal electric field averaged over the beam cross section, r is the major radius of the beam, n is the harmonic number of the density modulation and N_n is the number of particles.

The data required to simulate this geometry will not be described, nor will the operating instructions be given since the status of this program is still

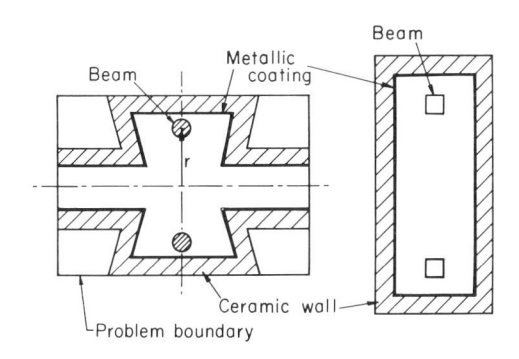

FIG. 112. Typical compressor-type cavities for electron ring accelerators.

```
5X23CM. PILLBOX (INFIN. CONDUCTING WALLS). BEAM AT 18CM.

MEAN RADIUS OF BEAM       =    .180000 METERS
KINETIC ENERGY OF BEAM    =   2.300000 MEV
XSECT. AREA OF BEAM       =    .000040 SQ.METERS
HARMONIC NUMBER           =   1
GAMMA                     =   5.500926
BETA                      =    .983338
K0                        =   5.462988
FREQUENCY                 = 260.666468 MHZ
KMAX                      =  11
LMAX                      =  62
THIS PROBLEM HAS SYMMETRY ABOUT THE PLANE K=1

MOD(GN) =  5.87388E+00    MOD(ZN/N) =  3.72102E+01    ZN/N = (   0.     ,  3.72102E+01)

  K   L    RE(R*EPHI)    IM(R*EPHI)    RE(R*HPHI)    IM(R*HPHI)
  1   2    6.82769E-04   0.            0.            0.
  2   2    6.74361E-04   0.           -1.16689E-04   0.
  3   2    6.49364E-04   0.           -2.30503E-04   0.
  4   2    6.08335E-04   0.           -3.38637E-04   0.
  5   2    5.52347E-04   0.           -4.38429E-04   0.
  6   2    4.82760E-04   0.           -5.27421E-04   0.
  7   2    4.01288E-04   0.           -6.03424E-04   0.
  8   2    3.09939E-04   0.           -6.64567E-04   0.
  9   2    2.10963E-04   0.           -7.09346E-04   0.
 10   2    1.06795E-04   0.           -7.36662E-04   0.
 11   2    0.            0.           -7.45842E-04   0.
  1   3    1.52013E-03   0.            0.            0.
  2   3    1.50141E-03   0.           -2.59860E-04   0.
  3   3    1.44572E-03   0.           -5.13317E-04   0.
  4   3    1.35443E-03   0.           -7.54127E-04   0.
  5   3    1.22976E-03   0.           -9.76357E-04   0.
  6   3    1.07483E-03   0.           -1.17454E-03   0.
                    ⋮                           ⋮
 11  59    0.            0.           -8.30397E-02   0.
  1  60   -1.24371E-02   0.            0.            0.
  2  60   -1.22573E-02   0.           -1.32545E-02   0.
  3  60   -1.17267E-02   0.           -2.60481E-02   0.
  4  60   -1.08733E-02   0.           -3.79700E-02   0.
  5  60   -9.73174E-03   0.           -4.86950E-02   0.
  6  60   -8.36081E-03   0.           -5.79944E-02   0.
  7  60   -6.81905E-03   0.           -6.57242E-02   0.
  8  60   -5.16788E-03   0.           -7.18004E-02   0.
  9  60   -3.45024E-03   0.           -7.61731E-02   0.
 10  60   -1.73205E-03   0.           -7.88103E-02   0.
 11  60    0.            0.           -7.96917E-02   0.
  1  61   -7.10501E-03   0.            0.            0.
  2  61   -7.00227E-03   0.           -1.27302E-02   0.
  3  61   -6.69940E-03   0.           -2.50469E-02   0.
  4  61   -5.21189E-03   0.           -3.65742E-02   0.
  5  61   -5.56372E-03   0.           -4.70013E-02   0.
  6  61   -4.78502E-03   0.           -5.50936E-02   0.
  7  61   -3.90915E-03   0.           -6.36881E-02   0.
  8  61   -2.96389E-03   0.           -6.96785E-02   0.
  9  61   -1.99220E-03   0.           -7.39982E-02   0.
 10  61   -9.98856E-04   0.           -7.66055E-02   0.
 11  61    0.            0.           -7.74771E-02   0.
  1  62    0.            0.            0.            0.
  2  62    0.            0.           -1.25409E-02   0.
  3  62    0.            0.           -2.46844E-02   0.
  4  62    0.            0.           -3.60665E-02   0.
  5  62    0.            0.           -4.63816E-02   0.
```

FIG. 113. Partial output printed by program AZTEC; the quantities listed are explained in the text.

experimental; the code is undergoing various modifications at this time to increase its generality and usefulness.

Figure 113 shows a partial printout obtained with this code; some of the input parameters are printed (kinetic energy, harmonic number, etc.) along with various important calculated parameters such as frequency, mean radius of the beam, etc. The amount of output is controlled by the user's properly assigning the limits (K,L) of the grid he desires to print. The quantity MOD(GN) specifies the modulus of g_n defined by Eq. (40.14), the quantity MOD(ZN/N) specifies the modulus of the impedance divided by the harmonic number; and ZN/N shows the complex components of the impedance. The quantities are the average across the beam. The output also gives the scaled field values (EPHI, HPHI) at each point of the mesh.

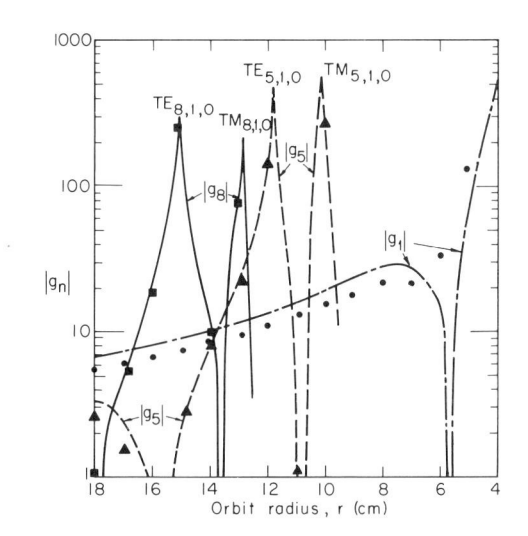

FIG. 114. Plot of $|g_n|$, $n = 1, 5$, and 8 versus r obtained for program AZTEC (dots) and that obtained from Zotter [110] (dashed line).

Figure 114 shows a plot of $|g_n|$ versus r for the pillbox model as obtained by using AZTEC and the program mentioned by Zotter [110]. The agreement of both programs is quite good, indicating a certain degree of confidence in the results. However, further experimentation is needed to establish the usefulness of AZTEC in more sophisticated geometries.

APPENDIXES

I | Summary of Computer Programs That Solve the Nonlinear Partial Differential Equation for Magnetostatics in 2 Dimensions

Date	Name Center	Authors	Method	Discretization	Remarks
1963	SIBYL MURA-LBL (Other variants include LINDA)	Christian [16, 24] Dahl [24] Parzen [24] Dorst [13, 24b] Colonias [17]	Two potential finite difference. SOR.	Fixed topology and geometry. Rectangular meshes in the two regions.	Design tool for HEP apparatus at LBL. Good agreement with measurements.
1964		Trut [111]	Scalar potential.	Fixed topology and fixed geometry. Rectangular mesh interfaces have to coincide with mesh lines.	
1964	TRIM/LBL Modification.	Winslow [6] Colonias [10] Diserens [9] Lari [26] Khoe [112]	Vector potential. Variational and finite difference. SOR.	Fixed topology and variable geometry. Triangular mesh.	Design tool for all kinds of magnets. In regular international use at LBL, RHEL, NAL, etc. Good agreement with measurements over wide range of fields into saturation.

I. Summary of Computer Programs

Date	Name Center	Authors	Method	Discretization	Remarks
1965	NUTCRACKER/ SLAC	Burfine [21] Anderson [21] Brechna [21] Anderson [22]	Vector potential finite difference SOR.	Fixed topology and geometry. Rectangular mesh. Later version appears to have variable geometry for currents.	At SLAC used for HEP; apparatus. Good agreement with measurement.
1966	MARE/CERN	Perin [18, 20] Van der Meer [19, 20]	Two potential. Extensive use made of Hornsby [21] program for solving elliptic PDE. Finite difference and SOR.	Fixed topology and geometry.	Powerful design tool at CERN. Used for design of ISR and numerous HEP magnets. Good agreement with measurement.
1966	Colorado Univ.	Erdelyi [29a, 30] Ahamed [29a, 30] Hopkins	Vector potential. Finite difference. Novel method of SOR.	Rectangular and polar meshes. Variable geometry.	Design tool for electromagnetic machinery.
1968		Reichert [113]	Vector potential. SOR.	Rectangular and polar lattices.	Applied to MDH magnets and to various rotationally symmetric magnets. Good agreement with experiment.
1970	McGill Univ.	Silvester [114] Chari [114]	Vector potential. Finite element (Variational method.) Gaussian for the linear part, Newton-Raphson for the non-linear variable part.	Variable topology and geometry. Triangular mesh.	One of the first attempts in applying the finite element method. Good agreement with measurements.
1970	JINR Dubna, USSR.	Vorozhtsov [123] Zakamskaya [123] Zaplatin [123]	Vector potential. Adaptation of Program TRIM.	Fixed topology and variable geometry. Triangular mesh.	Seem to be successfully utilized on the BESM-6 Computer.

Date	Name Center	Authors	Method	Discretization	Remarks
1970	FATIMA/CERN	Iselin [115]	Vector potential. Finite element (with first second order versions). SOR.	Fixed geometry and variable topology. Triangular mesh.	Treats anisotropy effects. Good agreement with measurement.
1970	GRACY	Parzen [28] Jellett [28]	Vector potential. Finite difference. SOR.	Fixed geometry and topology. Rectangular mesh.	Used in study of saturated yokes in superconducting dipoles.

II | Summary of Computer Programs That Solve the Partial Differential Equations for Magnetostatics in 3 Dimensions

Date	Name Center	Authors	Method	Discretization	Remarks
1967	Carnegie	Foss [39]	Scalar potential. Finite difference.	Three dimensional lattice.	Written for CDC G21 in local software. Little information available.
1970	MIMI-A CERN	Caeymaex [43]	Modified scalar potential. Air-iron only at present infinite μ boundaries. Finite difference SOR.	Three dimensional lattice with variable spacing in Z direction.	Used for assessing the performance of beam handling magnets with end effects. Good agreement with measurements.
1971	AEG-Telefunken Germany	Müller [116] Wolff [116]	Scalar potential Finite difference. SOR.	Cylindrical polar lattice. Variable geometry.	To be used in the design of turbine generators.
1971	LBL	Backett [37] Colonias [36]	Vector potential. Finite element. SOR.	Triangular prism elements irregular in xy plane.	Under development.
1971	RHEL	Diserens Trowbridge Zienkiewiczs	Scalar potential. Finite element. Frontal solution.	Isoparametric elements.	Under development.
1971	LAMPF	Christian [117] Vogel [117]	Scalar potential. Infinite boundaries. Use made of analytic functions for corner regions.	Three dimensional lattice.	Used for design of C magnets. Reasonably good agreement with test models.

III | Summary of Computer Programs That Use the Integral Equation Method

Date	Name Center	Authors	Method	Discretization	Remarks
1967	TAMI	Sackett [118]	Vector potential. Dipole magnetisation. SOR.	Rectangular. Blocks.	Experimental only. Reported instabilities in the relaxation process
1967/ 1970	RENO	Halacsy [41, 42] Schneider [119]	Scalar potential. Dipole magnetisation. Non relaxation method. Iterative loop for variable permeability.	Regular cubes.	Well established program but owing to limitations in the size of memory result only available for simple cases.
1970	University of Toronto	Zaky [120] Robertson [121] Karmarker [121]	Three methods used. Scalar potential magnetisation currents, and vector potential.	Variety of meshes used in 2 D. Regular blocks in 3 D.	Under development. Three-D version tested on constant μ problem. Good agreement with measurements.
1971	GFUN RHEL	Newman [122] Trowbridge [122] Turner [122]	Dipole magnetisations. Equation solved for magnetisation field vector by Gaussian elimination. Non linear loop for variable by simple iterative scheme.	Current and iron elements for a wide variety of shape for 2 D.	2 D version fully operational with established reliability. 3 D version under advanced development with good agreement with measurement.

References

1. J. A. Stratton, "Electromagnetic Theory." McGraw-Hill, New York, 1941.
2. R. S. Varga, "Matrix Iterative Analysis," p. 272. Prentice-Hall, Englewood Cliffs, Englewood Cliffs, New Jersey, 1962.
3. P. Concus, Numerical solution of the minimal surface equation. *Math. Comput.* **21**, 340–350 (1967).
4. J. S. Colonias and A. C. Paul, Magnetic field calculations for the 184-inch synchrocyclotron. UCRL-18882. Lawrence Berkeley Lab., Berkeley, California, 1970.
5. J. S. Colonias and J. H. Dorst, Magnet design applications of the magnetostatic program TRIM. *Proc. Int. Conf. Magnet Technol., Stanford Univ., Stanford, California, 1965*, pp. 188–195. Clearinghouse for Fed. and Tech. Inform., Nat. Bur. Stand., US Dept. of Comm., Springfield, Virginia.
6. A. M. Winslow, Numerical solution of the quasilinear Poisson equation in a nonuniform triangular mesh. *J. Comput. Phys.* **1**, 149–172 (1966).
7. A. M. Winslow, An irregular triangular mesh generator. UCRL-7880. Lawrence Livermore Lab., Livermore, California, 1964.
8. A. M. Winslow, Lawrence Livermore Lab., Livermore, California, private communication, 1971.
9. N. J. Diserens, Further development of the magnetostatic computer programme TRIM at the Rutherford Laboratory. RHEL/R 171. Rutherford Lab., Chilton, Didcot, Berkshire, England, 1969.
10. J. S. Colonias, TRIM – A magnetostatic computer program for the CDC 6600. UCRL-18439. Lawrence Berkeley Lab., Berkeley, California, 1968.
11. J. S. Colonias, Calculation of two-dimensional magnetic fields by digital display techniques. *Proc. Int. Conf. Magnet Technol., 2nd*, pp. 79–86. Rutherford Lab., Chilton, Didcot, Berkshire, England, 1967.
12. C. Besev, J. S. Colonias, and B. Olsen, Computer design and optimization of iron cored particle focusing devices. UUIP-586. Inst. of Phys., Uppsala Univ., Uppsala, Sweden, 1968.
13. J. H. Dorst, Lawrence Berkeley Lab., Berkeley, California, private communication, 1969.
14. H. Brown, ETH, Zurich, Switzerland, private communication, 1969.
15. S. T. Lin, Brookhaven Nat. Lab., Upton, New York, private communication, 1971.

16. R. S. Christian, Purdue Univ., Lafayette, Indiana, private communication, 1969.
17. J. S. Colonias, Operating instructions for SIBYL magnetostatic program. UCID-2264. Lawrence Berkeley Lab., Berkeley, California, 1964.
18. R. Perin, Preliminary note about MARE, a computer program for the calculation of static magnetic fields. ISR-MAG/66-8. CERN, Geneva, Switzerland, 1966.
19. S. Van der Meer, A computer program for calculating the flux distribution in two-dimensional saturated magnet cores. NPA/Int. 66-9. CERN, Geneva, Switzerland, 1966.
20. R. Perin and S. Van der Meer, The program MARE for the computation of two-dimensional static magnetic fields. CERN 67-7/20. CERN, Geneva, Switzerland, 1967.
21. L. R. Anderson, H. Brechna, and E. A. Burfine, A computer code for variable permeability magnetostatic field problems. SLAC-56. Stanford Univ., Stanford, California, 1966.
22. L. R. Anderson, Numerical methods for variable permeability magnetostatic field problems. Ph. D. Thesis, Univ. Microfilms, Ann Arbor, Michigan, 1970.
23. S. Saint-Lorent, SLAC, Stanford Univ., Stanford, California, private communication, 1971.
24. R. S. Christian, P. F. Dahl, and G. Parzen, Computer calculation of the magnetic field of alternating gradient synchrotron magnets. *IEEE Trans. Nucl. Sci.* **12**, 408 (1965).
24a. L. J. Laslett, Use of scalar potential in two-dimensional magnetostatic computations with distributed currents. MURA 211, *Midwestern Universities Research Assoc.*, Stoughton, Wisconsin, 1956.
24b. J. H. Dorst, Magnet design applications of the magnetostatic program SIBYL. *Proc. Int. Conf. Magnet Technol., Stanford Univ., Stanford, California, 1965,* pp. 182–187. Clearinghouse for Fed. and Tech. Inform., Nat. Bur. Stand., US Dept. of Comm., Springfield, Virginia.
25. J. P. Blewett and M. S. Livingston, "Particle Accelerators," p. 253. McGraw-Hill, New York, 1962.
26. R. Lari, Argonne Nat. Lab., Argonne, Illinois, private communication, 1971.
27. Measurements on the prototype magnet unit. CERN-PS/Int. MM59-5. CERN, Geneva, Switzerland, 1959.
28. K. Jellett and G. Parzen, Computation of high field magnets. AADD-165. Brookhaven Nat. Lab., Upton, New York, 1970.
29. S. V. Ahamed, R. D. Burtness, and E. A. Erdelyi, Flux distribution in saturated d.c. machines at no-load. *IEEE Proc. Power Group, Winter Meeting, 31 TP 65-113.* Institute of Electrical and Electronics Engineers, New York, New York, 1965.
29a. S. V. Ahmed and E. A. Erdelyi, Nonlinear theory of salient pole machines. Companion paper to Ahmed *et al.* [29].
30. S. V. Ahamed, E. A. Erdelyi, and R. E. Hopkins, Non-linear theory of synchronous machines on-load. Companion paper to Ahamed *et al.* [29].
31. M. W. Garrett, An elliptic integral computer package for magnetic fields, forces, and mutual inductances of axisymmetric systems and a versatile line-tracking routine. ORNL-3575. Oak Ridge Nat. Lab., Oak Ridge, Tennessee, 1965.
32. D. L. Mumaugh, COILS – Magnetic properties of current systems using elliptic integrals. UCID-2465. Lawrence Berkeley Lab., Berkeley, California, 1963.
33. G. T. Ewan, J. S. Geiger, and R. L. Graham, A one-meter-radius iron-free double-focusing $\pi\sqrt{2}$ spectrometer for β-ray spectroscopy with a precision of $1:10^5$. *Nucl. Instrum. Methods* **9**, 245–286 (1960).
34. S. J. Sackett and R. N. Healey, JASON – A Digital computer program for the numerical solution of the linear Poisson equation $\nabla \cdot (\varkappa\nabla\phi) + \varrho = 0$. UCRL-18721. Lawrence Berkeley Lab., Berkeley, California, 1969.

35. J. P. Meulders, Inst. De Phys. Corpusculaire, Ottignies, Belgium, private communication, 1971.
36. J. S. Colonias, Report on field calculations. *Proc. Discuss. Anal. Magnet Fields*, Engineering Report 18, pp. 35–37. Univ. of Nevada, Reno, Nevada, 1968.
37. S. J. Sackett, An algorithm for the numerical solution of the nonlinear magnetostatic-field equation in three-dimensions. UCID-3435. Lawrence Berkeley Lab., Berkeley, California, 1970.
38. P. Concus, Numerical solution of the nonlinear magnetostatic field equation in two-dimensions. *J. Comput. Phys.* **1**, 330–342 (1967).
39. M. H. Foss, Scalar potential magnet programs. *Proc. Discuss. Anal. Magnet Fields*, Eng. Rpt. 18, pp. 80–95. Univ. of Nevada, Reno, Nevada, 1968.
40. M. H. Foss, Carnegie-Mellon Univ., Pittsburg, Pennsylvania, private communication, 1969.
41. A. A. Halacsy, Three dimensional calculation of magnetic fields and accuracy of the calculation of H, in a three dimensional space. *Proc. Discuss. Anal. Magnet Fields*, Engineering Report 18, pp. 108–112, 126–135. Univ. of Nevada, Reno, Nevada, 1968.
42. A. A. Halacsy, Univ. of Nevada, Reno, Nevada, private communication, 1969.
43. S. Caeymaex, MIMI – A computer program for three-dimensional magnetic field problems. ISR-MA/70-19. CERN, Geneva, Switzerland, 1970.
44. J. C. Brown and W. A. Perkins, MAFCO – A magnetic field code for handling general current elements in three-dimensions. UCRL-7744-Rev II. Lawrence Livermore Lab., Livermore, California, 1966.
45. C. D. Henning, FORCE – A computer program for calculating magnetic forces developed in electromagnets. UCRL-14917. Lawrence Livermore Lab., Livermore, California, 1966.
46. A. K. Chargin and C. D. Henning, Magnetic field and force by finite element techniques. UCRL-71538. Lawrence Livermore Laboratory, Livermore, California, 1969.
47. H. S. Butler, S. K. Howry, and C. H. Moore, TRANSPORT – A computer program for designing beam transport systems. SLAC. Stanford Univ., Stanford, California, unpublished, 1963.
48. A. C. Paul, LBL Berkeley version of TRANSPORT. UCID-3525. Lawrence Berkeley Lab., Berkeley, California, 1971.
48a. S. Kowalsky, MIT, Cambridge, Massachusetts, private communication, 1969.
49. C. H. Moore and B. A. Norman, Theory and design of beam transport systems. AGSCD-26. Brookhaven Nat. Lab., Upton, New York, 1967.
50. K. L. Brown, A first- and second order matrix theory for the design of beam transport systems and charged particle spectrometers. SLAC-75. Stanford Univ., Stanford, California, 1967.
50a. S. Penner, Calculations of properties of magnetic deflection systems. *Review of Scientific Instrum.* **32**, 150–160 (1961).
51. A. C. Paul, User guide for LBL teletype and vista TRANSPORT. LBL-951. Lawrence Berkeley Lab., Berkeley, California, 1972.
51a. J. S. Colonias, TRACE – An on line beam transport design program. *Nucl. Instrum. Methods* **76**, 13–22 (1969).
52. J. S. Colonias and A. C. Paul, On-line calculation of beam transport systems to first- and second-order. *Proc. Int. Conf. Magnet Technol.*, DESY, Group -L-, 2 Hamburg 52, Notkestieg, Germany, 1970.
53. T. J. Devlin, OPTIK – An IBM 709 computer program for the optics of high-energy particle beams. UCRL-9727. Lawrence Berkeley Lab., Berkeley, California, 1961.

54. R. B. Chaffee, Lawrence Berkeley Lab., Berkeley, California, private communication, 1970.
55. D. F. Kane, Lawrence Berkeley Lab., Berkeley, California, private communication, 1971.
56. J. W. Gardner and D. Whiteside, TRAMP – Tracking and matching program. NIRL/M/21. Rutherford Lab., Chilton, Didcot, Berkshire, England, 1961.
57. J. W. Gardner and D. Whiteside, A FORTRAN version of TRAMP. NIRL/M/41. Rutherford Lab., Chilton, Didcot, Berkshire, England, 1963.
58. W. R. Butler, Lawrence Berkeley Lab., Berkeley, California, private communication, 1970.
59. P. F. Meads, An IBM 7094 beam, quadrupole aberration program, *Nucl. Instrum. Methods* **40**, 166–168 (1966).
60. P. Shapiro, FORTRAN programs for first-order calculation of properties of beam transport systems. NRL-6710/No. 4. Naval Res. Lab., Washington, D.C., 1968.
61. G. R. Dangerfield and R. L. Walsh, BOPTIC – A beam transport program in FORTRAN IV. AAEC/TM-544. Aust. AEC Res. Establishment, Lucas Heights, Australia, 1970.
62. G. Burton, M. Hanney, and P. Strolin, BEATCH – A FORTRAN programme for the particle optics of beam transfer channels. ISR-BT-TH/69-27. CERN, Geneva, Switzerland, 1969.
63. M. Bassetti, R. M. Buonanni, and M. Placidi, Beam optics computation for particle transport systems by means of an improved New-Raphson method. *Nucl. Instrum. Methods* **45**, 93–101 (1966).
64. J. Fronteau and J. Hornsby, Etude de la transmission par un systeme magnetique lineaire d'un faisceau de particules primaires et des secondaires emis en vol. programme FORTRAN. (CERN-06008B50), CERN 62-36. CERN, Geneva, Switzerland, 1962.
65. A. Citron, J. Fronteau, and J. Hornsby, Calcul automatique des faisceaux de π et des faisceaux de μ programme FORTRAN, (CERN 6008 B56/57). Application au projet de faisceau $\pi - \mu$ du ps, CERN 63-30. CERN, Geneva, Switzerland, 1963.
66. J. Banaigs, A. Citron, J. Fronteau, and J. Hornsby, Diffusion coulombienne en milieu focalisant, programmes FORTRAN. (CERN-W101-B51 et B58), CERN 64-29. CERN, Geneva, Switzerland, 1964.
67. E. R. Close, SOTRM – A programs to generate first- and second-order matrix elements by tracking charged particles in a specified magnetic field. UCRL-19823. Lawrence Berkeley Lab., Berkeley, California, 1970.
68. E. R. Close, Generation of first- and second-order transformation elements from a given magnetic field. *Nucl. Instrum. Methods* **89**, 205–214 (1970).
69. A. C. Paul, TRAJECTORY – An orbit and ion optic matrix program for the 184-inch cyclotron. UCRL-19407. Lawrence Berkeley Lab., Berkeley, California, 1969.
70. M. M Gordon and T. A. Welton, Computation methods for AVF cyclotron design studies. ORNL-2765. Oak Ridge Nat. Lab., Oak Ridge, Tennessee, 1959.
71. D. I. Hopp, A general orbit code for cyclotron analysis. Tech. Rep. P-67. UCLA, Los Angeles, California, 1965.
72. L. Shaw, A general orbit program for sector focused cyclotrons. USNRDL-TR-936. U.S. Naval Radiol. Defense Lab., San Francisco, California, 1965.
73. A. C. Paul, GOC3D – A general orbit code for cyclic accelerator development. LBL-589. Lawrence Berkeley Lab., Berkeley, California, 1971.
74. L. J. Laslett and W. A. Perkins, Betatron amplitude growth upon traversing resonances during the compression cycle of an electron ring accelerator. UCRL-20143. Lawrence Berkeley Lab., Berkeley, California, 1971.

75. V. O. Brady, MAGOP – A program to generate first-order matrix elements by tracking charged particles in a specified two-dimensional magnetic field. LBL-752. Lawrence Berkeley Lab., Berkeley, California, 1972.
76. W. A. Perkins and S. J. Sackett, MAFCOIII – A code for calculating particle trajectories in magnetic and electric fields. LBL-765. Lawrence Berkeley Lab., Berkeley, California, 1972.
77. L. P. Meissner, Lawrence Berkeley Lab., Berkeley, California, private communication, 1972.
78. R. I. Arnette, M. M. Gordon, H. C. Owens, and T. A. Welton, An IBM 704 code for determining equilibrium orbits and properties of small-amplitude oscillations in cyclotron fields. ORNL/59-11-2. Oak Ridge Nat. Lab., Oak Ridge, Tennessee, 1959.
79. A. A. Garren, Calculation of cyclotron trim coil currents for field optimization by linear programming methods. *Nucl. Instrum. Methods* **18, 19**, 309–322 (1962).
80. K. L. Kelly, General description and operating characteristics of the Berkeley 88-inch cyclotron, *Nucl. Instrum. Methods* **18, 19**, 33–40 (1962).
81. E. R. Close, CYDE – A collection of FORTRAN II cyclotron development codes for the use on the IBM 7094. UCID-2869. Lawrence Berkeley Lab., Berkeley, California, 1968.
82. R. E. Berg, MSU cyclotron. Proj. Rep. MSUCP-24. Ph. D. thesis, Michigan State Univ., East Lansing, Michigan, 1966.
83. R. E. Berg, H. G. Blosser, and M. M. Gordon, Theoretical and experimental beam studies for the Michigan State University cyclotron. *Nucl. Instrum. Methods* **58**, 327–341 (1968).
84. D. H. Nelson, The 88-inch cyclotron main-coil magnetic field as a continuous function of current. UCID-2464. Lawrence Berkeley Lab., Berkeley, California, 1965.
85. H. C. Owens and T. A. Welton, An IBM 704 code for studying particle orbits in cyclotron fields. ORNL 59-11-3. Oak Ridge Nat. Lab., Oak Ridge, Tennessee, 1959.
86. M. M. Gordon and T. A. Welton, Computation methods for AVF cyclotron design studies. ORNL-2765. Oak Ridge Nat. Lab., Oak Ridge, Tennessee, 1959.
87. A. A. Garren, D. L. Judd, L. Smith, and H. A. Willax, Electrostatic deflector calculations for the Berkeley 88-inch cyclotron. *Nucl. Instrum. Methods* **18, 19**, 525–547 (1962).
87a. H. C. Owens, Three computer programs for calculating cyclotron orbits. *Nucl. Instrum. Methods* **18, 19**, 295–349 (1962).
88. A. A. Garren and H. C. Owens, An IBM 704 code for studying effects of electric and magnetic deflectors in cyclotron fields. Lawrence Berkeley Lab., Berkeley, California, unpublished, 1962.
89. D. J. Clark, Lawrence Berkeley Lab. Berkeley, California, private communication, 1971.
90. V. O. Brady, Lawrence Berkeley Lab., Berkeley, California, private communication, 1971.
91. M. Reiser and J. Kopf, Electrolytic tank facility and computer program for central region studies for the MSU cyclotron. MSUCP-19. Michigan State Univ., East Lansing, Michigan. 1964.
92. E. R. Close, PINWHEEL – Orbit tracking in combined electric and magnetic fields. T1-BKY-PINWEL, Lawrence Berkeley Lab., Berkeley, California, 1966.
93. A. A. Garren and A. S. Kenney, SYNCH – A computer system for synchrotron design and orbit analysis. Lawrence Berkeley Lab., Berkeley, California, unpublished, 1968.
94. E. D. Courant and H. S. Snyder, Theory of the Alternating Gradient Synchrotron. *Annals of Physics* **3**, No. 1, 1–48 (1958).

95. P. L. Morton and J. R. Rees, The design of low-beta insertions for storage rings. *IEEE Trans. Nucl. Sci.* **14**, 630–634 (1967).
96. E. Keil and P. Strolin, AGS – ISR computer system for synchrotron design and orbit analysis. CERN 69-5. CERN, Geneve, Switzerland, 1969.
97. G. Parzen, A mesh-iteration program for field calculations in linac cavities. *Proc. Proton Linear Accelerator Conf., Brookhaven Nat. Lab., Upton, New York, 1968*, BNL 50120 (C-54), pp. 537–542. Clearinghouse for Fed. and Tech. Inform., Nat. Bur. Stand., US Dept. of Comm., Springfield, Virginia.
98. M. D. J. MacRoberts and W. F. Rich, Numeric solution of the fundamental mode of cylindrically symmetrical resonant cavities. LA-4219. Los Alamos Sci. Lab., Los Alamos, New Mexico, 1969.
99. H. C. Hoyt, Numerical studies of the shapes of drift tubes and linac cavities. *IEEE Trans. Nucl. Sci.* **12**, 153–155 (1965).
100. H. C. Hoyt, W. F. Rich, and D. D. Simmons, Computer designed 805-MHz proton linac cavities. *Rev. Sci. Instrum.* **37**, 755–762 (1966).
101. H. C. Hoyt, Designing resonant cavities with the LALA computer program. *Proc. Linear Accelerator Conf., Los Alamos, New Mexico, 1966*, LA-3609, pp. 119–124. Clearinghouse for Fed. and Tech. Inform., Nat. Bur. Stand., US Dept. of Comm., Springfield, Virginia.
102. G. Parzen, Perturbations and tolerances in a 200-MeV proton linac. *Proc. Int. Conf. High Energy Accelerators, Cambridge, Massachusetts, 1967*, CEAL-2000, pp. A34, 35. Clearinghouse for Fed. and Tech. Inform. Nat. Bur. Stand., US Dept. of Comm., Springfield, Virginia.
103. P. F. Dahl, S. Giordano, J. P. Hannwacker, K. Jellett, and G. Parzen, Linac cavity field calculations. *Proc. Int. Conf. High Energy Accelerators, Cambridge, Massachusetts, 1966*, pp. 115–118. Clearinghouse for Fed. and Tech. Inform., Nat. Bur. Stand., US Dept. of Comm., Springfield, Virginia.
104. T. W. Edwards, MESSYMESH—Programs for calculations for linear accelerator cavities. No. 642, MURA, *Midwestern Universities Research Assoc.*, Stoughton, Wisconsin, 1962.
105. T. W. Edwards, Proton linear accelerator cavity calculations. No. 622, MURA, *Midwestern Universities Research Assoc.*, Stoughton, Wisconsin, 1961.
106. B. Austin, T. W. Edwards, J. E. O'Meara, M. L. Palmer, D. A. Swenson, and D. E. Young, The design of proton linear accelerators for energies up to 200-MeV. No. 713, MURA, *Midwestern Universities Research Assoc.*, Stoughton, Wisconsin, 1965.
107. S. J. Sackett and A. A. Garren, AZTEC – A code for calculating the impedance of an azimuthally bunched beam in a cylindrically symmetric structure. LBL-774. Lawrence Berkeley Lab., Berkeley, California, 1972.
108. A. C. Entis, A. A. Garren, and L. Smith, Negative mass instabilities in an electron-ring compressor. *IEEE Trans. Nucl. Sci.* **18**, 1092–1096 (1971).
109. C. E. Neilsen, A. M. Sessler, and K. R. Symon, Longitudinal instabilities in intense relativistic beams. *Proc. CERN Symp. High Energy Accelerators, 1959*, pp. 239–252. CERN, Geneva, Switzerland.
110. B. Zotter, Longitudinal instability of relativistic particle beams in laminated vacuum chambers. CERN Rep. ISR-TH/69-35. CERN, Geneva, Switzerland, 1969.
111. F. C. Trut, Analysis of Homopolar inductor alternators. Ph. D. Thesis. Univ. of Delaware, 1964.
112. T. K. Khoe and R. J. Lari, FORGY—A companion computer program of TRIM to calculate forces and energy in electromagnets. *Proc. Int. Conf. Magnet Technol.*, *4th*,

Brookhaven Nat. Lab. 1972. Nat. Tech. Inform. Service, US Dept. of Comm., Springfield, Virginia.
113. K. Reichert, *Arch. Elektrotech. (Berlin)* **52**, 176–195 (1968).
114. P. Silvester and M. V. K. Chari, *IEEE Trans. Power App. Syst.*, **89**, p. 1642 (1970).
115. C. Iselin, Some recent developments in computation of two-dimensional fields in iron magnets. *Proc. Int. Conf. Magnet Technol., 3rd.* DESY, Group -L-, 2 Hamburg 52, Notkestieg, Germany, 1970.
116. W. Müller and W. Wolff, Calculation of three dimensional magnetic field with field dependent permeability and given current density using a scalar potential. *Proc. Conf. Magnet. Fields, 3rd.* Engineering Report No. 46, pp. 94–106, Univ. of Nevada, Reno, Nevada, 1971.
117. R. Christian and H. Vogel, Three dimensional field computations for C-magnets and electrostatic separators. *Proc. Conf. Magnet. Fields, 3rd.* Engineering Report No. 46, pp. 290–292, Univ. of Nevada, Reno, Nevada, 1971.
118. S. J. Sackett, Solution of the magnetostatics problem in three space dimensions—preliminary investigations at LBL Berkeley. *Proc. Conf. Magnet. Fields, 2nd.* Engineering Report No. 26, pp. 67–76, Univ. of Nevada, Reno, Nevada, 1969.
119. J. E. Schneider, A matrix method for calculating magnetic field intensity. *Proc. Conf. Magnet. Fields, 3rd.* Engineering Report No. 46, pp. 219–261. Univ. of Nevada, Reno, Nevada, 1971.
120. S. G. Zaky and S. D. T. Robertson, Integral equation formulation for the solution of magnetic field problems–dipole and current models. *Proc. Conf. Magnet. Fields, 3rd.* Engineering Report No. 46, pp. 142–158, Univ. of Nevada, Reno, Nevada, 1971.
121. S. D. T. Robertson and H. C. Karmaker, Integral equation formulation for the solution of magnetic field problems–magnetization vector approach. *Proc. Conf. Magnet. Fields, 3rd.* Engineering Report No. 46, pp. 159–171, Univ. of Nevada, Reno, Nevada, 1971.
122. M. J. Newman, C. W. Trowbridge, and L. R. Turner, GFUN: An interactive program as an aid to magnet design. *Proc. Int. Conf. Magnet Technol., 4th. Brookhaven Nat. Lab., 1972.* Clearinghouse for Fed. and Tech. Inform., Nat. Bur. Stand., US Dept. of Comm., Springfield, Virginia.
123. S. B. Vorozhtsov, L. T. Zakamskaya, and N. L. Zaplatin, Calculation of a two dimensional magnetic field using irregular triangular coordinate mesh. Rep. No. P9-5013. Joint Inst. for Nucl. Res., Dubna, USSR, 1970.

Author Index

Numbers in italics refer to the pages on which the complete references are listed. Numbers in parentheses are reference numbers and are inserted to enable the reader to locate those cross references where the author's name does not appear at the point of reference in the text.

A

Ahamed, S. V., 93, 289, *294*
Anderson, L. R., 63(21, 22), 289, *294*
Arnette, R. I., 233, 237, *297*
Austin, B., 280, *298*

B

Banaigs, J., 190(66), 191, 192, *296*
Bassetti, M., 190, *296*
Berg, R. E., 234, 238, *297*
Besev, C., 37(12), *293*
Blewett, J. P., 79(25), *294*
Blosser, H. G., 234(83), 238(83), *297*
Brady, V. O., 223, 238, 246, *297*
Brechna, H., 63(21), 289, *294*
Brown, H., 39, *293*
Brown, J. C., 119, 120, 124, 128, 227, *295*
Brown, K. L., 140(50), *295*
Buonanni, R. M., 190(63), *296*
Burfine, E. A., 63(21), 289, *294*
Burtness, R. D., 93(29), *294*
Burton, G., 189, *296*
Butler, H. S., 140, *295*
Butler, W. R., 173, 186, *296*

C

Caeymaex, S., 119, 291, *295*
Chaffee, R. B., 165, *296*
Chargin, A. K., 129(46), *295*
Chari, M. V. K., 289, *299*
Christian, R. S., 39, 76(24), 79(24), 288, 291, *294*, *299*

Citron, A., 190(66), 191(66), 192(66), *296*
Clark, D. J., 238, 251, *297*
Close E. R., 195(67, 68), 195, 233, 238, 246, *296*, *297*
Colonias, J. S., 16(4, 5), 21(10), 27, 35, 37(11, 12), 39(17), 89, 115(36), 160 (51a, 52), 288, 291, *293*, *294*, *295*
Concus, P., 10(3), 117, *293*, *295*
Courant, E. D., 254, *297*

D

Dahl, P. F., 76(24), 79(24), 279, 280, 288, *294*, *298*
Dangerfield, G. R., 189, *296*
Devlin, T. J., 164, *295*
Diserens, N. J., 21, 288, *293*
Dorst, J. H., 16(5), 39(13), 79(24b), 288, *293*, *294*

E

Edwards, T. W., 280, *298*
Entis, A. C., 281(108), 283, *298*
Erdelyi, E. A., 93(29, 29a, 30), 289, *294*
Ewan, G. T., 97(33), 99, 100, *294*

F

Foss, M. H., 118, 291, *295*
Fronteau, J., 190(64, 65, 66), 191, 192 (66), *296*

G

Gardner, J. W., 172, *296*
Garren, A. A., 233, 237(87, 88), 253, 255, 281, 283, *297, 298*
Garrett, M. W., 94, *294*
Geiger, J. S., 97(33), 99(33), 100(33), *294*
Giordano, S., 279(103), 280(103), *298*
Gordon, M. M., 210, 233(78), 234(83), 237(86), 238(83), *296, 297*
Graham, R. L., 97(33), 99(33), 100(33), *294*

H

Halacsy, A. A., 118, 119(42), 292, *295*
Hanney, M., 189(62), *296*
Hannwacker, J. P., 279(103), 280(103), *298*
Healey, R. N., 102, 103, *294*
Henning, C. D., 129, 129(46), 130, *295*
Hopkins, R. E., 93(30), 289, *294*
Hopp, D. I., 210, *296*
Hornsby, J., 190(64, 65, 66), 191, 192(66), *296*
Howry, S. K., 140(47), *295*
Hoyt, H. C., 270(99, 100, 101), 270, 274, *298*

I

Iselin, C., 290, *299*

J

Jellett, K., 91, 279(103), 280(103), 290, *294, 298*
Judd, D. L., 237(87), *297*

K

Kane, D. F., 165, 173, 186, *296*
Karmaker, H. C., 292, *299*
Keil, E., 264, *298*
Kelly, K. L., 233(80), *297*
Kenney, A. S., 253, 255(93), *297*
Khoe, T. K., 288, *298*
Kopf, J., 246, *297*
Kowalsky, S., 140(48a), *295*

L

Lari, R. J., 189, 288, *294, 298*
Laslett, L. J., 76(24a), 79(24a), 222, 223, *294, 296*
Lin, S. T., 39, *293*
Livingston, M. S., 79(25), *294*

M

MacRoberts, M. D. J., 270, *298*
Meads, P. F., 188(59), *296*
Meissner, L. P., 227(77), *297*
Meulders, J. P., 112(35), *295*
Moore, C. H., 140(47, 69), *295*
Morton, P. L., 261(95), *298*
Müller, W., 291, *299*
Mumaugh, D. L., 94, *294*

N

Neilsen, C. E., 283, *298*
Nelson, D. H., 236(84), 236, *297*
Newman, M. J., 292, *299*
Norman, B. A., 140(49), *295*

O

Olsen, B., 37(12), *293*
O'Meara, J. E., 280(106), *298*
Owens, H. C., 233(78), 237(78, 85, 87a, 88), *297*

P

Palmer, M. L., 280(106), *298*
Parzen, G., 76(24), 79(24), 91, 269(97), 279, 280(103), 288, 290, *294, 298*
Paul, A. C., 16(4), 140(48), 153, 160(51, 52), 203, 210, *293, 295, 296*
Penner, S., 143, *295*
Perkins, W. A., 119, 120, 124, 128, 222, 223, 227, *295, 296, 297*
Perin, R., 39(18, 20), 74, 75, 289, *294*
Placidi, M., 190(63), *296*

R

Rees, J. R., 261(95), *298*
Reichert, K., 289, *299*
Reiser, M., 246, *297*
Rich, W. F., 270(100), 270, *298*
Robertson, S. D. T., 292, *299*

S

Sackett, S. J., 102, 103, 116, 227, 281, 291, 292, *294, 295, 297, 298, 299*
Saint-Lorent, S., 68(23), *294*
Schneider, J. E., 292, *299*
Sessler, A. M., 283(109), *298*
Shapiro, P., 188(60), 188, *296*
Shaw, L., 210, *296*
Silvester, P., 289, *299*
Simmons, D. D., 270(100), *298*
Smith, L., 237(87), 283(108), *297, 298*
Snyder, H. S., 254(94), *297*
Stratton, J. A., 5(1), *293*
Strolin, P., 189(62), 264, *296, 298*
Swenson, D. A., 280(106), *298*
Symon, K. R., 283(109), *298*

T

Trowbridge, C. W., 292, *299*
Trut, F. C., 288, *298*
Turner, L. R., 292, *299*

V

Vand der Meer, S., 39(19, 20), 74, 75, 289, *294*
Varga, R. S., 8(2), *293*
Vogel, H., 291, *299*
Vorozhtsov, S. B., 289, *299*

W

Walsh, R. L., 189, *296*
Welton, T. A., 210, 233(78), 237(78, 85, 86), *296, 297*
Whiteside, D., 172, *296*
Willax, H. A., 237(87), *297*
Winslow, A. M., 17, 17(7), 19, 288, *293*
Wolff, W., 291, *299*

Y

Young, D. E., 280(106), *298*

Z

Zakamskaya, L. T., 289, *299*
Zaky, S. G., 292, *299*
Zaplatin, N. L., 289, *299*
Zotter, B., 285, *298*

Subject Index

A

Aberration properties, third-order, 188
Accelerator, linear cavity
 design of, 269, 270, 280
 electromagnetic fields of, 279
 resonant frequency of, 279
Accuracy, increase of, 89
Accuracy of solution, 11
AGS, 264–267
Alternating direction implicit (ADI)
  method of solution, 9
AZTEC, 281–285

B

Beam, alpha-particle, 242
 phase history comparison, 239
Beam conditions
 dispersion and separation, 179–181
 matching, 178
Beam elements, 174–178
Beam, particle, optical design of, 164
Beam profile, 173
Beam transfer channels, particle motion in, 189
Beam transport, 138, 140, 164, 186, 190
 design of, 173, 190, 194
 programs, 188
 first-order properties calculation, 188
BEATCH, 189–190

Betatron, amplitude growth, 220
Betatron functions, 254, 257, 264
BMAP, 61–63
BOPTIC, 189
Boundary conditions, 107
Boundary requirements, 64

C

COILS, 6, 11, 94–101, 124
Coil, trim
 current adjustment, 238
 current calculation, 233
 effects of, 237
Convergence, rate of, 89
Coulomb scattering deviation, 191
CURE, 270–279
Cyclotron
 design of, 233
 88-inch, 112, 233, 242
 184-inch, 198, 208
CYDE, 136, 193, 233–246
CYDEB, 237

D

Direct method of solution, 8
Dirichlet boundaries, 107, 112–113
Discretization, 4, 7
Discretization error, 11
DORO, 236

Subject Index

E

Equations of motion, 211, 247
Euler equation, 117

F

FATIMA, 290
FIELD, 17, 27–34
Field,
  electric, 102–113, 227
    programs for, 102 134
  electromagnetic, 270
    of linear accelerator, 279
  magnetic, 93–94, 119, 227, 233
    three-dimensional, 114
    programs for, 1–101
    of rotating electrical machinery, 93–94
    solution of, 27–34
    cyclotron, orbit properties of, 236
Fields
  from azimuthally bunched beam, 281
  magnetic and electric, 246
  time-varying, 227
FORCE, 129–133
Forces, calculation of, 94
Frequency, calculation of, 270

G

Gauss-Seidel, 8
GFUN, 292
GOC, 136, 193
GOC3D, 101, 210–223
GRACY, 91–92, 290
Graphic display, 37, 160–164, 170

I

Inductance, calculation, 94
Integral equation, 292
Integration of equations of motion, 193–267
Interface, iron–air, 39, 63
Ion-optic matrix, first-order, 195, 203, 210
Ion optics, 147

J

JASON, 2, 9, 102–113
JESSY, 275, 279–280

L

LALA, 270, 274
Laplace's equation, 93
Lattice, magnetic, design of, 261
LESCO, 236
LINDA, 6, 39–63, 84, 86–91, 288
Lorentz equation, 211

M

MAFCO, 11–9128, 128, 129, 227, 228
MAFCOIII, 227–233
Magnet, CERN-PS, 22–26, 37, 59, 84–91
  magnetic field in air region, 63
  184-inch synchrocyclotron, simulation of, 16
Magnet types, 43–46, 83
Magnetic force, 129
Magnetostatic programs, two-dimensional, 15–101
MAGOP, 223–227
MARE, 289
MAREC, 6, 39, 74–91
Matrix formalism, 135, 136, 138, 139
Matrix transformation, 139
Maxwell's equations, 2, 281
MESH, 17, 21–27, 103
Mesh, 89
  irregular, 13
    irregular triangle, 17, 21–27
  nonuniform quadrilateral, 103
  quadrilateral, 112
  rectangular, 13
  triangular, 13, 17, 21–27
  triangular prism, 117
  uniform, 39–63
Mesh spacing, 11
MESSYMESH, 275, 280
MIMI, 119
MIMI-A, 291
MIMI-I, 119

N

Neumann boundaries, 107, 113
Numerical integration, 135, 136, 193
Numerical solution, 3–4
NUTCRACKER, 6, 63–74, 86–91, 289

O

OPTIK, 136, 164–176, 184, 186, 203
Orbit, closed, 257

Orbit, particle, 173, 189
  calculation of, 135, 203, 210, 223, 247
  tracking of, 246
  tracking facilities, 181–183
Orbit properties, 254
  isochronous equilibrium, 233
Orbit, transverse, linear properties of, 253

P

Partial differential equations
  solved in two dimensions, 288
  solved in three dimensions, 291
Particle motion, 189
Particle orbit, *see* Orbit
Particles, charged, motion in electromagnetic field, 136
Particle transmission, 190–192
Phase history, 233, 246
Phase space, 140, 155, 161, 176, 189
PINWHEEL, 136, 246–254
Plot, equipotential and flux lines, 56
Plotting capability, 35
PLUTO, 56–61
Point successive overrelaxation (PSOR), 8–11, 116–117, 269, 280,
  linearized, 9
Poisson's equation, 5, 17, 64, 77, 93, 102
Potential
  scalar, 5, 6, 39, 76, 115, 118
  vector, 5, 6, 39, 64, 91, 93, 94, 115, 116
Program
  beam transport 138, 140
  electric field, 102–134
  linear accelerator, 269–285
  magnetic field, 1–101
  orbit calculation, 135–267
  summary, 288–292
  three-dimensional magnetostatic, 114–137
  two-dimensional, magnetostatic, 15–113
  using matrix formalism, 138–192
Program B-50, 191
Program B-51, 191
Program B-52, 192
Program B-56/57, 192
Program 4P, 136, 188
Programs, two-dimensional magnetostatic, comparison, 68–91
PSOR, *see* Point successive overrelaxation

R

Ray tracing, 112, 160–164
RENO, 118–119, 292
Resonances, 211, 220
Resonant cavity, *see* Accelerator, linear cavity

S

SIBYL, 39, 288
SIMONE, 115
Solution, *see* individual method
Solution method, 3–4
Solution, numerical, 6–11
SOR, *see* Point successive overrelaxation
SOTRM, 135, 136, 139, 193, 194–203, 210
SYNCH, 136, 254–264
Synchrotron, alternating gradient, design of, 264
SYTRAN, 190

T

TAMI, 115–116, 292
THOR, 10, 116–118
TRAJ, 136
TRAJECTORY, 203–210
TRAMP, 136, 172–187
TRAMTIK, 186
Transfer matrix
  first-order, 143, 151, 153, 162, 165, 195, 198, 203, 208, 210, 223, 253, 264
  second-order, 140, 195
TRANSPORT, 136, 138, 140–164, 167, 184, 186, 190, 203, 210, 261
TRED, 37–39
TRIM, 6, 11, 15–39, 84, 86–91, 103, 117, 288
TRIMCO, 238
TRIP, 35–37

W

Wave equation, 269, 271

Z

ZAM, 227
ZIG, 37
ZORO, 237